Mental Health and Illness Worldwide

Series Editors
Norman Sartorius
Association for the Improvement of Mental Health Programmes (AMH)
Geneva, Switzerland

Ee Heok Kua
Department of Psychological Medicine
National University of Singapore
Singapore, Singapore

Most books on mental health and illness are published for readers in North America and Europe, and not much is known about psychiatric practice, services and research in Asia, Africa, and South America. This series will include contributions of clinicians and researchers worldwide. Each volume will cover broad issues including epidemiology, cross-cultural comparison, clinical research, stigma of mental illness, cultural issues in mental healthcare, health economics, innovative services, preventive programs and health service outcome research. The volumes will find a wide readership among psychiatrists, psychologists, sociologists, health policy makers, social workers, health economists, anthropologists and philosophers. It will provide the readers a broader perspective of mental health and illness worldwide and also future research initiatives.

More information about this series at http://www.springer.com/series/14178

Santosh Kumar Chaturvedi
Editor

Mental Health and Illness in the Rural World

With 6 Figures and 30 Tables

🐎 Springer

Editor
Santosh Kumar Chaturvedi
Psychiatric Rehabilitation
Services, Department of Psychiatry
National Institute of Mental Health and
Neurosciences
Bangalore, Karnataka, India

Department of Mental Health Education
National Institute of Mental Health and Neurosciences
Bangalore, Karnataka, India

ISSN 2511-8323   ISSN 2511-8315 (electronic)
ISBN 978-981-10-2343-9   ISBN 978-981-10-2345-3 (eBook)
ISBN 978-981-10-2344-6 (print and electronic bundle)
https://doi.org/10.1007/978-981-10-2345-3

This Springer imprint is published by the registered company Springer Nature Singapore Pte Ltd.
The registered company address is: 152 Beach Road, #21-01/04 Gateway East, Singapore 189721, Singapore

Series Preface

Psychiatry lives exciting and challenging times. Advances of knowledge stemming from basic sciences and epidemiological and clinical research have provided a better understanding of the etiopathogenesis, psychopathology, and natural history of mental disorders. Improved methods of treatment have changed clinical practice and prolonged the life of people with mental illness. Economic consideration and the emphasis on human rights of people with mental illness made a profound impact on the way in which psychiatry is to be practiced.

Regrettably, however, psychiatry is not practiced in the same manner around the world. Undergraduate and postgraduate education in psychiatry varies in content and duration from country to country. Psychiatrists use different doses of medication for the same disorders. The systems of care for people with mental illnesses differ in the organization and content of their interventions. Support to scientific investigations of matters related to psychiatry fluctuates and in many countries amounts to very little.

Information about the function of psychiatric services varies in quantity and quality. The series of seven books on Mental Health and Illness Worldwide aims to help in reducing these differences and facilitate international collaboration in psychiatry. We have invited top experts from different countries to edit the volumes, and they have in turn selected authors from different parts of the world. We have also decided to approach the body of psychiatry from a public health and epidemiological perspective rather than have books dealing with different groups of diseases. The series includes books examining and presenting knowledge assembled according to social and public health variables – gender, urbanity, migratory status, age, and education. Each of the volumes has adopted a wide perspective and included chapters based on knowledge stemming from epidemiology, on results of the investigation of cultural issues, on the best of psychopathology, on the results of the investigation of biological factors, mental health care and its innovations, health economics, and experience gained in preventive programs. The volume editors have agreed to aim at producing volumes marked by the balance of information and knowledge from basic social and behavioral sciences and from clinical practice.

The seven volumes of this opus are:

1. Mental Health and Illness of the Elderly
   Editors: Helen Chiu (Hong Kong) and Ken Shulman (Canada)

2. Mental Health and Illness in the City
   Editors (Denmark): Povl Munk-Jorgensen, Niels Okkels, and Christina Kristiansen
3. Mental Health and Illness of Women
   Editors: Prabha S. Chandra (India), Helen Herrman (Australia), Jane Fisher (Australia) and Anita Riecher-Rössler (Switzerland)
4. Mental Health and Illness in the Rural World
   Editor: S. Chaturvedi (India)
5. Mental Health, Mental Illness and Migration
   Editors: Driss Moussaoui (Africa), Dinesh Bhugra (United Kingdom), and Antonio Ventriglio (Italy)
6. Mental Health and Illness of Children and Adolescents
   Editors: Eric Taylor (United Kingdom), John Wong (Singapore), Frank Verhulst (Netherlands), and Keiko Yoshida (Japan)
7. Education About Mental Health and Illness
   Editors: Marc H.M. Hermans (Belgium), Tan Chay Hoon (Singapore), and Edmond Pi (USA)

We were delighted to see that the volume editors have succeeded in recruiting outstandingly knowledgeable authors for the chapters of their books. Most of them have received worldwide recognition for their contributions in their fields of specialization, and all of them have written their texts with authority and excellent judgment concerning the materials to be included.

We believe that these series of books demonstrate the importance and value of interdisciplinary and international collaboration and that it will provide readers a global perspective of mental health and mental illness. We also hope that it will help to make our discipline more homogenous and bring its practitioners worldwide closer together in the pursuit of helping people with mental illness worldwide.

It is our pleasure and a privilege to thank Professor Helen Chiu and Professor Ken Shulman, editors of this volume dealing with Mental Health and Illness of the Elderly – the first of the series – for their hard work and for their insights and dedication to excellence.

January 2020   Norman Sartorius
                                                                Ee Heok Kua

Volume Preface

For the sake of better mental health and well-being people express their desire to go to a "village" or a rural setting! Somehow, this concept is quite popular; people say life is simple in the villages and more relaxing. There is no hustle-bustle and urgency of city life, no traffic woes, and no great demands from the society and comparatively low cost of living. Mental health is expected to be better in the rural settings. While this concept is perhaps popular among general public, there are also other beliefs about lack of adequate health infrastructure, facilities, and services. People from villages head toward towns and cities to seek better health services. Hence, mental health may not get enough attention and care in the rural settings. Rural people look for modern amenities like domestic equipment in the shopping malls of the cities. Rural living is considered idyllic, confined to a small community with a traditional lifestyle. Once I met a person who wanted to live in a village after retirement, provided the "hut" had all modern luxuries like air conditioning, refrigerator, and a car!

Rural areas are often considered backward, both economically and in terms of development. Due to availability of only basic health services, health problems are expected to go untreated. There is also difficulty in getting standard medications, as rural areas may not have big pharmacies and clinical laboratories for investigations. The main reason for this may be lack of business opportunities for the laboratories or pharmacies. Rural areas are likely to have traditional healing services available, which do not have enough scope in cities or towns. Low cost and affordable treatments, such as yoga, Ayurveda, and herbal and related therapies, are likely to be popular due to their availability and affordability.

Thus, there are certain benefits but also some drawbacks with regard to mental health care in rural areas. However, mental health care in rural areas has not received direct attention from policy makers and health professionals. The rural background has been dealt with remotely as one of the demographic variables in research studies, to compare with the urban population. It is a key social determinant, only to the advantage of the urban habitat. The other points of focus have been certain conditions considered a challenge in rural areas, like farmer suicides, dissociative disorders, acute psychosis, and alcoholism.

This volume on rural mental health is thus a unique contribution to the literature on the subject. It covers the conceptual issues about what is rural and the differences between rurality and urbanity. More than three decades ago I visited a health center in a rural area on the outskirts of New York, called Lafayette; I was amazed to see that at that time it looked more modern and developed than any city in India. The chapters in the is volume have attempted to highlight specific disorders and their variations in the rural background. Some of these are, as expected, the common mental disorders, dissociative and somatoform disorders, and stress-related disorders. Substance use disorders and alcoholism need special mention in the light of use of cannabis and opium in traditional societies and locally brewed alcohol. The challenges of living with intellectual disabilities in rural areas have been discussed in detail.

Among specific themes, mental health issues in women in rural areas and their challenges, stigma toward mental illness in rural areas, and its relationship with mental health literacy are discussed in detail. Suicide and self-harm in rural areas, with specific reference to farmer suicides, have been well elaborated. On the one hand, there are constraints and challenges in conducting epidemiological research on mental health in rural settings; on the other, it is equally challenging to organize mental health care services in rural areas. Many times, it is a camp approach on a periodic basis.

Rural mental health scenarios in Bangladesh and Sri Lanka have been described in two chapters; and in the Asia Pacific rural communities with a focus on cultural factors.

Management of mental health problems in rural settings, both pharmacological and non-pharmacological, and the use of yoga and traditional healing methods have been described in separate chapters, including challenges in rehabilitation. The challenges of mental health professionals in rural areas are probably being highlighted for the first time, in this book. The chapter on the future of rural living and impact on mental health does not only summarize the key messages conveyed by the chapters in this volume but also provides guidelines and steps for better mental health care delivery in rural areas in future.

This volume aims to be a valuable addition to the literature on rural mental health, which would take forward research and development in this field and enthuse more professionals to dedicate services in rural areas. Prevention, promotion, and rehabilitation are as equally important for rural areas as the management of mental health problems. These appear complex, distant, and hazy, but with advancements in health care delivery and modernization, these should be possible. Would modernized villages still be considered rural is another debatable question. The success of this volume should not be measured by the number of views, downloads, and citations of this book or its chapters but from the inspiration it may provide for furthering the cause of services and research on rural mental health globally.

Psychiatric Rehabilitation   Santosh Kumar Chaturvedi
Services, Department of Psychiatry   MD, FRCPE, FRCPsych
National Institute of Mental Health and   Editor
Neurosciences
Bangalore, Karnataka, India
Department of Mental Health Education
National Institute of Mental Health and
Neurosciences
Bangalore, Karnataka, India

Contents

About the Series Editors

Professor Norman Sartorius, obtained his M.D. in Zagreb (Croatia). He specialized in neurology and psychiatry and subsequently obtained a Master's Degree and a Doctorate in Psychology (Ph.D.). He carried out clinical work and research and taught at graduate and postgraduate levels at the University of Zagreb, at the Institute of Psychiatry in London, at the University of Geneva, and elsewhere. Professor Sartorius joined the World Health Organization (WHO) in 1967 and soon assumed charge of the program of epidemiology and social psychiatry. In 1977, Professor Sartorius was appointed Director of the Division of Mental Health of WHO. He was the Principal Investigator of several major international studies on schizophrenia, depression, and of mental and neurological disorders. In 1993, Professor Sartorius was elected President of the World Psychiatric Association (WPA) and served as President-elect and then President until August 1999, after which he was elected President of the European Psychiatric Association. Professor Sartorius is currently the President of the Association for the Improvement of Mental Health Programmes, and he is a member of the Geneva Prize Foundation, having been its President from 2004 to 2008. Professor Sartorius holds professorial appointments at universities in different countries including China, UK, and USA.

Professor Sartorius has published more than 400 articles in scientific journals, authored or coauthored 12 books, and edited more than 80 others. He is the coeditor of three scientific journals and is a member of editorial and advisory boards of many scientific journals. Professor Sartorius is also a corresponding member and fellow of a large number of international organizations and advisory boards. He has several honorary doctorates and is a member of academies of science and of medicine in different countries. He speaks Croatian, English, French, German, Russian, and Spanish.

Dr. Ee Heok Kua is the Tan Geok Yin Professor of Psychiatry and Neuroscience at the National University of Singapore (NUS) and Senior Consultant Psychiatrist at the National University Hospital, Singapore.

He was trained as a doctor at the University of Malaya and received postgraduate training in psychiatry at Oxford University and geriatric psychiatry at Harvard University.

A member of the World Health Organization team for the global study of dementia, he is the previous Head of the Department of Psychological Medicine and Vice Dean, Faculty of Medicine, at NUS, and the Chief Executive Officer and Medical Director at the Institute of Mental Health, Singapore.

His research interest includes depression, dementia, and alcoholism, and he has written 23 books on psychiatry, aging, and addiction. A novel he wrote, *Listening to Letter from America,* is used in a module on anthropology at Harvard University.

The former President of the Pacific-Rim College of Psychiatrists and President of the Gerontological Society of Singapore, he was Editor of the *Singapore Medical Journal* and *Asia-Pacific Psychiatry* journal.

About the Editor

Dr. Santosh Kumar Chaturvedi, MD, FRCPE (Edinburgh, UK), FRCPsych (UK), is a Senior Professor of Psychiatry and former Dean of Behavioural Sciences at the National Institute of Mental Health and Neurosciences, Bangalore, India. He is Head of the Department of Mental Health Education and has been the Head of the Department of Psychiatry and Psychiatry Rehabilitation Services in the past.

Dr. Chaturvedi is an Honorary Member of the World Psychiatric Association and member of many international and national societies and associations. His main areas of research and work have been consultation-liaison psychiatry, chronic pain and somatization, neuropsychiatry, quality of life research, psycho-oncology and palliative care, and cultural psychiatry.

He has received national awards by the Indian Council of Medical Research on three occasions and awards by the Indian Psychiatric Society – DLN Murthy Rao Oration and Tilak Venkoba Rao Oration – and Poona Psychiatric Association on many occasions.

He is the Editor in Chief of the *Journal of Psychosocial Rehabilitation and Mental Health*, published by Springer Nature.

He has numerous publications in various international and national journals. He has edited and/or published 20 books, manuals, and journals.

For more information, visit https://santoshchaturvedi. in/.

Contributors

Preeti Pansari Agarwal Department of Psychiatry, NIMHANS, Bangalore, India

Hargun Ahluwalia Department of Clinical Psychology, National Institute of Mental Health and Neurosciences, Bangalore, Karnataka, India

Chittaranjan Andrade Department of Psychopharmacology, National Institute of Mental Health and Neurosciences, Bangalore, Karnataka, India

Abhilash Balakrishnan Department of Psychiatry, National Institute of Mental Health and Neurosciences, Bangalore, Karnataka, India

Chethan Basavarajappa Psychiatric Rehabilitation Services, Department of Psychiatry, National Institute of Mental Health and Neurosciences, Bangalore, Karnataka, India

Prakash B. Behere Department of Psychiatry, Jawaharlal Nehru Medical College, Wardha, Maharashtra, India

Aniruddh P. Behere Helen DeVos Children's Hospital, Michigan State University, Grand Rapids, MI, USA

Vishal Bhavsar Section of Women's Mental Health, Department of Health Services and Population Research, Institute of Psychiatry, Psychology and Neurosciences, De Crespigny Park, London, UK

Dinesh Bhugra Mental Health and Cultural Psychiatry, Institute of Psychiatry (KCL), London, UK

Department of Psychosis Studies, King's College London, London, UK

Prabha S. Chandra Department of Psychiatry, National Institute of Mental Health and Neurosciences, Bangalore, Karnataka, India

Santosh Kumar Chaturvedi Psychiatric Rehabilitation Services, Department of Psychiatry, National Institute of Mental Health and Neurosciences, Bangalore, Karnataka, India

Department of Mental Health Education, National Institute of Mental Health and Neurosciences, Bangalore, Karnataka, India

Madhuporna Dasgupta Department of Mental Health Education, National Institute of Mental Health and Neurosciences, Bangalore, Karnataka, India

Geetha Desai Department of Psychiatry, National Institute of Mental Health and Neurosciences, Bangalore, Karnataka, India

B. N. Gangadhar NIMHANS Integrated Centre for Yoga, Department of Psychiatry, National Institute of Mental Health and Neurosciences (NIMHANS), Bangalore, Karnataka, India

Shiv Dutt Gupta IIHMR University, Jaipur, India

Nutan Jain IIHMR University, Jaipur, India

Sujeet Jaydeokar Learning Disability, Neurodevelopmental Disorders, and Acquired Brain Injury Services, Centre for Autism, Neurodevelopmental Disorders and Intellectual Disability (CANDDID), Cheshire and Wirral Partnership NHS Foundation Trust, Chester, UK

Arun Kandasamy Centre for Addiction Medicine, Department of Psychiatry, National Institute of Mental Health and Neurosciences, Bangalore, Karnataka, India

Matthew Kelly Research Department of Clinical, Educational and Health Psychology, University College London, London, UK

Muhammad Zillur Rahman Khan Department of Psychiatry, Patuakhali Medical College, Patuakhali, Bangladesh

Ee Heok Kua Department of Psychological Medicine, National University of Singapore, Singapore, Singapore

Dhanuja Mahesh Ministry of Health, Nutrition and Indigenous Medicine, Colombo, Sri Lanka

Himanshu Mansharamani Department of Psychiatry, Jawaharlal Nehru Medical College, Wardha, Maharashtra, India

Meena Kolar Sridara Murthy Department of Mental Health Education, National Institute of Mental Health and Neurosciences, Bangalore, Karnataka, India

Abhinav Nahar Department of Psychiatry, National Institute of Mental Health and Neurosciences, Bangalore, Karnataka, India

Venkata Lakshmi Narasimha Centre for Addiction Medicine, Department of Psychiatry, National Institute of Mental Health and Neurosciences, Bangalore, Karnataka, India

Mahesh M. Odiyoor Learning Disability, Neurodevelopmental Disorders, and Acquired Brain Injury Services, Centre for Autism, Neurodevelopmental Disorders and Intellectual Disability (CANDDID), Cheshire and Wirral Partnership NHS Foundation Trust, Chester, UK

Samir Kumar Praharaj Department of Psychiatry, Kasturba Medical College, Manipal, Manipal Academy of Higher Education, Manipal, Karnataka, India

Shiva Shankar Reddy Department of Psychiatry, National Institute of Mental Health and Neurosciences, Bangalore, Karnataka, India

P. S. Reddy Department of Psychiatry, Queens University, Barrie, ON, Canada

Diana Ross Department of Psychiatry, NIMHANS, Bangalore, India

Eesha Sharma Department of Child and Adolescent Psychiatry, National Institute of Mental Health and Neurosciences, Bangalore, Karnataka, India

Nikhil Sharma South London and Maudsley NHS Foundation Trust, London, UK

Vimal Kumar Sharma University of Chester, Chester, UK
Cheshire and Wirral Partnership NHS Foundation Trust, Chester, UK

Thanapal Sivakumar Psychiatric Rehabilitation Services, Department of Psychiatry, National Institute of Mental Health and Neurosciences, Bangalore, Karnataka, India

Bettahalasoor Somashekar Adult Community Mental Health Services, Swanswell Point, Coventry and Warwickshire Partnership NHS Trust, Coventry, UK

Athula Sumathipala Research Institute for Primary Care and Health Sciences, Faculty of Medicine and Health Sciences, Keele University, Stoke on Trent, UK

Paola Tejada Universidad El Bosque, Bogotá, Colombia

Jagadisha Thirthalli Psychiatric Rehabilitation Services, Department of Psychiatry, National Institute of Mental Health and Neurosciences, Bangalore, Karnataka, India

Shivarama Varambally NIMHANS Integrated Centre for Yoga, Department of Psychiatry, National Institute of Mental Health and Neurosciences (NIMHANS), Bangalore, Karnataka, India

Antonio Ventriglio University of Foggia, Foggia, Italy

Hiranya Wijesundara Ministry of Health, Nutrition and Indigenous Medicine, Colombo, Sri Lanka

Balaji Wuntakal Solent NHS Trust, Langston Centre, St James Hospital, Portsmouth, UK

Richa Yadav Elmhurst Hospital Center, Ichan School of Medicine at Mount Sinai, New York, NY, USA

Rural Living and Mental Health: Concepts and Contexts

Eesha Sharma and Santosh Kumar Chaturvedi

Contents

Abstract

Urban and rural areas differ in epidemiological rates of physical and mental morbidity. Possibly, rural and urban habitats differentially impact health. It is important to decipher spatial, socioeconomic, and sociological components that differentiate urban and rural habitats, in order to study their etiopathological role in health-related morbidity. Operational definitions for rural areas are typically based on accessibility of healthcare, e.g., distance from the nearest health facility.

E. Sharma
Department of Child and Adolescent Psychiatry, National Institute of Mental Health and Neurosciences, Bangalore, Karnataka, India
e-mail: eesha.250@gmail.com

S. K. Chaturvedi (✉)
Psychiatric Rehabilitation Services, Department of Psychiatry, National Institute of Mental Health and Neurosciences, Bangalore, Karnataka, India

Department of Mental Health Education, National Institute of Mental Health and Neurosciences, Bangalore, Karnataka, India
e-mail: skchatur@gmail.com

© Springer Nature Singapore Pte Ltd. 2020
S. Chaturvedi (ed.), *Mental Health and Illness in the Rural World*, Mental Health and Illness Worldwide, https://doi.org/10.1007/978-981-10-2345-3_1

1

Rural living, however, is much more than mere physical accessibility of modern healthcare. Rural populations have unique environmental profile and lifestyle choices about occupation, social norms, beliefs about health, and expressions of illness and help-seeking behavior. This introduction to rural living and mental health looks at (a) conceptual issues around defining "rural" and "urban," (b) urban and rural characteristics and their association with health, and (c) implications of these findings on rural healthcare systems and policy.

Keywords
Rural · Urban · Mental health · Health services · Mental health policy

Introduction

Epidemiological studies have repeatedly shown differences in morbidity rates of mental health disorders across urban and rural settings. The urban-rural dichotomy has been associated with differences not only in prevalence of mental ill-health but also differences in beliefs and practices that affect mental ill-health, mental well-being, and help-seeking behavior. It is hypothesized that the transactional impact of environmental determinants on genetic and psychological structures underlies these differences. This chapter will touch upon conceptual clarifications about what constitutes "urban" and "rural" and their interactions and influences with mental health.

What Is "Rural"?

To understand the *impact* of rural, it is critical to have conceptual clarity on *what the rural entails*. Rural has usually been defined by the context and not by the construct. Lexicographers, sociologists, economists, demographers, and health scientists, among others, have studied the impacts of rural based on dichotomies that emerged from parameters of their interest, which do not always overlap. Perhaps the biggest debate among these varying conceptualizations is whether rural is a geographical or a social representation, i.e., is it about "where" people live or "how" people live? Rural is most often identified with a specific pattern of land use and population density. Historically, "cities" developed by the coming together of many people, living in an area for the purposes of business and governance. Important religious or educational institutions further characterized or led to the establishment of cities around them. Towns also served similar purposes but with lesser population density and possibly a lower engagement with governance and other matters of state interest. All areas outside of cities and towns, i.e., the countryside, came to be regarded as rural.

State and national governments have operational definitions for "urban"; and rural has been defined as that which is "not urban," i.e., any land area/population that does not meet the criteria for urban is considered rural. Definitions encompass different characteristics (Nicholson 2008):

- Spatial: population density, population numbers, and distance to urban centers
- Socioeconomic: principal employment in the area
- Sociological: subjective aspects and experiences of rurality/urbanicity

The criteria for urban/rural are country- and service-specific. Following from some of the national criteria – the population density identifying a rural area could vary from <150 persons per square kilometer in Canada (Directorate for Public Governance and Territorial Development 2011) to 999 persons per square kilometer in the United States (What is rural? [Online] 2016). However, actual population density in a rural area could be as low as one person per square kilometer. Rural areas have also been identified as those that do not come under a municipal board's urban development program, the limits of which are based on resource use and administrative feasibility. Another common characteristic of rural areas is that a majority of the local population works in the agricultural and allied fields such as fisheries, poultry and forest, or small-scale industry. In fact, the National Sample Survey Organisation in India defined urban and rural areas based on the proportion of population engaged in agriculture-related work (>75% = rural) (Concepts and Definitions Used in NSS [Online] 2001).

Rural is often linked with an agrarian community that is predominantly engaged in agriculture. In the last few decades, this association is changing due to mechanization and urbanization. With an increased usage of agricultural tools and scientific progress resulting in reduced requirement for manual labor, rural communities are shifting toward other occupations. Moreover, urban facilities are increasingly available in rural areas. So, even though population size and resource access stay "rural," there is an urbanization trend in consumption categories like transport services, media, and personal products. Further, several areas that were earlier classified rural are now being identified as towns on the basis of increased population density and employment characteristics, even though they may not have municipal bodies. The move to urban areas, even in developed economies, has been increasing exponentially from the taking away of rural land for construction and industrial purposes (Fraser 2006). Urbanization has progressed in several stops and starts resulting in rural areas with varying shades of urbanicity – rural area with high/moderate/low urban influence and highly rural/remote areas (Fraser 2006). Therefore, it is possibly more meaningful to talk about "degree" of urbanization rather than a categorical urban/rural classification. Even with an equivocal cost-benefit situation, there has been increasing migration and urbanization in the last several decades. It is estimated that over the next couple of decades, more than 70% of the world's population will be living in urban areas (Urban Population Growth 2018). In a developing country like India, more than 30% of the population was urban in 2014 and is likely to go up to at least 50% by 2050 (India: Urban Health Profile 2016). Moreover, almost 25% of the urban population in India resides in slums, where the negative consequences of urbanicity are probably the densest.

The "urbanism" versus "ruralism" debate has guided policy decisions the world over due to the obvious ecological and economic impacts of these ideologies. The idea of "new-urbanism" is committed to sustaining urban areas and developing cities, with industry-based high-density centers with "passive" urban residents as the main consumers (Azadi et al. 2012). On the other hand, "new-ruralism" is committed to sustaining rural areas, agriculture-based agro-ecosystems with low-density peripherals and "active" farmers as the main target group (Azadi et al. 2012). While there is no consensus about which of these two ideologies would serve the ever-changing complicated needs of humankind better, it appears that a co-existence of the two frameworks may be essential for socioeconomic and psychological well-being.

"Rural" Depends on the Context

The varying definitions of "urban," and consequently "rural," have important implications. For one, research findings and service planning cannot be directly transferred from one definition to another. Different definitions generate different proportions of rural populations. The definitions are useful in that they help define the population of interest for a given purpose in a given location at a given time. For instance, in the United Kingdom, the National Health Services categorize a person as rural if he/she lives $>/=1$ mile from a doctor or chemist (Rurality, controlled localities and the provision of pharmaceutical services by doctors – an explanation of the history 2010). Access to a health facility appears to be the most crucial determinant of "rurality" where healthcare is considered (Steinhaeuser et al. 2014). Absolute population size is also given consideration while planning rural health services. In India, services are classified by population size – a subcenter serves a population of 3000–5000, primary health center a population of 20,000–30,000, and a community health center a population of 80,000–120,000 (Indian Public Health Standards – Ministry of Health and Family Welfare, Government of India 2012). The healthcare staff in these centers is accordingly allocated per population requirements. Thus, in the context of healthcare, rural has largely been conceptualized from the service planning perspective, with the goal to ensure adequate coverage and access for every person to a health facility. This, of course, is not enough to understand the multi-dimensional details of rural health. It is also pertinent that service planning would have limited effectiveness if the nuances of rural health were ignored. In the following sections, we discuss social and environmental characteristics of rural living that are pertinent to a reconceptualization of rural health services.

Rural Living: A Social Construct and an Environmental Construct (Table 1)

Understanding rural is to appreciate rural living and lifestyle. When one imagines "city life," one is immediately swarmed by images of dazzling lights, traffic, and "no time." This is in contrast to images of rural living characterized by a laissez faire attitude, large open spaces, and "a lot of time." Rural lifestyle, the social and environmental fabric of rural living, is very relevant in a discussion to understand

Table 1 What is rural? Frames of reference and relevance to mental health

Frame of reference	Description	Relevance to mental health
Historical	Countryside area, outside of cities and towns, which were centers of governance and trade	Self-identities of people as "rural" or "urban"
Spatial	Population density Absolute population size Distance to "urban centers"	High population densities (urban) associated with social adversities; low population densities (rural) associated with limited care access
Socioeconomic	Principal employment characteristics – Agriculture-related Areas outside of municipality's development programs	Unique stressors such as agrarian distress have direct impact on mental Well-being
Sociological	Subjective experiences of people residing in rural areas	Ethnographic influences that can impact on several facets of health beliefs and help-seeking behaviors
Health services	Rural = living more than specified distance away from health facility Understaffed health systems Governmental institutions complemented by non-government-run services and programs	Service planning and accessibility
Habitats	Large open green spaces Less night-time light Proximity to livestock and animals	Possibly influence health via direct neurobiological mechanisms
Lifestyle	Independence, self-reliance, creativity: Key problem-solving strategies Cohesion, stoicism self-sufficiency More physical activity Less "stress eating" Earlier assumption of adult roles and responsibilities Greater pro-environmental concern and behaviors	Affect risk for health problems and help-seeking behaviors
Population profile	Higher proportion of elderly people Low income Low cost of living Low education level Self-employment Large family size Early age at marriage (multiple pregnancies, health problems) Low divorce rates	Affect risk for health problems, help-seeking behaviors, and service planning
Health perspectives	Health = ability to work Low risk perception Low belief in safety of interventions Low need for preventive healthcare Stoic and fatalistic acceptance of the inevitability of death	Affect risk for health problems, help-seeking behaviors, and service planning

(continued)

Table 1 (continued)

Frame of reference	Description	Relevance to mental health
	Stigma about mental illnesses Cultural idioms of distress Folk illnesses Belief in traditional systems of medicine	

rural perspectives on health. A larger proportion of elderly individuals, lower mean levels of education, lower mean levels of income, and a larger proportion of self-employed individuals characterize socio-demographics in rural areas. In addition, there are a lower cost of living, a larger family size, and a lower divorce rate. In essence, rural populations illustrate simple community living and rely heavily on shared resources, knowing each other well, and having a large social support network. Independence, self-reliance, and creativity are highly valued when it comes to handling mundane "life crises" and health issues (Sav et al. 2015). From a sociological perspective, rural attributes include community cohesion, stoicism, and self-sufficiency (Nicholson 2008). Closeness to natural habitats in rural areas also results in closer engagement with animals, especially livestock. It is well known that pets and some animal species can positively impact mental health. This realm of enquiry has generated interest recently with the suggestion that closeness to animal life in rural areas could buffer the effects of trauma and emotional conflicts (Glass et al. 2014).

Lifestyle characteristics in rural areas extend to factors affecting health such as eating habits and physical exercise. A study from Poland on 600 18-year-olds found that greater amount of physical activity and less "stress eating" characterized rural youth (Suliburska et al. 2012). Interestingly, food consumption did not differ across urban and rural youth, pushing the authors to ponder if urbanization and media exposure lead to similarities in food choices and prevalence of disorders like obesity that was not different between urban and rural youth in this study. Similar findings were echoed by Regis et al. (2016) in their study on more than 6000 14–19-year-olds. They found that rural youth not only had better activity levels but were also less engaged with television and computer/video games. Rural living also differs when it comes to family life and choices. There is a tendency for girls to marry early, earlier assumption of adult roles and responsibilities, multiple pregnancies and health concerns thereof, quite often accompanied by attendant negative consequences (Coward 2006).

Urbanicity as a Risk Factor for Physical and Mental Health

There has been a flurry of literature on the detrimental effects of urban living on physical and mental health. There has been interest in identifying characteristics of the urban environment that could influence mental health. Urban and rural environments differ in their physical features. Natural habitats, more characteristic of rural areas, could influence health and well-being by several mechanisms – better quality air and water, more opportunities for engaging with nature, more opportunities for social interaction, and the visual and aromatic effects of natural surroundings (Ulrich 1999). Long-term exposure to urban environment has been associated with sub-clinical airway narrowing and reduction in lung vital capacity (Priftis et al. 2007). Exposure to night-time light has also been studied as an important physical determinant of mental health by virtue of its potential impact on circadian rhythms (Bedrosian and Nelson 2013) that could have a causal role in the development of mood and anxiety disorders, as also other non-communicable diseases (Fonken and Nelson 2011).

The exact nature of and the mechanistic bases for the influence of physical characteristics of urban environments on health are challenging to assess. Some interesting work in recent decades has looked at the attitudes of urban and rural residents toward the environment. A Spanish study found that while urban residents expressed concern for the environment, they reported lesser pro-environmental behaviors (Berenguer et al. 2005). Further, the pro-environmental concern and behavior may be modified by the socio-occupational involvements of people, with agriculturists scoring higher than rural residents employed in other kinds of work (Freudenburg 1991). Related to these findings is the interesting concept of *ecological grief*, which refers to mental health disturbances arising from ecological loss such as climate change, loss of valued species, and ecosystems (Cunsolo and Ellis 2018). It is possible that a combination of physiological and psychological mechanisms plays a role in the impact of ecological change on mental health and, possibly, vice versa.

Social stress and exposure to social adversity, which are also constituents of urban living, are more amenable to investigation, in terms of their possible impact on psychological and social well-being and health. Neighborhood characteristics (vandalism, litter or trash, vacant housing, teenagers hanging out, burglary, drug selling, and robbery) affect perceptions of social organization, support, and cohesion, and these were inversely associated with depressive symptoms at a 9-month follow-up in a community sample of 800 drug users, even after controlling for baseline depressive symptoms (Latkin and Curry 2003). A stratified, multistage, probability-based sample survey on >3000 adults as part of the Chicago Community Adult Health Survey systematically assessed neighborhood characteristics on the following measures – perceived neighborhood disorder and violence, reciprocal exchange, social ties, and social cohesion (Mair et al. 2010). Less physical disorder and decay, lower perceived violence and disorder, greater reciprocal exchange and social cohesion, greater proportions of married couples/fewer single mothers, and more residential stability were associated with lower levels of depressive symptoms. An interesting observation was that individual perception played a seemingly more important role than actual "neighborhood" effects, suggesting that neighborhood stressors probably operate through individuals' perceptions of their environments.

It is evident from these research findings that while land development may improve resource availability and service access in urban areas, it compromises exposure to natural habitats and increases negative social consequences such as poverty, overcrowding, pollution, health problems (Urban Habitats | National Geographic 2017), and greater social stress. These findings also reiterate the point that understanding "rural" must take into account the gamut of environmental and social constructs of rural living and not merely demographical administrative divisions.

A recurrent phenomenon that has been seen in rural areas is *agrarian distress*. With the changing face of rural living, there is an overall increase in stress levels of agrarian communities. Agrarian communities have been facing losses on multiple accounts – loss of agricultural lands from urbanization, loss of wages from mechanization, loss of crop yields from erratic weather patterns, loss of earnings from low crop yield and debts, and even loss of family structure from members migrating to cities. Farmers land up in a state of heightened stress from the chronic nature, and psychological impact, of these losses. There have been several reports of farmer suicides from both developing (Kurian 2007) and developed (Brumby et al. 2011) countries. Thus, rural should not by default be equated with protective factors. The characteristics of the rural and the urban have to be understood in the context of the scope of the health/mental health enquiry. Subjective processing and experience of stressors would play a greater role in understanding the mechanistic links to mental health.

Rural Perspectives on Health and Mental Health (Table 1)

In the rural areas, health is often equated with the "ability to work" (Weinert and Long 1987). In comparison to urban communities, death is viewed as more neutral, beneficent, and preventive healthcare is not prioritized (Gessert et al. 2015). Traditional and *natural*, simple practices – eat well, sleep well, work hard, exercise, stay busy, meet people, trust God, availability of wide open spaces, sense of safety and freedom – are seen as key to maintaining a good health status (Gessert et al. 2015). Delays in help-seeking may be compounded by a low risk perception and low confidence in the utility of safety interventions along with a stoic and fatalistic acceptance of the inevitability of death (Gessert et al. 2015). Rates of help-seeking lower further where mental health issues are concerned. Common mental health disorders like depression and anxiety are most often kept hidden and considered issues that other's needn't be troubled with (Johnson et al. 2011).

Several factors could affect the urban-rural differences in mental health including socioeconomic status of the population studied, cultural differences, knowledge, attitude, and practices related to mental health. Low awareness and help-seeking make mental ill-health a highly stigmatized subject. Small communities preclude the possibility of confidentiality and anonymity and the perceived self-stigma becomes a greater problem for the individual than the primary mental health problem. Governments all over the world have made efforts to improve service access for both physical and mental health needs. Setting up of a rural health network, mobile clinics, and even tele-health services are endeavors to bring health services to people's doorsteps. However, alongside the provision of needs and services lies the concept of "realized access" (Fraser 2006). Realized access is an indicator of actual service utilization and not merely the presence of services. Realized access is moderated by felt needs, attitudes toward health, ideas about possible benefits from service use, and, of course, the "potential access," i.e., actual provision and feasibility of accessing services. Evidently, efforts to increasing realized access must focus on understanding and addressing attitudes such as stoicism that could be perpetual barriers to help-seeking (Judd et al. 2006).

Understanding of bodily functions and disease influences people's experience and reporting of ill-health. Several psychiatric presentations have been described as "culture-bound," i.e., certain symptom clusters are seen in specific cultures, e.g., *dhat syndrome* in India, *jhum-jhum syndrome* in Nepal and parts of North India (Desai and Chaturvedi 2017), *koro-like syndromes* in West Africa (Aina and Morakinyo 2011), and *ataques de nervios* in Latin America (Nogueira et al. 2015). *Dhat syndrome* is a preoccupation with beliefs about semen discharge in urine, which the sufferer interprets as harmful by way of causing "weakness" and impotence. *Jhum-jhum syndrome* manifests with experience of tingling and numbness in limbs without any apparent neurological cause. Koro-like syndromes are characterized by perceptual, cognitive, and emotional disturbances arising from anxiety that the penis will recede into the body and may even cause death. Ataques de nervios are culturally acceptable responses to acute stressful experiences, especially the loss of loved ones, family conflict, or threat. These syndromes typically arise from misbeliefs about physiological processes and from cultural idioms of distress. These syndromes are a means of communicating distress. To the sufferer, acknowledgment of the "reality" of these experiences plays a huge role in future course of intervention and outcome. "Popular hidden illnesses" refers to a locally acceptable way of communicating an illness; for example, in several developing countries, depression presents with a greater emphasis on somatic symptoms rather than mood and cognition symptoms. "Folk illnesses" are specific patterns of presentations found in specific groups. They also respond typically to specific forms of folkloric treatments. These presentations are more culture-bound rather than purely a representation of the urban-rural divide. However, rural living is laden with traditional beliefs and practices that determine what "symptoms" people report and seek help for. Better awareness in urban areas reduces the occurrence of these presentations.

Even with ever-advancing modern medicine systems, people's faith and interest in traditional medicine systems persists. Chinese medicine, Ayurveda, and other systems have existed for centuries and usually have high acceptability rates, in view of the perspective that these treatments are naturally based and have fewer side effects. Unfortunately, scientific research into these practices has lagged far behind modern medicine. The World Health Organization (WHO) has taken cognizance of the scenario of traditional and complementary medicine practices in different countries and has come up with the WHO Traditional Medicine Strategy 2014–2023 (Traditional medicine 2009). The WHO sees merit in recognizing alternative modes of healthcare, understanding their overall role in health and well-being, and seeks to encourage safe and effective use through regulation, research, and integration into primary healthcare systems.

Urban-Rural Differences in Mental Health

Urbanicity, or the quality or fact of being urban, has been associated with a higher prevalence of mental health problems (Peen et al. 2010). The recently concluded National Mental Health Survey of India, 2016, found that the prevalence of mental health problems was greatest in the urban metros as compared to rural areas (Gururaj et al. 2016). This was true for both common and severe mental health disorders. Epidemiological studies (Paykel et al. 2000; Vassos et al. 2016) from different parts of the world echo these findings. A meta-analysis (Peen et al. 2010) of epidemiological studies also found that the urban-rural difference stays even after controlling for various confounders, such as socioeconomic status, diagnostic methods, diagnostic categories used, and socio-cultural background. Unlike other psychiatric disorders, substance-use disorders have a unique profile when comparing urban and rural areas. The prevalence of substance-related mental health problems is comparable across urban and rural areas (Gfroerer et al. 2007). The patient profiles show greater differences in the context of these disorders. Rural patients are younger, more likely to be educated and employed, have a younger age at onset of substance use, are less likely to have sought help for substance use, and are more likely to have alcohol as the primary substance used (The TEDS report: a comparison of rural and urban substance abuse treatment admissions 2012). Urban patients on the other hand are more likely to seek help voluntarily and are more likely than rural subjects to be homeless and unemployed and use drugs other than alcohol, such as injectable drugs (The TEDS report: a comparison of rural and urban substance abuse treatment admissions 2012). This suggests that rather than the urban-rural factor, it may be social adversity combined with biological risk profiles that perpetuate substance-use disorders.

The epidemiological differences in mental health in urban and rural environments are supported by studies looking at neurobiological differences. In a study on healthy volunteers from Germany, urban living and urban upbringing showed specific associations with fMRI activity in amygdala and perigenual anterior cingulate cortex (pACC) during social stress tasks (Lederbogen et al. 2011). Another study on healthy volunteers looked for structural differences in the pACC, and hippocampus. Urbanicity scores were calculated for each participant derived from place of residence (city/town/village) and years of stay till the age of 15; the first 15 years of life were considered since there is evidence to show that urban exposure during the developmental period may have a larger role to play than urban residence in adulthood. They found reduced gray matter volumes in the dorsolateral prefrontal cortex and pACC, in males>females, associated with urban upbringing (Haddad et al. 2015).

Urban-rural differences in mental health statistics have awoken much interest in the mechanistic underpinnings and identification of "risk" and "protective" factors in urban and rural areas, respectively. There are a few considerations, however, in the validity of these findings. The epidemiological differences exist for cases identified by available diagnostic/assessment tools. It was earlier discussed that the communication of distress and ill-health is influenced by the socio-cultural background, medical/biological models of health and disease, and traditional practices. The "caseness" and identification by epidemiological tools is, thus, called into question in rural areas that are more laden with socio-cultural symbolism.

Rural Mental Health Services

Rural areas are underserved where any kind of health services are concerned. Reports and publications from all parts of the world are also rampant with discussions on the shortage in mental health services in rural areas. As discussed earlier, healthcare systems are typically planned for population sizes, with limited attention to the local needs and scope of care. A "one size fits all" approach has been taken so far in medical teaching and planning of services. Challenges, thus, lie on both sides – the local population's felt needs and acceptability of a care system based on modern medicine practice and the health service providers' practical and psychological readiness to serve in a rural area.

Rural mental health services are frequently understaffed. Unfamiliarity of the rural area and lack of supportive resources at work are most often to blame for the reluctance health professionals show for serving in these areas. Small populations mean that social contact is high and the typical professional etiquettes relevant in the urban model – confidentiality and boundaries – are difficult to implement (Hastings and Cohn 2013; Policy Implications for Rural Mental health (A Report) [Online] 2016). Challenges in practice of psychiatry abound. Rural folk, especially in the developing world, are keener on being given directive advice rather than a reflective process of working through problems in psychotherapy, or on being an active participant in treatment selection. At several levels such as these, the training of the mental health professional falls short of addressing these disparate circumstances. Staff shortages also mean that the rural health professional has to be engaged not only in the medical evaluation and treatment but also administrative paperwork. Even with a much lower patient load, clinicians report a burnout from the atypical professional circumstances and little contact with colleagues. Besides, occupational and educational opportunities for family members of health professionals may be limited, resulting in the professional having to live away from family. It always helps to suit the professional's personality to the task at hand. Some find working in rural areas an opportunity. They appreciate working with "families rather than individuals," establishing their niche away from the competitive scenario of urban medicine, and being able to have a flexible work routine without the fast-paced lifestyle of urban areas (Hastings and Cohn 2013).

While state-run health services have not been able to adequately cover mental health needs in rural areas, NGOs and public-private partnerships have provided some solace (Kumar 2011). They run outreach services, including school mental health programs and short- and long-stay homes for the mentally ill. Several governmental and non-governmental centers have initiated novel service provision methods such as telemental health services for rural areas. These services include client-clinician interaction via use of video-conferencing or telephones, web-based interventions, and social media and group discussion forums. A systematic review found that there appears to be clear support for the use of these technology-based approaches to service planning, with benefits in terms of patient satisfaction, reduced cost of care, and flexibility of care access (Langarizadeh et al. 2017). With the support of local health workers, trained in primary health assessments and care, the urban models of collaborative, specialist care can be delivered to rural areas.

There has been concern from the clinicians'/therapists' perspective – privacy, data security, work overload, and novelty of the work scenario with unpredictable problems. Efforts are underway to organize these services with attention to these concerns (Ybarra and Eaton 2005).

The Way Forward: Need for a Rural Mental Health Policy

Rural mental health has been researched and talked about for several decades now. A dedicated *Journal of Rural Mental Health*, quarterly published by the American Psychological Association, since 1977, best exemplifies this (http://www.apa.org/pubs/journals/rmh/index.aspx). The changing structure, organization, population, and mental health needs in rural areas have posed a challenge for systematic research and, possibly, policy efforts. The majority of medical research is designed and carried out in urban areas, and rural is identified as that which is "not urban," i.e., if the research participant does not come from the city/town, they come from a rural area. As would be clear by now that this hardly represents the truth. The study of environmental influences goes beyond demographic descriptors. Novel ethnographic methods that take into consideration various environmental stressors, "stressor"-specific experiences of individuals, "respondent-driven sampling" to access minority/hidden populations, and, also, existing epidemiological definitions of urbanicity/rurality could lead to conceptualizing the "rural" as a comprehensive dimensional construct (Manning 2017).

On the service planning side, a rural mental health policy is needed, driven by healthcare needs of rural areas, and not merely extensions of health policies serving urban areas. Mental health meets more challenges, and runs a higher risk of being ignored altogether, in countries where infectious and nutritional diseases are still the priority health problems. Mental healthcare needs an independent policy, rather than just being latched upon primary healthcare services. The WHO's dictum "No health without mental health" reiterates this point. A comprehensive framework (Kelly et al. 2006) for rural mental health should be developed driven by principles such as (a) physical and psychological safety of health seekers and providers; (b) remote or on-site access to care and supervision; (c) appropriateness to local needs and existing healthcare practices and beliefs; (d) inclusion into mainstream healthcare services, so as to reduce stigma and acknowledge the importance and relevance of the global health concept wherein physical and mental health are complementary to each other; and (e) self-sustained by empowerment of local health workers.

Conclusion

Epidemiological research has identified differences in mental morbidity rates and help-seeking patterns between urban and rural areas. The etiopathological mechanisms that could help explain these differences, in terms of epigenetic and socio-ecological influences of the urban and rural environments, are unclear. More insights in this area could add greatly to our understanding of specific, operational, risk, and protective factors. However, the varying conceptualizations of what constitutes rural and what constitutes urban have mitigated progress in these endeavors, as research is typically limited to using spatial definitions. There is a need to approach this area of enquiry with a multi-dimensional framework that combines spatial, sociological, socioeconomic, and ethnographic characteristics of the rural.

References

Aina O, Morakinyo O (2011) Culture–bound syndromes and the neglect of cultural factors in psychopathologies among Africans. Afr J Psychiatry 14. https://doi.org/10.4314/ajpsy.v14i4.4

Azadi H, Van Acker V, Zarafshani K, Witlox F (2012) Food systems: New-Ruralism versus New-Urbanism. J Sci Food Agric 92:2224–2226. https://doi.org/10.1002/jsfa.5694

Bedrosian TA, Nelson RJ (2013) Influence of the modern light environment on mood. Mol Psychiatry 18:751–757. https://doi.org/10.1038/mp.2013.70

Berenguer J, Corraliza JA, Martín R (2005) Rural-urban differences in environmental concern, attitudes, and actions. Eur J Psychol Assess 21:128–138. https://doi.org/10.1027/1015-5759.21.2.128

Brumby S, Chandrasekara A, McCoombe S, Kremer P, Lewandowski P (2011) Farming fit? Dispelling the Australian agrarian myth. BMC Res Notes 4(89). https://doi.org/10.1186/1756-0500-4-89

Concepts and Definitions Used in NSS [Online] (2001) National Sample Survey Organisation. Ministry of Statistics & Programme Implementation, Government of India

Coward RT (2006) Rural Women's health: mental, behavioral, and physical issues. Springer, New York

Cunsolo A, Ellis NR (2018) Ecological grief as a mental health response to climate change-related loss. Nat Clim Chang 8:275–281. https://doi.org/10.1038/s41558-018-0092-2

Desai G, Chaturvedi S (2017) Idioms of distress. J Neurosci Rural Pract 8:94. https://doi.org/10.4103/jnrp.jnrp_235_17

Directorate for Public Governance and Territorial Development (2011) OECD Regional Typology [Online]

Fonken LK, Nelson RJ (2011) Illuminating the deleterious effects of light at night. F1000 Med Rep 3. https://doi.org/10.3410/M3-18

Fraser J (2006) Rural health: a literature review for the National Health Committee. National Health Committee, Wellington

Freudenburg WR (1991) Rural-urban differences in environmental concern: a closer look. Sociol Inq 61:167–198. https://doi.org/10.1111/j.1475-682X.1991.tb00274.x

Gessert C, Waring S, Bailey-Davis L, Conway P, Roberts M, VanWormer J (2015) Rural definition of health: a systematic literature review. BMC Public Health 15. https://doi.org/10.1186/s12889-015-1658-9

Gfroerer JC, Larson SL, Colliver JD (2007) Drug use patterns and trends in rural communities. J Rural Health 23:10–15. https://doi.org/10.1111/j.1748-0361.2007.00118.x

Glass N, Perrin NA, Kohli A, Remy MM (2014) Livestock/animal assets buffer the impact of conflict-related traumatic events on mental health symptoms for rural women. PLoS One 9:e111708. https://doi.org/10.1371/journal.pone.0111708

Gururaj G, Varghese M, Benegal V, Rao GN, Pathal K, Singh LK, Mehta RY, Ram D, Shibukumar TM, Kokane A, Lenin Singh RK, Chavan BS, Sharma P, Ramasubramanian C, Dalal PK, Saha PK, Deuri SP, Giri AK, Kavishvar AB, Sinha VK, Thavody J, Chatterji R, Akoijam BS, Das S, Kashyap A, Ragavan VS, Singh SK, Misra R, NMHS Collaborators Group (2016) National Mental Health Survey of India, 2015–16: summary report. National Institute of Mental Health and Neurosciences, Bengaluru

Haddad L, Schäfer A, Streit F, Lederbogen F, Grimm O, Wüst S, Deuschle M, Kirsch P, Tost H, Meyer-Lindenberg A (2015) Brain structure correlates of urban upbringing, an environmental risk factor for schizophrenia. Schizophr Bull 41:115–122. https://doi.org/10.1093/schbul/sbu072

Hastings SL, Cohn TJ (2013) Challenges and opportunities associated with rural mental health practice. J Rural Ment Health 37:37–49. https://doi.org/10.1037/rmh0000002

India: Urban Health Profile (2016) World Health Organization

Indian Public Health Standards – Ministry of Health and Family Welfare, Government of India [WWW Document] (2012). http://nhm.gov.in/nhm/nrhm/guidelines/indian-public-health-standards.html. Accessed 18 Oct 2018

Johnson IR, McDonnell C, O'Connell AM, Glynn LG (2011) Patient perspectives on health, health needs, and health care services in a rural Irish community: a qualitative study. Rural Remote Health 11:1659

Judd F, Jackson H, Komiti A, Murray G, Fraser C, Grieve A, Gomez R (2006) Help-seeking by rural residents for mental health problems: the importance of agrarian values. Aust N Z J Psychiatry 40:769–776. https://doi.org/10.1080/j.1440-1614.2006.01882.x

Kelly B, Tonna A, Paton M, Wolfenden K, Nielsen B, Hudson B, Lines K, Marshall C, Whitwell P (2006) The rural mental health emergency and critical care access plan. NSW Health, Sydney, Australia

Kumar A (2011) Mental health services in rural India: challenges and prospects. Health 3:757–761. https://doi.org/10.4236/health.2011.312126

Kurian NJ (2007) Widening economic & social disparities: implications for India. Indian J Med Res 126:374–380

Langarizadeh M, Tabatabaei MS, Tavakol K, Naghipour M, Rostami A, Moghbeli F (2017) Telemental health care, an effective alternative to conventional mental care: a systematic review. Acta Inform Med 25:240–246. https://doi.org/10.5455/aim.2017.25.240-246

Latkin CA, Curry AD (2003) Stressful neighborhoods and depression: a prospective study of the impact of neighborhood disorder. J Health Soc Behav 44:34–44

Lederbogen F, Kirsch P, Haddad L, Streit F, Tost H, Schuch P, Wüst S, Pruessner JC, Rietschel M, Deuschle M, Meyer-Lindenberg A (2011) City living and urban upbringing affect neural social stress processing in humans. Nature 474:498–501. https://doi.org/10.1038/nature10190

Mair C, Diez Roux AV, Morenoff JD (2010) Neighborhood stressors and social support as predictors of depressive symptoms in the Chicago community adult health study. Health Place 16:811–819. https://doi.org/10.1016/j.healthplace.2010.04.006

Manning N (2017) Mechanisms in urban mental health disorders

Nicholson LA (2008) Rural mental health. Adv Psychiatr Treat 14:302–311. https://doi.org/10.1192/apt.bp.107.005009

Nogueira BL, Mari J d J, Razzouk D (2015) Culture-bound syndromes in Spanish speaking Latin America: the case of Nervios, Susto and Ataques de Nervios. Arch Clin Psychiatry (São Paulo) 42:171–178. https://doi.org/10.1590/0101-60830000000070

Paykel ES, Abbott R, Jenkins R, Brugha TS, Meltzer H (2000) Urban-rural mental health differences in great Britain: findings from the national morbidity survey. Psychol Med 30:269–280

Peen J, Schoevers RA, Beekman AT, Dekker J (2010) The current status of urban-rural differences in psychiatric disorders. Acta Psychiatr Scand 121:84–93. https://doi.org/10.1111/j.1600-0447.2009.01438.x

Policy Implications for Rural Mental health (A Report) [Online] (2016) National Association for Rural Mental Health, Washington, DC

Priftis KN, Anthracopoulos MB, Paliatsos AG, Tzavelas G, Nikolaou-Papanagiotou A, Douridas P, Nicolaidou P, Mantzouranis E (2007) Different effects of urban and rural environments in the respiratory status of Greek schoolchildren. Respir Med 101:98–106. https://doi.org/10.1016/j.rmed.2006.04.008

Regis MF, de Oliveira LMFT, dos Santos ARM, da Leonidio ACR, Diniz PRB, de Freitas CMSM, Universidade de Pernambuco, Brazil, Centro Universitário Asces-Unita, Brazil, Faculdade Boa Viagem, Brazil, Universidade Federal de Pernambuco, Brazil (2016) Urban versus rural lifestyle in adolescents: associations between environment, physical activity levels and sedentary behavior. Einstein (São Paulo) 14:461–467. https://doi.org/10.1590/s1679-45082016ao3788

Rurality, controlled localities and the provision of pharmaceutical services by doctors – an explanation of the history (2010) National Health Services

Sav A, King MA, Kelly F, McMillan SS, Kendall E, Whitty JA, Wheeler AJ (2015) Self-management of chronic conditions in a rural and remote context. Aust J Prim Health 21:90–95. https://doi.org/10.1071/PY13084

Steinhaeuser J, Otto P, Goetz K, Szecsenyi J, Joos S (2014) Rural area in a European country from a health care point of view: an adaption of the rural ranking scale. BMC Health Serv Res 14. https://doi.org/10.1186/1472-6963-14-147

Suliburska J, Bogdański P, Pupek-Musialik D, Głód-Nawrocka M, Krauss H, Piątek J (2012) Analysis of lifestyle of young adults in the rural and urban areas. Ann Agric Environ Med 19:135–139

The TEDS report: a comparison of rural and urban substance abuse treatment admissions (2012) Center for Behavioral Health Statistics and Quality. Substance Use and Mental Health Services Administration, Rockville

Traditional medicine (2009) World Health Organization; Sixty-second World Health Assembly, Resolutions and decisions, annexes, Geneva

Ulrich RS (1999) Effects of gardens on health outcomes: theory and research. In: Healing gardens. Therapeutic benefits and design recommendations. Wiley, New York

Urban Habitats | National Geographic [WWW Document] (2017). https://www.nationalgeographic.com/environment/habitats/urban/. Accessed 18 Oct 2018

Urban Population Growth (2018) World Health Organization. Global Health Observatory

Vassos E, Agerbo E, Mors O, Pedersen CB (2016) Urban-rural differences in incidence rates of psychiatric disorders in Denmark. Br J Psychiatry J Ment Sci 208:435–440. https://doi.org/10.1192/bjp.bp.114.161091

Weinert C, Long KA (1987) Understanding the health care needs of rural families. Fam Relat 36:450. https://doi.org/10.2307/584499

What is rural? [Online] (2016) United States Department of Agriculture, national agricultural library. Rural Information Center, Beltsville

Ybarra ML, Eaton WW (2005) Internet-based mental health interventions. Ment Health Serv Res 7:75–87

Mental Health Aspects of Rural Living

2

Matthew Kelly, Antonio Ventriglio, and Dinesh Bhugra

Contents

Abstract

With increased globalization and movement of people, more people are moving to urban areas, but rural mental health remains of paramount importance due to specific issues and needs related to that. In this chapter, we will explore the mental health aspects of individuals living in rural settings. In this chapter, we describe the epidemiological rates of mental distress and psychiatric illnesses in rural communities. We also explore etiological differences in presentation in comparison to urban psychiatry. We will take a specific focus on rural farming communities who across the world face far higher rates of suicide than the general population. While looking at the factors that contribute this increased

M. Kelly (✉)
Research Department of Clinical, Educational and Health Psychology, University College London, London, UK
e-mail: mattkelly94@hotmail.com

A. Ventriglio
University of Foggia, Foggia, Italy

D. Bhugra
Mental Health and Cultural Psychiatry, Institute of Psychiatry (KCL), London, UK

Department of Psychosis Studies, King's College London, London, UK

© Springer Nature Singapore Pte Ltd. 2020
S. Chaturvedi (ed.), *Mental Health and Illness in the Rural World*, Mental Health and Illness Worldwide, https://doi.org/10.1007/978-981-10-2345-3_32

risk of suicidality, we will consider interventions that can help to reduce risk including social protective factors, healthcare, and policy changes and in doing so discuss the challenges and changes that need to be made to health provisions in rural settings to adequately provide mental health care for this at-risk population.

Keywords
Rural mental health · Suicidality · Suicidal behavior · Pesticides

Introduction

There is no doubt that in the past few decades there have been major migrations from rural to urban areas. The reasons for such a huge migration are many, and these can be related to globalization and its impact on increased industrialization in many parts of the world leading to increased urbanization. It is evident that rates of various (though not all) psychiatric disorders are higher in urban areas. It is also worth recognizing that virtually all of psychiatric services whether they are clinical or research activities are more likely to be established in urban areas. It is important to recognize differences between rural areas in different parts of the world. For example, rural areas in sub-Saharan Africa or in low-income countries will experience somewhat different issues when compared with rural Denmark or rural USA. Thus the epidemiological findings and certainly etiological factors may well differ. It is helpful to recognize that there are likely to be variation in rates of various psychiatric disorders between rural areas, urban conurbations, and urban areas. In this chapter, we describe some of these differences and illustrate these with examples of suicidality as well as predictive protective or contributory factors.

Differences Between Rural and Urban Mental Health

Within psychiatric literature, it is usually stated that urbanization correlates with higher prevalence for certain psychiatric disorders. When examining differences in British rural, semi-rural, and urban areas through psychiatric morbidity surveys, urban subjects were found to have higher rates of mental illness and substance misuse. However, interestingly when controlling for levels of social deprivation, the difference in prevalence was non-significant. The findings suggest the difference in urban-rural psychopathology may be largely attributable to the more adverse social environments associated with urbanization (Paykel et al. 2000).

However, it is when observing suicide rates along with the prevalence of depression, anxiety, and substance-use disorders and use of health professionals for mental health problems that a difference in epidemiology for psychiatric disorders arises. When looking at differences in Australian metropolitan, rural and remote areas researchers found higher suicide rates were evident for men, particularly young men in rural and remote populations compared with metropolitan. Interestingly in Australia, the proportion of young men who were reporting mental health disorders did not differ significantly between rural and remote areas compared with urban metropolitan areas. What did significantly differ between rural and urban areas was the proportion of young men with a mental health disorder from non-metropolitan areas were significantly less likely than those from metropolitan areas to seek professional help for a mental health disorder (Caldwell et al. 2004). When examining the mental health difficulties of rural dwellers, we must start by comparing the differences in etiology between rural and urban areas.

Suicide and Self-Harm

A key observation has been that various occupational groups whether in urban areas or in rural areas tend to have varying rates of psychiatric illnesses. Within the elevated risk of suicide for rural dwellers, it is well documented that in the profession of farming, rates of suicide are both far higher in incidence compared to the general population (rural and urban) and also carry the highest suicide rate in comparison to any other occupation. It is therefore unsurprising that suicide is the second most common cause of death in young farmers after accidents (Booth et al. 2000). The question which most researchers and clinicians in the area want to answer is what aspects of both rural and farming lifestyles contribute to this increased likelihood of suicidality.

One of the main reasons proposed in the literature is the lifestyle of a being unavoidably linked with their work and community, not only because of the long hours which may override one's work-life balance and increase psychological stress but also because the majority of farmers live on their farms. Geographical and social isolation is frequently cited as a major psychosocial risk factor affecting the overall health of rural dwellers and in particular farmers. There is also some evidence that due to the changes in the profession the social networks of farmers are in fact shrinking over time (Raine 1999). The literature notes that these social networks even when present can also be enmeshed, judgmental, exclusionary to outsiders, demanding, and difficult to withdraw from (Judd et al. 2006a).

An example of how harmful such societal attitudes can be for rural mental health can be seen in minority groups such as western rural lesbian, gay, bisexual, and transgender (LGBT) citizens. The risk of mental illness is already far greater for LGBT people in what is thought to due to "minority stress" compounded by repeated exposure to psychosocial stressors from anti-LGBT attitudes and behaviors such as discrimination, stigmatization, and on occasion violence and hate crime (Meyer 2003). Exposure to such stressors is likely to be exacerbated living in small rural communities. Interviews with providers of mental health-related care including psychology, social work, substance abuse counselors, and HIV-AIDS outreach workers serving rural New Mexico communities have shown both individual and institutional forms of anti-LGBT bias within health services. Consistent experiences in outpatient mental health clinics, residential treatment centers, and inpatient hospitals serving rural LGBT people showed that members of staff were untrained to deal with LGBT issues. This lack of awareness created a culture where therapists encouraged not disclosing sexuality or gender identity to the community and group facilitators prevented LGBT integration to group therapy for substance misuse by suppressing discussion of the sexuality and gender issues as not being directly relevant. As a result, patients discussed remaining socially isolated, depressed, reluctant to seek care locally, and not being able to afford to travel to urban areas for more specialized services (Willging et al. 2006).

Such inability to access rural specialized rural services could explain why rural trans communities show significant differences in mental health between rural and non-rural transmen and higher levels of drug use and unprotected sex among transwomen (Horvath et al. 2014). These findings demonstrate both how a lack of knowledge of LGBT mental health issues and on a wider level a lack of individualized patient-centered care for specific mental health needs in rural areas prevent access and openness to discussing and help-seeking for mental health issues. These then have wider implications not just on mental health but physical health and overall well-being.

In addition to social isolation, the typical day-to-day working life in rural areas can be grueling; few rural workers, in particular farmers, are in the position to rest, take holidays, or break away. This is due to having to find someone both willing and competent enough to take over responsibilities in their absence. Farmer's lives are governed by often unpredictable forces which they may not be able to control, such as weather, disease, and other issues with livestock, which can cause a great amount of uncertainty and stress. While it should also be noted that farmers are unique as a group whose work is intimately tied with every aspect of their lives, the lives of their families, and often work that has spanned across several generations of ancestors (Gregoire 2002), such factors therefore play a major role in being provocative or protective factors.

The effect of being intimately tied to one's farming livelihood has been shown in research looking at the possibility of regional variation in farmer suicides within England and Wales. When examining suicide rates both by county and by type of farm no significant heterogeneity emerged. However, it did appear that farm size had a significant impact with data showing of those in the sample 92% of farmers who committed suicide had farms of 300 acres or less. This is consistent with other findings which suggest that farmers with smaller holdings suffer more stresses with fewer supports (Hawton et al. 1999). This observation has major implications for creating public health approaches.

When work becomes so bidirectionally tied to one's way of life, it only takes the small setbacks in work or health and well-being to have a large impact on the other. There is a high prevalence of physical ill-health within the farming community with research suggesting around one-third of the farmers having physical health problems that interfered with their work. Unsurprisingly, as in many other groups, physical health problems were more likely to be a problem for older farmers, but surprisingly, it also affected a quarter of farmers under the age of 50. The most common symptoms described were back pain and arthritis. These physical health concerns are clearly linked to the farming lifestyle both due to being occupational-related injury and because of the large impact physical disability could have on work (Hawton 1998).

When investigating risk factors present in British farmers prior to suicide, as expected, the most common single factor was the presence of mental health problems which were reported in 82% of farmer suicides. In most cases, there was evidence of a previous depressive illness with one-third receiving treatment with antidepressants. While two-thirds of the sample had seen their general practitioner (GP) in the previous 3 months, most of the presentations had been with physical symptoms. It was noted by the researchers that farmers and their families often lacked knowledge about mental health problems and the ways in which their symptoms might be expressed (Hawton 1998). This is a significant observation which requires further detailed focused study.

This presentation of comorbid physical and mental health issues has also been found cross culturally with a higher rate of stress symptoms present in Canadian farmers compared with non-farmers, including not only psychological symptoms, such as tension, poor sleep, and irritability, but also higher rates of physical symptoms, in particular tiredness and back pain (Walker and Walker 1988). It has been suggested by some in the field that the reason for these findings may be that farmers are more likely to present to primary care with somatic or misinterpreted bodily symptoms rather than the usual psychological presentation of depression and anxiety (Booth et al. 2000). It is also entirely possible that people see the role of the doctor as dealing with physical illness hence their emphasis on somatic features in their presentation to the GP. It is also entirely possible that the GP may miss serious mental illness or investigate these unnecessarily. An additional way of describing the potential reasons for this comorbidity could be attributed to issues with isolation and stigma leaving mental health issues untreated leading to poorer physical health.

Interestingly in at least one study, researchers found no evidence to support the proposition that farmers experience higher rates of mental health problems than non-farmer rural residents. It makes sense that individual personality differences, gender, and community attitudes which limit an individual's ability to acknowledge or express mental health problems were the most significant risk factors for suicide in farmers. These differences included higher levels of conscientiousness among farmers and levels of neuroticism (which may drive help-seeking) being significantly lower. In addition, there appeared to be significant attitudinal barriers to seeking help for male farmers who may have mental health problems. Attitudes of being one's own boss and not needing support were prevalent in the community with farming success or failure seen to fall to the responsibility of the individual farmer (Judd et al. 2006b).

Prevention

When starting to introduce ways of preventing rural farmer suicides one must first look at the methods farmers chose when attempting to end one's life. When looking at methods of suicide used by farmers in developed countries, farmers differ markedly from the general population as they are much more likely to use firearms (usually shotguns) and less likely to take drug overdoses or use car exhaust fumes in comparison to the general population (Booth et al. 2000; Hawton 1998).

While in developing farming communities such as India, pesticide ingestion and subsequent poisoning has been found the most common method of deliberate self-harm and suicide in both men and women (Chowdhury et al. 2007), pesticide suicide also accounts for 60% of suicides in China and 71% in Sri Lanka and is also a major concern in many other countries including Brazil, Trinidad, Malaysia, and Malawi. Studies from Asia suggest that pesticide suicides are impulsive acts, undertaken by individuals during stressful life events, and majority do not suffer from prediagnosed mental disorders at the time (Vijaykumar 2007). Data from Chinese psychological autopsy designs have found that even after controlling for social and psychiatric risk factors, such as education level, living situation, marital status, income, and mental disorder, increased risk of suicide was most significantly accounted for by direct access to insecticides specifically over any other types of pesticides (Kong and Zhang 2009). Organophosphorus compounds are readily available as insecticide pesticides and are among the most common poisons used with Indian reports estimating the rates of poisoning as suicidal method ranging from 20.6% (10.3% organophosphorus) to 56.3% (43.8% organo-phosphorus). With regard to lethality, organophosphates have neurotoxic effects so stands to reason they are an extremely lethal method when ingested leading to respiratory failure and death if taken in a significant amount (Kar 2006).

With suicide methods varying in their lethality, it is vital to understand the relationship between the suicidal ideations, impulsivity, attempts, and suicide completions. Shotguns, for example, are statistically a much more reliably lethal method of suicide than a drug overdose. The use of more lethal methods may in part explain the higher rate of successful suicide attempts among farmers and rural communities across the world. This possibility is also supported by the oppositely low rate of non-fatal deliberate self-harm reported among farmers in comparison to the general population. This explanation in itself would suggest that suicide among farmers differs and it is more likely to result from impulsive acts which are more violent, lethal, and less planned than suicide in the general population (Lewis et al. 1997). There is an additional factor related to rates of suicide which is gender. Men are much more likely to use violent methods for committing suicide and are also less likely to attempt self-harm.

Yet another complicating factor is alcohol; again, men are much more likely to use alcohol (Grant et al. 2004), thereby increasing the likelihood of completing suicide. Alcohol abuse and dependence prevalent in the last 6 months along with current drug abuse and higher levels of impulsivity and aggression have all been found to be associated with higher rates of suicide in males (Dumais et al. 2005). Such findings add to the complexity of comorbid substance issues that may contribute to increased suicidality. Over the last decade, research has shown that while among all ages substance abuse is slightly lower in rural areas than in urban areas, what was not known was how substance use varied across different levels of rurality. More recent research suggests rural American youth and young people have higher rates of alcohol and illicit drug use. Type of drug use also varies;

while marijuana use is higher in urban areas, those living in the most rural areas have nearly twice the rate of methamphetamine use as urban young adults. Individuals were also more likely than urban youth to have engaged in the high-risk behaviors such as driving and other self-destructive and harming behavior while under the influence of alcohol or other illicit drugs (Lambert et al. 2008). By hypothesizing these risk factors being responsible for higher suicide rates in rural areas, we can begin to place preventative methods to reduce this impulsive immediate access to lethal methods.

In general, with regard to suicidal intent, a reduction in access to a particular means of suicide usually leads to an increase in the use of alternative methods but to an overall decrease in rates. In many farming communities in particular in the US, shotguns are a daily tool and access to this method is largely unregulated and readily available (Gunderson et al. 1993). In 1989, the political fallout from Dunblane massacre saw large changes to the law in England and Wales covering the registration, ownership, and storage of shotguns. This led to a marked decrease in the use of shotguns as a means of suicide among farmers. As a result, the use of shotguns has now been overtaken by hanging as the principal suicide method. Hanging, while still serious, carries less chance of mortality than the use of a firearm. This is a perfect example of the influence that reducing access to means has on impulsive suicidal behavior and suicide rates (Cummings et al. 1997) as has been shown in reducing access to pesticides.

Rates of alcohol dependence and abuse are also like in western rural cultures relatively high in Indian farming communities with around 30–50% of male suicides thought to be under the influence of alcohol at the time of suicide; a huge impulsive factor driving suicide attempts also plays a role. While changes at policy levels influence rates of suicide as demonstrated in the case of gun regulation, policies can also have a negative impact. For example, attempts to make attempted suicide a punishable offense under Section 309 of the Indian Penal Code have been counterproductive. Rather than acting as a deterrent, it has prevented individuals seeking help for fear from both patients and healthcare departments of legal complications (Vijaykumar 2007). Looking at the storage of pesticides in most households, found to be mostly unsafe, and exploring the inadequate knowledge among the community concerning the adverse effects of pesticides have been more productive (Chowdhury et al. 2007). As with firearms, it has been reducing the availability of and access to pesticides and reducing alcohol as an impulsive factor in both availability and consumption in priority areas that has had the most impact on lowering suicide rates (Vijaykumar 2007). When implementing a community randomized controlled feasibility study using villages randomized to intervention sites and control sites, Vijaykumar demonstrated how using centralized storage facilities constructed with local government involvement could help lower suicide rates. While providing lockable storage boxes at government buildings only accessible under supervision at certain times in the day, household surveys were conducted at baseline and one and a half years after intervention documented to provide sociodemographic data on pesticide usage, storage, and suicides. This method was chosen partly due to the level of underreporting of suicide attempts to government as discussed above. Results found 248 households utilized the storage facility with most reporting they found the storage useful, safe, and convenient. While the control group had less suicides at baseline, at follow-up, there was a significant drop in suicides in the intervention arm of trial with more deaths in the control site observed during the follow-up period. Importantly, anyone who had attempted or completed suicide in the intervention sites had not utilized the central storage facility (Vijaykumar et al. 2013).

While reducing access to methods and putting restrictions in place to prevent impulsive self-harming suicidal behaviors, we can also look at protective factors which appear to act as a preventative variable for reducing suicide. Some of these factors can be existing factors already present in farming lifestyle, factors whose presence is perhaps underutilized or their importance unrecognized. When looking at farmers in Northumberland, UK, men appeared to be relatively protected by being married or by having some form of a confidant at home. Other protective factors identified in the Northumberland study included frequent leisure activities and having close friends or support network. Good physical health was a protective factor for both men and women (Paxton and Sutherland 2000).

In addition to the individual factors that may elevate a person's risk of suicide, collective and contextual factors must also be considered as possible contributors to the elevated rate of suicide among rural males. The literature describes a number of factors including rural socioeconomic decline leading to increased stress and finical uncertainty, barriers to service utilization such as service availability and accessibility, and rural culture including both community attitudes to mental illness help-seeking and the accessibility to firearms (Judd et al. 2006a). Previous findings suggest that the presence of these community attitudes regarding stigma to mental health interventions and services has a much more negative impact on help-seeking behavior in rural areas than in urban areas (Rost et al. 1993). With contextual risk factors in mind, clinicians and service directors must consider how best to design treatment pathways that reduce the barriers preventing rural-based patients accessing mental health services.

Treatment and Pathways to Care

When trying to develop treatment pathways to reach rural communities, the obvious challenge at a primary care level in tackling rural mental health is firstly to encourage individuals to be knowledgeable about mental health and mental illnesses and then provide accessible services so that people can choose the right strategy for seeking help for mental health as well as for physical health problems. Not only would this involve training GPs and practice nurses to better recognize underlying mental health problems presenting as somatic symptoms but will also depend on patient knowledge to identify potential mental health symptoms (Booth et al. 2000). The patients will choose pathways depending upon what their explanatory models are and where they see the most appropriate and suitable help is available matching their models. These explanatory models are strongly influenced by socio-economic status and educational status.

At a broader community level and context, cultural factors within a small rural community however have been shown to likely work against this, affecting how mental illness is perceived and accepted and potentially preventing help-seeking. This effect is also more significant in smaller communities than in larger urban populations (Rost et al. 1993). In addition, the isolation of remote rural communities may reflect a closed social network which appears to produce a culture of self-reliance and stoicism toward health problems, thereby resulting in less likelihood in help-seeking behaviors. This means that mental healthcare professionals must be aware that residents of rural areas will frequently present with more severe symptoms as a result of seeking help later into the course of a mental disorder (Smalley et al. 2010). These social networks preventing help-seeking may also limit its members from taking up opportunities from outside the network that lead to a better work-life balance (Judd et al. 2006b) and thereby changing the expected and perhaps comfortable ways of living and functioning. There have been suggestions that governments provide specific practical social care support for farmers in areas of financial, retirement, housing, and retraining for those who wish to leave the farming profession. It is suggested that such policies would have the potential for reducing the hopelessness and reduce the need for increased mental healthcare support (Nicholson 2008).

To firstly counteract the lack of knowledge and negative social attitudes to mental health support and treatment in rural areas, low-level psychoeducation interventions have been proposed and implemented in certain areas. Targeted psychoeducation aims to increase awareness and educate the population about mental health problems and coping strategies in the hope of contributing to the de-stigmatization of mental illness and promoting help-seeking. Such materials have been suggested to take the form of articles in the farming media publications and easy read self-help materials distributed as leaflets and posters in appropriate locations. Another suggested intervention is the education of younger generations through programs on the syllabus in rural schools and that mental health services remain in close connections with schools. This is particularly appropriate given the multigenerational nature of farming. Many children will inherit farms from their parents and subsequently the responsibilities and stresses that come with a rural farming lifestyle (Gregoire 2002; Nicholson 2008; Smalley et al. 2010).

However, regardless of an improvement in attitudes to mental health, the issue of confidential access to care remains a huge issue in small rural communities. Especially as noted above, often, specialist services are likely to be based in urban areas. Everything from the location of mental health facilities being easily identifiable or a mental health worker being well-known or living locally (immersed in local community), this can cause further difficulties in patients seeking help. There is also often limited choice of healthcare professionals in rural areas with a single professional covering a large geographical area. A healthcare professional may be considered an integral part of the rural community, living and working in the same small area, and because of this, people may be embarrassed to admit to mental illness. This also causes obvious role conflicts and boundary issues with GPs, psychiatrists, nurses, and psychologists increasingly likely to bump into their patients outside of a clinical setting (Nicholson 2008). The literature often refers to this inability to hide mental health issues from a tight-knit community as "rural gossip networks." Qualitative research investigating rural gossip networks and social visibility within rural communities found that these observations only further heighten the experience of stigma and exclusion for young people seeking help for mental ill-health. Participants explained how these experiences negatively impacted on service utilization, both access and engagement, and hindered progress toward recovery. The results suggest a complexity of barriers to mental health service utilization for rural patients which must be further considered (Aisbett et al. 2007).

In addition to these attitude changes with rural communities on mental health, there is also a need for changing attitudes to firearm restriction. In most countries, farming communities are very resistant to any restrictions on their access to shotguns. Guns in many parts of the world are often perceived by most farmers as essential tools for their work and by some an essential part of their heritage or even an inalienable constitutional right as seen in the United States (Gunderson et al. 1993). It is noted for this reason formal restrictions may be difficult to implement. Informal methods of restricting access to firearms by farmers such as temporary voluntary restriction for those thought to be at risk such as adapting Vijaykumar's (2013) model of attempting to restrict pesticides in rural India could prove more promising. In order to implement a gun locker scheme, it would be vital to raise awareness with the direct support network of the individual at risk such as family members and primary healthcare workers who may oversee this safeguarding.

When considering treatment approaches, it cannot be underestimated the large geographical and logistical issues with running mental health services in rural areas. Overall, rural areas are less deprived than urban areas (Paykel et al. 2000) but within them exist distinct pockets of deprivation. Rural deprivation can often go unrecognized as a consequence of economically averaging out measures of deprivation over larger areas and population bases. Elements of cultural living can be missed, for example, car use is an essential in a rural area, whereas in urban areas, public transport can be relied on. Unemployment may appear to be less, but part-time, seasonal, or casual work with a friend is far more common with average earnings also being far less than in urban areas (Philo et al. 2003). These areas of deprivation including lack of independent transport, distance to travel a long way for treatment, and potential missing of employment impacting on uncertain wages all impact on service engagement (Nicholson 2008).

Suggestions to counteract travel barriers to engagement have with the improvement of technology and greater Internet connectivity in rural areas fallen to telemedicine interventions such as therapy and psychoeducation delivered via telephone and Internet/smartphone applications. Early comparison studies found anxious and/or depressed patients in primary care settings have shown greater improvement to telephone-based intervention than treatment-as-usual in large randomized control trials (Proudfoot et al. 2003). Investigations into modernized Internet-based applications have also been found to be effective in reducing both depressive symptoms and stigmatizing attitudes to depression and improving depression literacy in rural areas of Australia. The results indicate that mental health self-help and information programs can be delivered effectively by means of using the Internet as a platform, especially important in rural areas where accessibility of face-to-face mental health services is poor (Griffiths and Christensen 2007). More recent meta-analysis has found slightly favoring of telehealth interventions compared with face-to-face interventions but has not shown significant differences in outcomes. Even similar outcomes are encouraging however given the potential benefits of telehealth in rural and remote areas with regard to healthcare access, time, and cost savings. It also provides another means to accessing support discreetly, important where there is a strong culture of self-reliance and self-managing health problems among rural residents (Speyer et al. 2018).

Due to the above issues with addressing rural deprivation financially, it costs more to provide services in rural areas. Low population numbers make providing 24/7 cover and emergency or crisis services more expensive, and as a result, rural services remain relatively underfunded. Due to this underfunding, rural post can struggle with the recruitment and retention of staff (Nicholson 2008). Previous investigations into rural staffing levels have found historic understaffing with nearly three-fourths of counties in the US with populations between 2,500 and 20,000 lacking a psychiatrist and approximately half without a master's-level or doctoral-level social worker or psychologist (Holzer et al. 1998). These levels have remained low making this issue of chronic staff shortage one that has lasted decades (Thomas et al. 2009).

Mental health care professionals working in rural areas have extended services roles compared with their urban counterparts traveling large distances. Simple inconveniences such as transport expenses for staff and patients end up having both direct financial costs and indirect costs in terms of efficiency with travel time. In addition, clinicians can suffer professional isolation including the inability to attend centralized further training (Nicholson 2008). One suggestion to encourage chose to relocate and work in more rural areas is to extend the responsibility and opportunities available for those who chose to do so. One example is the extension of psychologists' role in rural US areas to the prescribing of certain psychotropic medications. Giving prescription privileges to psychologists has been the subject of heated debate in the field but has been introduced to address lack of psychiatrists in rural areas and is currently implemented in five US states: Iowa, Idaho, Illinois, New Mexico, and Louisiana (APA 2011). While there are concerns of the use of medication by professionals who do not have a medical training and the implications for protected psychology sessions over shorter medication consultations, arguments for point out that the level of psychopharmacology training in clinical psychology doctorates (and subsequent post docs) far outweighs what is present in the training of general practitioners. It is also argued that other allied healthcare professionals such as nurses already have prescription rights in some countries.

The move is seen by some as adding greater competence to the overall mental health system by adding an additional resource, especially for general practitioners in rural areas who need professional consultation regarding psychological disorders and the use of psychotropic medications alongside talking therapy (Jameson and Blank 2007).

Conclusion

Overall, it is clear from both high- and low-income countries that rural communities are uniquely at risk of suffering mental ill-health and an increased incidence of suicidality. This is particularly true for those in the farming profession. The interconnectivity with one's lifestyle and work, social isolation, lack of community mental health awareness and specific services, access to impulsive lethal methods, dispositional personality differences, and social stigma preventing help-seeking all contribute to this increased risk of suicide. Research has shown the interventions focusing on the restriction of methods such as firearms and pesticides along with increased awareness of mental health issues and the promotion of protective factors significantly lower these rates.

Government policy promoting the restriction of lethal methods and providing social care to farmers struggling in today's economic climate are vital to providing support to mental health workers treating this population. When developing rural mental health services, developers must consider the influence of rural culture on mental health service utilization. While community- and school-based interventions aimed at reducing the social stigma are positive steps, the co-location of mental health services and general health services is suggested as one way to reduce the fear associated with social visibility. With rural patients often presenting with physical symptoms, the integration of physical and mental health services could prove useful while also reducing costly and inefficient traveling of patients and staff to different geographical locations. Further exploring the use of telehealth interventions and future research should evaluate the efficacy and effectiveness of such programs in rural settings which could be hugely beneficial in both reducing social visibility of mental health services and costly travel for staff and patients.

What is evident is that clinicians tasked with working in rural areas must work proactively to undertake the role of potentially being the soul mental health advocate in their local area to both counteract the burden of stigma in rural areas and balance ethical dilemmas due to potential dual personal and clinical relationships with patients that occur in small communities. They must be adequately trained for the mental health difficulties that disproportionately affect rural communities including suicide, substance misuse, and worsened presentation due to delays in help-seeking. Working to further develop rural mental health services will involve a large amount of clinical creativity and versatility guiding the interaction with and referrals from physical health services and providing support despite a lack of local specialized services such as support for the farming occupation or rural LGBT communities. It will involve exploring a range of nontraditional treatment modalities such as the use of telehealth and potentially embracing roles usually taken by other members of the multidisciplinary team such as the prescription of medication and safeguarding of individuals against easy access to suicide methods.

References

Aisbett DL, Boyd CP, Francis KJ, Newnham K, Newnham K (2007) Understanding barriers to mental health service utilization for adolescents in rural Australia. Rural Remote Health 7(624):1–0

Booth N, Briscoe M, Powell R (2000) Suicide in the farming community: methods used and contact with health services. Occup Environ Med 57:642–644

Caldwell TM, Jorm AF, Dear KB (2004) Suicide and mental health in rural, remote and metropolitan areas in Australia. Med J Aust 181(7):S10

Chowdhury AN, Banerjee S, Brahma A, Weiss MG (2007) Pesticide practices and suicide among farmers of the Sundarban region in India. Food Nutr Bull 28(2 Suppl):S381–S391

Cummings P, Koepsell TD, Grossman DC, Savarino J, Thompson RS (1997) The association between the purchase of a handgun and homicide or suicide. Am J Public Health 87(6):974–978

Dumais A, Lesage AD, Alda M, Rouleau G, Dumont M, Chawky N, Roy M, Mann JJ, Benkelfat C, Turecki G (2005) Risk factors for suicide completion in major depression: a case–control study of impulsive and aggressive behaviors in men. Am J Psychiatr 162(11):2116–2124

Grant BF, Dawson DA, Stinson FS, Chou SP, Dufour MC, Pickering RP (2004) The 12-month prevalence and trends in DSM-IV alcohol abuse and dependence: United States, 1991–1992 and 2001–2002. Drug Alcohol Depend 74(3):223–234

Gregoire A (2002) The mental health of farmers. Occup Med 52(8):471–476

Griffiths KM, Christensen H (2007) Internet-based mental health programs: a powerful tool in the rural medical kit. Aust J Rural Health 15(2):81–87

Gunderson P, Donner D, Nashold R, Salkowicz L, Sperry S, Wittman B (1993) The epidemiology of suicide among farm residents or workers in five north-central states, 1980–1988. Am J Prev Med 9(3):26–32

Hawton K (1998) Suicide and stress in farmers. Stationery Office Books (TSO), London

Hawton K, Fagg J, Simkin S, Harriss L, Malmberg A, Smith D (1999) The geographical distribution of suicides in farmers in England and Wales. Soc Psychiatry Psychiatr Epidemiol 34(3):122–127

Horvath KJ, Iantaffi A, Swinburne-Romine R, Bockting W (2014) A comparison of mental health, substance use, and sexual risk behaviors between rural and non-rural transgender persons. J Homosex 61(8):1117–1130

Jameson JP, Blank MB (2007) The role of clinical psychology in rural mental health services: defining problems and developing solutions. Clin Psychol Sci Pract 14(3):283–298

Judd F, Cooper AM, Fraser C, Davis J (2006a) Rural suicide—people or place effects? Aust N Z J Psychiatry 40(3):208–216

Judd F, Jackson H, Fraser C, Murray G, Robins G, Komiti A (2006b) Understanding suicide in Australian farmers. Soc Psychiatry Psychiatr Epidemiol 41(1):1–0

Kar N (2006) Lethality of suicidal organophosphorus poisoning in an Indian population: exploring preventability. Ann General Psychiatry 5(1):17

Lambert D, Gale JA, Hartley D (2008) Substance abuse by youth and young adults in rural America. J Rural Health 24(3):221–228

Lewis G, Hawton K, Jones P (1997) Strategies for preventing suicide. Br J Psychiatry 171(4):351–354

Meyer IH (2003) Prejudice, social stress, and mental health in lesbian, gay, and bisexual populations: conceptual issues and research evidence. Psychol Bull 129(5):674

Nicholson LA (2008) Rural mental health. Adv Psychiatr Treat 14(4):302–311

Paxton R, Sutherland R (2000) Stress in farming communities: making best use of existing help. North Tyneside e Northumberland NHS Trust, Newcastle

Paykel ES, Abbott R, Jenkins R, Brugha TS, Meltzer H (2000) Urban–rural mental health differences in Great Britain: findings from the National Morbidity Survey. Psychol Med 30(2):269–280

Philo C, Parr H, Burns N (2003) Rural madness: a geographical reading and critique of the rural mental health literature. J Rural Stud 19(3):259–281

Proudfoot J, Goldberg D, Mann A, Everitt B, Marks I, Gray JA (2003) Computerized, interactive, multimedia cognitive-behavioural program for anxiety and depression in general practice. Psychol Med 33(2):217–227

Raine G (1999) Causes and effects of stress on farmers: a qualitative study. Health Educ J 58(3):259–270

Rost K, Smith GR, Taylor JL (1993) Rural-urban differences in stigma and the use of care for depressive disorders. J Rural Health 9(1):57–62

Smalley KB, Yancey CT, Warren JC, Naufel K, Ryan R, Pugh JL (2010) Rural mental health and psychological treatment: a review for practitioners. J Clin Psychol 66(5):479–489

Speyer R, Denman D, Wilkes-Gillan S, Chen YW, Bogaardt H, Kim JH, Heckathorn DE, Cordier R (2018) Effects of telehealth by allied health professionals and nurses in rural and remote areas: a systematic review and meta-analysis. J Rehabil Med 50(3):225–235

Thomas KC, Ellis AR, Konrad TR, Holzer CE, Morrissey JP (2009) County-level estimates of mental health professional shortage in the United States. Psychiatr Serv 60(10):1323–1328

Vijaykumar L (2007) Suicide and its prevention: the urgent need in India. Indian J Psychiatry 49(2):81

Walker JL, Walker LJ (1988) Self-reported stress symptoms in farmers. J Clin Psychol 44(1):10–16

Willging CE, Salvador M, Kano M (2006) Unequal treatment: mental health care for sexual and gender minority groups in a rural state. Psychiatr Serv 57(6):867–870

Common Mental Disorders and Folk Mental Illnesses

3

Abhilash Balakrishnan, Hargun Ahluwalia, and Geetha Desai

Contents

A. Balakrishnan · G. Desai (✉)
Department of Psychiatry, National Institute of Mental Health and Neurosciences, Bangalore, Karnataka, India
e-mail: abhi.gilli@gmail.com; desaigeetha@gmail.com

H. Ahluwalia
Department of Clinical Psychology, National Institute of Mental Health and Neurosciences, Bangalore, Karnataka, India
e-mail: gunn1802@gmail.com

© Springer Nature Singapore Pte Ltd. 2020
S. Chaturvedi (ed.), *Mental Health and Illness in the Rural World*, Mental Health and Illness Worldwide, https://doi.org/10.1007/978-981-10-2345-3_8

Abstract

Common mental disorders encompass spectrum of anxiety and depressive disorders. They are associated with significant disability and might not be recognized in the primary health settings. CMDs are associated with multiple social factors which influence the prevalence, presentation, and health seeking. Clinical presentation of CMDs varies across cultures, and it is necessary for health professionals to be aware of the myriad of presentations. Treatment of CMDs needs to incorporate the existing explanatory models in managing these common conditions.

Keywords

Common mental disorders · Anxiety · Depression · Folk illness

Common mental disorders (CMDs), a term introduced by Goldberg and Huxley (1992), includes depression, anxiety disorders and other neurotic disorders. CMDs are important for public health implications as they are often unrecognized in the primary care and medical settings. CMDs are especially important for their public health implications, since these often remain unrecognized in the primary care and medical settings, and are responsible for increased healthcare burden and costs (World Health Organization 2001). About one in six people (15.7%) was found to have CMDs. Although less disabling than the severe mental disorders, the higher prevalence implies that the cumulative cost of CMDs is very high.

Cross-national epidemiologic research has confirmed that while major depression and anxiety disorders occur worldwide, the symptomatic expression, interpretation, and social response to these syndromes vary widely. It is understood that illnesses arise in a context and may be characterized by symptoms unique to that context. In literature, there are several terms for such manifestations of CMDs – folk illnesses, culture-bound syndromes, popular hidden illnesses, etc. While the differences between these terms were often vague, an attempt has been made to elucidate some of these terms in the section below.

Terminology

Folk illnesses are set of symptoms that co-occur in a community, and the individuals respond in similarly patterned ways (Rubel 1964). A contextual meaning is made of the set of symptoms and the traditional healing practices that are adopted. An examination of folk illnesses involves a cultural construction of sickness and medicine among laypeople and folk healers. It employs folklore to look at the meaning of sickness and health (Rubel 1964).

Idioms of distress were alternate ways to express distress and are rooted in the person's unique and cultural meaning. They help the person cope when there is no other way to express the psychological distress. Thus, the somatic expression serves an important communicative function. Some examples of these are medicine-taking behavior, the use of diagnostic tests, and increased smoking (Nichter 2010).

Popular hidden illness is an ethnomedicine category and is parallel to professional disease. "Popular" implies that it is recognized in the community, and "hidden" implies that it can only be known through a culture-sensitive exploration and is not easily detectable. Popular hidden illness may be understood as an acceptable way of being ill in that society, and often help is sought from the traditional healing systems. In this way, idioms of distress, when they become common modes of presentation, manifest as popular hidden illness.

The term "cultural concept of distress" is a new addition to the *Diagnostic and Statistical Manual of Mental Disorders* (DSM) series with the publication of DSM-V (American Psychiatric Association 2013):

> Cultural concepts of distress refers to ways that cultural groups experience, understand, and communicate suffering, behavioral problems, or troubling thoughts and emotions.

The term is a recent advance in the history of attempts to categorize psychological distress with demonstrable cultural influence that lacks one-to-one unity with biomedical psychiatric diagnoses. Its predecessor "culture-bound syndromes" in the DSM-IV had several limitations such as lack of cohesive sets of symptoms for a syndromal diagnosis, similar presentations in different cultural settings, and high variability in the etiology, patient groups, and symptoms. The DSM-V recognizes that all forms of distress are shaped by context. The current notion of "cultural concept of distress" is an aggregate of several labels used to describe the role of culture on pathology (Kohrt et al. 2013).

In this chapter we would like to discuss CMDs, their cultural presentations, and other cultural idioms of distress. Social, cultural, and religious beliefs greatly influence the phenomenological experience of illness, the presentation of symptoms, as well as diagnosis and management (Sheehan and Kroll 1990).

Social Factors in CMDs

The Developed Versus the Developing World

One major point of difference between the developed and the developing countries is that in the latter physical health has greater perceived importance than mental health. This is in context of problems of high infant mortality, infectious disorders, and population growth that plague the developing world (Saxena et al. 2003). In developing countries large majorities of the population dwell in rural parts and are below the poverty line. A small portion of the annual budget is spent on health of which mental health is a much smaller portion. These are traditional societies, often highly religious, and there is a preponderance of traditional healing practices (Khandelwal et al. 2004).

Poverty and its consequences have been associated consistently with CMDs. In high-income countries (HIC), unemployment, low income, low socioeconomic status (SES), and low education have been associated with negative mental health outcomes (Lund et al. 2010).

In low- and middle-income countries (LMIC), reviews of 5, 11, and 115 epidemiological studies have found several indicators of poverty to be associated with CMDs (Lund et al. 2010; Patel et al. 1999; Patel and Kleinman 2003). Education is a factor that has been consistently and strongly associated with CMDs (Lund et al. 2010; Patel et al. 1997; Gupta et al. 2010). A strong association has also been found of lower SES with CMDs (Lund et al. 2010; Patel et al. 1999) and with depression specifically (Patel et al. 1997; Gupta et al. 2010). Factors such as rapid social change, violence, and insecurity have also been found to have strong association with CMDs (Lund et al. 2010), particularly among women (Patel et al. 1999; Patel and Kleinman 2003). Other measures of poverty such as income, employment, and consumption have been found to have weaker associations (Lund et al. 2010). Patel and Kleinman (2003) discuss that in vulnerable persons, poverty and CMDs interact to create a vicious cycle of one condition maintaining the other.

Rural Versus Urban Areas

In the developed world, social problems and environmental stressors have been looked upon as phenomena of urbanization affecting the cities and from which the rural parts are largely shielded. The high population density of major metropoles harbors greater crime rates, mortality, social isolation, and pollution. Poorer mental health too has been associated with urbanization.

In a meta-analysis of rural–urban difference studies in HIC spanning over 20 years, it was found that mental disorder was 38% higher, with anxiety disorder being 21% higher in urban areas than rural area (Peen et al. 2010).

In a meta-analytic study in India, greater prevalence in the urban sector (79%) compared to 37% in rural was found (Reddy and Chandrashekar 1998). In another study the urban rate of mental disorders was 1.57 times that of the rural rate in India. The rates of specific disorders of depression and anxiety were also lower in rural areas, albeit marginally. Hysteria was found to be significantly more common in urban than in rural areas. The author explains these results by discussing that due to constant flux of the population between villages, towns, and cities, it was difficult to differentiate rural from urban precisely. Also in collectivistic cultures like India, unlike the urbanization in the West, extensive traditional kinships and caste relations continue to thrive in urban life, and the social isolation and lack of social support is limited (Ganguli 2000).

Of importance to the present discussion is also the explanatory model of mental illness in rural areas. Traditional healers, who practice in villages, are often the first point of contact in the pathway of care for people with CMDs. Illness categories understood by psychiatrists are often recognized as being unusual by traditional healers, and there are often no equivalent terms for syndromes in regional languages. Traditional healers in a focus group study felt that the presentations in CMDs did not reflect illnesses but rather psychological difficulties arising from marital problems, poverty, alcoholism, having several female children, poor nutrition, and bewitchment by others in the community (Shankar et al. 2006). Violence, fears, bizarre behaviors, and possessions in rural South Indian community were attributed to disturbed interpersonal relationships, spirits, and magic, and the locus of control was often placed outside the sufferer (Thara et al. 1998). Even though somatic complaints were found to be the commonest presentation, only a minority of patients believed that their illness was purely physical, and a majority had a psychological illness attribution (Shankar et al. 2006; Patel et al. 1998).

Gender

Among the multiple social determinants of CMDs, the most consistently reported and robustly supported is being a female. Women are one and a half to two times more likely to suffer from CMDs as compared by to men. Gender globally, but more so in the LMIC, determines social position, access to resources, and in turn mental health outcomes. Practices such as dowry in India, wherein the girl's family provides money, land, and other gifts to the groom's family at the time of marriage, are a cause of domestic violence, intimate partner violence, and dowry deaths (Shidhaye and Patel 2010). In a large-scale rural study on married women in India, it was found that lower education, interpersonal violence, husband's unsatisfactory reaction to dowry, and spousal substance use were associated with CMDs (Shidhaye and Patel 2010). Similarly a cohort study in Pakistan showed that lower SES was three times more likely to have postnatal depression than those from higher SES (Rahman and Creed 2007; Shidhaye and Patel 2010). A study collating data from four societies of low- or middle-income category – India, Zimbabwe, Brazil, and Chile – found that women's multiple roles, social position, failure to produce male children, postnatal depression, and violence make women especially vulnerable to CMDs (Patel et al. 1999).

Depression

Epidemiology

Depression is the commonest psychiatric diagnoses in patients attending psychiatric clinics, psychiatric outpatient departments, or mental health facilities. The lifetime prevalence of unipolar depression is about 15% in males and 25% in females in the first world countries with similar prevalence in the developing countries. According to the WHO, about one in four consultations to healthcare providers is depression related. The 12-month prevalence in the United States is about 7% with maximum prevalence in the 18–29-year-old individuals with females having a 1.5- to 3-fold higher rate than males (Sadock and Sadock 2011).

Clinical Features

The commonly described clinical features include depressed mood, loss of interest and enjoyment, and reduced energy leading to increased fatigability and diminished activity. Other common symptoms are reduced concentration and attention, reduced self-esteem and self-confidence, ideas of guilt and unworthiness, bleak and pessimistic views of the future, ideas or acts of self-harm or suicide, disturbed sleep, and diminished appetite (World Health Organization 1992). Somatic symptoms or bodily complaints are present in many of the cases. The common bodily complaints are headache, body ache, and back pain, as well as feelings of weakness, fatigue, and palpitations. Lack of appetite is mild, and there may be periods of overeating. Sleep is disturbed in many cases.

From a cross-cultural standpoint, it is difficult to define depressed mood either because the terms describing internal emotional states are not available in the culture or the behavioral signs and symptoms that describe manifestation of the internal state are not condoned in the culture. Depression may be manifest in physical symptoms rather than psychological ones such as aches and pain, appetite, and sleep disturbances. One such cultural manifestation is *shenjing shuairuo*, which is common in the Chinese population. This condition is associated with lesser stigma than depression and is characterized by feelings of mental and physical fatigue, poor attention and concentration, and biological disturbances (Chaturvedi and Desai 2011).

Management

The initial line of management is to identify any physical illnesses that may be associated with or leading to the symptoms of depression (e.g., anemia, hypothyroidism, etc.). If the symptoms are mild, then adequate control of the physical illness plus therapy (low-intensity psychosocial measures) would be started. If it is moderate to severe or if the above does not work, then drug treatment and/or psychological therapies would be warranted. Psychological interventions including cognitive behavior therapy (CBT) or interpersonal therapy (IPT) are recommended. However the psychosocial interventions need to take into the cultural setting where it has been administered.

Among the medications, selective serotonin reuptake inhibitors (SSRIs) are generally preferred as the first choice due to lesser side effects as compared to other classes of medication as well as due to its efficacy. Among the SSRIs, for persons with medical comorbidities, sertraline or citalopram is preferred due to lower propensity for interactions. Tricyclic antidepressants (TCAs) can also be used but need to be used with caution in the elderly and in people with a high suicidal risk (due to the risk of overdose). Serotonin–norepinephrine reuptake inhibitors (SNRIs) like duloxetine and venlafaxine can also be used (Sadock and Sadock 2011; National Institute for Clinical Excellence 2004a).

Anxiety Disorders

Anxiety disorders include disorders that have features of excessive fear and related behavioral manifestations. Fear is an emotional response to real or perceived imminent threat, and anxiety is anticipation of future threat. While in response to a threat, anxiety is a universal phenomenon, when this anxiety causes dysfunction in daily life and distress to the person, greater evaluation may be deemed necessary. Persons with anxiety disorders typically overestimate the threat, however clinicians determine if the reaction is out of proportion by taking into cultural and contextual factors into account. The threat interpretations and responses vary widely across cultures. Significant differences have been reported in rates of anxiety disorders across different ethnocultural groups with Mexican Americans having high rates of simple phobias (Brown et al. 1990). Contrary to this, a cross national study involving 7 countries (United states, Canda, Puerto Rico, Germany, Taiwan, Korea and New Zealand) have found comparable rates of OCD (Weissman et al. 1998). A variety of culture-related forms of anxiety disorders also have been identified including *koro* in South and East Asia, semen-loss anxiety syndrome (*dhat* and *jiryan* in India, *sukra prameha* in Sri Lanka, *shen-k'uei* in China, *taijin kyofusho* in Japan), "nervous fatigue" syndromes, including *shinkeishitsu* in Japan, *brain fag* in Nigeria, and *shenjing shuairuo* in China. Cultural influences are apparent in the content and focus on anxiety disorders (Chaturvedi and Desai 2011; Kirmayer et al. 1995).

The current classificatory systems like DSM-V categorize anxiety disorders into separation anxiety disorder, selective mutism, specific phobia, social anxiety disorder (social phobia), panic disorder, agoraphobia, and generalized anxiety disorder. The first two are anxiety disorders found in children and will not be discussed in this chapter. Obsessive–compulsive disorder, which in previous versions of the DSM featured within the anxiety disorders, is categorized in the obsessive–compulsive and related disorders subsection of the DSM-V (American Psychiatric Association 2013).

Specific (Simple) Phobia

Epidemiology

Prevalence rates in the United States and Europe are similar (about 6%), but the rates are lower in Asian, African, and Latin American countries (2–4%). Gender rates vary with different phobic stimulus (e.g., animal, natural environment, and situational specific phobias are predominantly experienced by females, whereas blood-injection-injury phobia is experienced nearly equally by both genders). Median age of onset is between 7 and 11 years, more commonly seen in nonmedical mental health settings if there is no comorbidity (American Psychiatric Association 2013).

Clinical Features

In this disorder the fear or anxiety is circumscribed to the presence of a particular object or situation. For a diagnosis of specific phobia, the response must differ from those to normal, transient fears and must be excessive. The phobic object or situation is actively avoided or endured with intense fear or anxiety, which is out of proportion to the actual danger posed by the object or situation. For its diagnosis, the phobia needs to be persistent, typically lasting for more than 6 months. There are phobias confined to highly specific situations such as certain animals, heights, thunder, darkness, flying, closed spaces, dentistry, sight of blood and fear of exposure to specific diseases, and many others (American Psychiatric Association 2013).

Management

Treatment of specific phobias is mainly psychological, involving graded exposure to increasingly higher anxiety-inducing stimulus and simultaneous relaxation training. This technique of behavioral therapy is known as systematic desensitization. The graded exposure can be real, imagined, or virtual. While therapy is ongoing, if the anxiety is debilitating, a short course of beta-adrenergic receptor antagonists may be useful, especially when associated with panic attacks (Sadock and Sadock 2011).

Social Anxiety Disorder (Social Phobia)

Epidemiology

The 1-year prevalence of social phobia has been estimated to be 7% with a median age around 13 years in the United States. In the rest of the world, a lower prevalence is seen ranging from 0.5% to 2%. Women have higher rates of social anxiety than men in epidemiological studies, but rates have been found to be more or less equal in the clinical population or slightly higher in males (American Psychiatric Association 2013).

Clinical Features

In social phobia, inappropriate anxiety is experienced in situations in which the person is observed and could be scrutinized. They tend to avoid such situations in restaurants, canteens, dinner parties, seminars, board meetings, etc. Japanese form of social phobia, *taijin kyofusho*, provides an example of the interaction of cultural beliefs and practices with anxiety. In taijin kyofusho, the person experiences fear that one will offend or make others uncomfortable through inappropriate social behavior and self presentation (e.g., having a physical blemish or emitting offensive odor). This is in accordance with Japanese preoccupation with proper public presentation of self (American Psychiatric Association 2013; Chaturvedi and Desai 2011; Kirmayer et al. 1995).

Management

Psychological and/or pharmacological management is warranted. SSRIs are the first line (citalopram/sertraline) of treatment. If there is a partial response at 10–12 weeks, then CBT can be added. If there is inadequate response after 10–12 weeks, then a switch to another SSRI (fluvoxamine or paroxetine) or an SNRI (venlafaxine) may be deemed necessary. If there is still no response, then MAOI can be given (National Institute for Health and Care Excellence 2013).

Panic Disorder (PD)

Epidemiology

Lifetime prevalence of panic disorder is 1–4%. Median age of onset is 20–24 years. Rates are similar among Hispanics, blacks, and whites. Women are two to three times more likely than men to be affected. Morbidity and impairment of quality of life in PD are comparable to that of depression (American Psychiatric Association 2013; Sadock and Sadock 2011).

Clinical Features

A panic attack is an abrupt surge of intense fear or intense discomfort that reaches a peak within minutes, and during which time, four (or more) of the following symptoms occur: palpitations, sweating, tremors, sensation of choking, shortness of breath, chest pain, nausea or abdominal distress, light-headedness, chills or heat, paresthesias, derealization or depersonalization, fear of losing control or going crazy, and fear of dying. Between attacks, patients may have anticipatory anxiety about having another attack. Hyperventilation may produce respiratory alkalosis and other symptoms. Comorbidities are very common in PD; around 30–90% of patients with

PD have comorbid anxiety disorders, and around 50% have major depression (American Psychiatric Association 2013).

PD needs to be distinguished from medical conditions causing panic attacks such as hyperthyroidism, hyperparathyroidism, pheochromocytoma, vestibular dysfunctions, seizure disorders, and cardiopulmonary conditions. An onset after the age of 45 and atypical features may be suggestive of medical conditions. PD also needs to be differentiated from substance or medication-induced anxiety disorder such as intoxication with cocaine, amphetamines, or caffeine or withdrawal from alcohol and barbiturates. Other psychiatric disorders in which panic attacks are a feature need to be ruled out. Situational-bound panic attacks may indicate conditions like phobia, OCD, and depressive disorder (American Psychiatric Association 2013).

Management

Pharmacological and psychological (CBT) therapies are useful. Among the medications, SSRIs are the first line, and if there is no adequate improvement after 12 weeks, then another SSRI/other classes (TCA) can be considered (Sadock and Sadock 2011; National Institute for Clinical Excellence 2004b).

Agoraphobia

Epidemiology

The 1-year prevalence of agoraphobia without panic disorder varies between 1.7% and 3.8%, and the lifetime prevalence is about 6–10%. Females are twice more likely to develop agoraphobia than men. Prevalence rates do not vary significantly across cultural or racial groups (American Psychiatric Association 2013).

Clinical Features

Agoraphobic patients avoid situations where help is not easily available. The term agoraphobia includes fears not only of open spaces but also situations like crowded stores, closed spaces, busy streets and wherever there is a difficulty of immediate or easy escape to a safe place. It is one of the most incapacitating of phobic disorders. Two groups of symptoms are described in agoraphobics, panic attacks and anxious cognitions about fainting and going crazy. Severely affected individuals become completely house-bound, especially women. Most patients are less anxious when accompanied by a trusted person or a family member. Depressive symptoms, depersonalization and obsessional thoughts may also be present (American Psychiatric Association 2013).

Agoraphobia may be difficult to differentiate from situational specific phobia, but the former applies to more than one situation, while the latter is limited to one specific situation. Also if the situation is feared for reasons other than a panic attack or fear of losing control/dying, a diagnosis of specific phobia is more likely. Similarly in social phobia, the fear is of being negatively evaluated which differentiates it from agoraphobia. In major depressive disorder, the avoidance of activities and going out of the house is not related to panic attacks or other anxious cognitions, as it is in agoraphobia. Also, agoraphobia is not diagnosed if the avoidance of situations is a physiological consequence of a medical condition (American Psychiatric Association 2013).

Management

Agoraphobia is generally comorbid with painc disorder. Treatment principles are the same as that of panic disorder (National Institute for Clinical Excellence 2004b).

Generalized Anxiety Disorders (GAD)

Epidemiology

The 1-year prevalence of GAD ranges from 3% to 8% with median age of 30 years. Women are twice as likely as men to be affected by the disorder. Persons of European descent and in developed parts of the world are more likely to have a diagnosis of GAD (American Psychiatric Association 2013; Sadock and Sadock 2011).

Clinical Features

The main symptoms of GAD are worry and apprehension, free-floating anxiety, motor tension like restlessness, inability to relax, headache, aching of the back and shoulders and stiffness of the muscles; autonomic hyperactivity, experienced as sweating, palpitation, dry mouth, epigastric discomfort, and giddiness. The anxiety is excessive and interferes with other aspects of the person's life. These symptoms must be present for more days than not for a period of at least 3 months. Patients often complain of difficulty in concentrating, poor memory, and heightened sensitivity to noise. Disturbances in sleep may be present along with tiredness, depressive symptoms, obsessional symptoms, and depersonalization (Sadock and Sadock 2011).

Etiology

Biological factors include gamma-aminobutyric acid (GABA) and serotonergic systems. The basal ganglia, the limbic system, the occipital lobe, and the frontal cortex have been implicated. Genetic factors have also been implicated. Psychodynamic theory hypothesizes that anxiety is a symptom of unconscious, unresolved conflicts. According to cognitive behavioral school, persons with GAD respond to incorrectly perceived dangers (Sadock and Sadock 2011).

Differential Diagnosis

GAD needs to be differentiated from anxiety disorder due to a medical condition such as pheochromocytoma or hyperthyroidism. Substance- or medication-induced anxiety disorder, social anxiety disorder, obsessive–compulsive disorder, post-traumatic stress disorder, adjustment disorder, depression, bipolar disorder, and psychotic disorders need to be ruled out (American Psychiatric Association 2013).

Management

Among the medications, SSRIs are the first-line agents, and among SSRIs, sertraline is preferred. If sertraline is ineffective, then a switch to another SSRI/SNRI can be made. Buspirone, which is a 5-HT 1A receptor partial agonist, has been found to be effective with GAD, especially in reducing the cognitive symptoms. Drugs with short half-lives such as venlafaxine and paroxetine should be avoided because they can cause withdrawal syndromes. Benzodiazepines may be used with caution on a short-term basis, due to their addiction potential. The effectiveness and side effects of the drugs must be reviewed every 2–4 weeks during the first 3 months of treatment and every 3 months thereafter (National Institute for Clinical Excellence 2004b).

Obsessive–Compulsive Disorder (OCD)

Epidemiology

The lifetime prevalence of OCD in the general population is estimated at 2–3%, and the median age of onset is 19.5 years. It is the fourth most common psychiatric diagnosis after phobias, substance-related disorders, and major depressive disorder. Females have slightly higher rates of prevalence than males. Studies have found consistent rates of OCD in various countries and diverse countries (American Psychiatric Association 2013; Sadock and Sadock 2011).

Clinical Features

Obsessive–compulsive disorder is a debilitating syndrome characterized by obsessions and compulsions. Obsessions are recurrent and persistent thoughts, impulses, or images that are experienced as intrusive and inappropriate and cause marked anxiety and distress. Compulsions are repetitive behaviors or mental acts that the person feels forced to perform in response to an obsession.

The common obsessions are about contamination, doubt, bodily symptoms, need for symmetry, aggressiveness, religion, blasphemy, and sex. Common compulsions are checking, washing, counting, and needing to ask or confess symmetry and precision and hoarding (American Psychiatric Association 2013; Sadock and Sadock 2011).

In studies of OCD patients seen in psychiatric clinics of Saudi Arabia and Egypt, the most common themes of obsessions and compulsions were religious (Okasha et al. 1994). The symptomatology of OCD then involves repetition and internal struggle with forbidden thoughts as these engender the greatest anxiety for the individual. Preponderance of obsessions concerning dirt and contamination was seen commonly in Indians. Obsessions with aggressive content were infrequent. The Hindu code of ethics provides for a great variety of purification rituals (Chaturvedi and Desai 2011).

Management

Psychological (CBT) and/or pharmacological treatment is warranted. SSRIs should be the first line of treatment, among which fluoxetine, fluvoxamine, paroxetine, sertraline, or citalopram can be tried. If there is a partial response on medication-only regimen, then CBT should be added. If there is no/inadequate response after 12 weeks, then another SSRI/clomipramine should be tried. If the above does not work, then augmentation of SSRI/clomipramine with antipsychotic can be tried. If the patient has comorbid body dysmorphic disorder, fluoxetine can be tried as it has more evidence compared to other SSRIs (National Institute for Clinical Excellence 2004c).

Implications of the Role of Culture on CMDs

It is clear that manifestations of CMDs vary greatly across cultures. The explanatory models adopted in a culture to explain behaviors also determine the modes of help seeking that are employed. Symptoms of depression, anxiety, and somatic concerns represent a cry for help and need to be viewed in the idiosyncratic cultural lens in which they thrive.

The concept of "cultural idioms of distress" was introduced to draw attention (Nichter 1981, 2010; Chaturvedi et al. 1995) to the fact that reports of bodily distress can serve a communicative function, when other modes of expression fail to communicate distress adequately or provide appropriate coping strategies.

Throughout the world, psychiatry is moving from the institution to the community. Understanding the community perception, explanations, and beliefs has become the need of the hour. Misrepresentations of these unique expressions of pathology may lead to unnecessary diagnostic procedures or inappropriate treatment. Among professionals, there is a need to develop cultural competence and sensitivity so as to meaningfully interpret, formulate, and treat symptoms. Recognizing idioms of distress helps build alliance between the doctor and the patient and creates an empathic bond that is essential in psychological treatment.

References

American Psychiatric Association (2013) Diagnostic and statistical manual of mental disorders (DSM-5®). American Psychiatric Publishing, Arlington

Brown DR, Eaton WW, Sussman L (1990) Racial differences in prevalence of phobic disorders. J Nerv Ment Dis

Chaturvedi SK, Desai G (2011) Neurosis. In: Bhugra D, Bhui K (eds) Textbook of cultural psychiatry. Cambridge University Press, Cambridge

Chaturvedi SK, Chandra PS, Sudarshan CY, Issac MK (1995) Popular hidden illness among women related to vaginal discharge. Indian J Soc Psychiatry 11:69–72

Ganguli HC (2000) Epidemiological findings on prevalence of mental disorders in India. Indian J Psychiatry 42(1):14

Goldberg DP, Huxley P (1992) Common mental disorders: a bio-social model. Tavistock/Routledge, London

Gupta R, Dandu M, Packel L, Rutherford G, Leiter K, Phaladze N, Percy-de Korte F, Iacopino V, Weiser SD (2010) Depression and HIV in Botswana: a population-based study on gender-specific socioeconomic and behavioral correlates. PLoS One 5(12):e14252

Khandelwal SK, Jhingan HP, Ramesh S, Gupta RK, Srivastava VK (2004) India mental health country profile. Int Rev Psychiatry 16(1–2):126–141

Kirmayer LJ, Young A, Hayton BC (1995) The cultural context of anxiety disorders. Psychiatr Clin North Am 18(3):503–521

Kohrt BA, Rasmussen A, Kaiser BN, Haroz EE, Maharjan SM, Mutamba BB, de Jong JT, Hinton DE (2013) Cultural concepts of distress and psychiatric disorders: literature review and research recommendations for global mental health epidemiology. Int J Epidemiol 43(2):365–406

Lund C, Breen A, Flisher AJ, Kakuma R, Corrigall J, Joska JA, Swartz L, Patel V (2010) Poverty and common mental disorders in low and middle income countries: a systematic review. Soc Sci Med 71(3):517–528

National Institute for Clinical Excellence (2004a) Depression: management of depression in primary and secondary care. Clinical guideline, 23. NICE, London

National Institute for Clinical Excellence (2004b) Anxiety: management of anxiety (panic disorder, with or without agoraphobia, and generalised anxiety disorder) in adults in primary, secondary and community care. Clinical guideline, 22. NICE, London

National Institute for Clinical Excellence (2004c) Obsessive- compulsive disorder and body dysmorphic disorder: treatment. Clinical guideline, 31. NICE, London

National Institute for Health and Care Excellence (2013) Social anxiety disorder: recognition, assessment and treatment of social anxiety disorder. Clinical guideline, 159. British Psychological Society, Leicester. https://www.guidance.nice.org.uk/CG159

Nichter M (1981) Idioms of distress: alternatives in the expression of psychosocial distress: a case study from South India. Cult Med Psychiatry 5(4):379–408

Nichter M (2010) Idioms of distress revisited. Cult Med Psychiatry 34(2):401–416

Okasha A, Saad A, Khalil AH, El Dawla AS, Yehia N (1994) Phenomenology of obsessive-compulsive disorder: A transcultural study. Compr psychiat 35(3):191–197

Patel V, Kleinman A (2003) Poverty and common mental disorders in developing countries. Bull World Health Organ 81(8):609–615

Patel V, Todd C, Winston M, Gwanzura F, Simunyu E, Acuda W, Mann A (1997) Common mental disorders in primary care in Harare, Zimbabwe: associations and risk factors. Br J Psychiatry 171(1):60–64

Patel V, Pereira J, Mann AH (1998) Somatic and psychological models of common mental disorder in primary care in India. Psychol Med 28(1):135–143

Patel V, Araya R, de Lima M, Ludermir A, Todd C (1999) Women, poverty and common mental disorders in four restructuring societies. Soc Sci Med 49:1461

Peen J, Schoevers RA, Beekman AT, Dekker J (2010) The current status of urban-rural differences in psychiatric disorders. Acta Psychiatr Scand 121(2):84–93

Rahman A, Creed F (2007) Outcome of prenatal depression and risk factors associated with persistence in the first postnatal year: prospective study from Rawalpindi, Pakistan. J Affect Disord 100(1):115–121

Reddy VM, Chandrashekar CR (1998) Prevalence of mental and behavioural disorders in India: a meta-analysis. Indian J Psychiatry 40(2):149

Rubel AJ (1964) The epidemiology of a folk illness: Susto in Hispanic America. Ethnology 3(3):268–283

Sadock BJ, Sadock VA (2011) Kaplan and Sadock's synopsis of psychiatry: behavioral sciences/clinical psychiatry. Lippincott Williams & Wilkins, Philadelphia

Saxena S, Sharan P, Saraceno B (2003) Budget and financing of mental health services: baseline information on 89 countries from WHO's project atlas. J Ment Health Policy Econ 6(3):135–143

Shankar BR, Saravanan B, Jacob KS (2006) Explanatory models of common mental disorders among traditional healers and their patients in rural South India. Int J Soc Psychiatry 52(3):221–233

Sheehan W, Kroll J (1990) Psychiatric patients' belief in general health factors and sin as causes of mental illness. Am J Psychiatry 147:112

Shidhaye R, Patel V (2010) Association of socio-economic, gender and health factors with common mental disorders in women: a population-based study of 5703 married rural women in India. Int J Epidemiol 39(6):1510–1521

Thara R, Islam A, Padmavati R (1998) Beliefs about mental illness: a study of a rural South-Indian community. Int J Ment Health 27(3):70–85

Weissman MM (1998) Cross-national epidemiology of obsessive-compulsive disorder. CNS Spectrums 3(S1):6–9

World Health Organization (1992) The ICD-10 classification of mental and behavioural disorders. World Health Organization, Geneva

World Health Organization (2001) The world health report 2001 – mental health: new understanding, new hope. World Health Organization, Geneva

Somatoform and Dissociative Disorders in Rural Settings

4

Abhinav Nahar, Shiva Shankar Reddy, and Geetha Desai

Contents

Abstract

Somatoform disorders and dissociative disorders are common presentations in rural health settings. They have been described conceptually as a method of expression of emotional distress or problems. In the rural health settings, presentations of somatic complaints to the health professionals are common. These symptoms might be due to underlying depression or anxiety and might be more acceptable forms for seeking help. Lack of availability of mental health services and stigma of mental illness label are some of the barriers for management of these conditions. Integrated care might be a better option of management where the health professionals are trained to detect and treat the conditions in the cultural milieu.

Keywords

Somatoform · Dissociation · Rural · Culture

A. Nahar · S. S. Reddy · G. Desai (⊠)
Department of Psychiatry, National Institute of Mental Health and Neurosciences, Bangalore, Karnataka, India
e-mail: abhinavnahar2002@gmail.com; shivakmc55@gmail.com; desaigeetha@gmail.com

© Springer Nature Singapore Pte Ltd. 2020   47
S. Chaturvedi (ed.), *Mental Health and Illness in the Rural World*, Mental Health and Illness Worldwide, https://doi.org/10.1007/978-981-10-2345-3_7

Introduction

Somatoform disorders are characterized by conditions with predominant physical symptoms, not directly attributable to any general medical condition or related to use of any substance and not due to any underlying psychiatric disorder. These symptoms can be severe enough to cause significant distress and impairment. The disorders that were included in DSM-IV were somatization disorder, undifferentiated somatoform disorder, conversion disorder, pain disorder, hypochondriasis, body dysmorphic disorder, and somatoform disorder not otherwise specified (NOS). There existed significant overlap between these different disorders as they were categorized quantitatively based on the number of symptoms rather than qualitatively. In addition, the primary care physicians who most frequently encounter these patients had difficulty comprehending the various somatoform disorder diagnoses. In view of the above shortcomings, DSM-V grouped four conditions, namely, somatization disorder, hypochondriasis, pain disorder, and undifferentiated somatoform disorder under a new category, somatic symptom disorder. Other significant changes was, presence of a single symptom, rather than requirement of a cluster of symptoms required in somatization disorder which should be significantly distressful and be associated with abnormal thoughts, behaviors and feelings in response to the symptom(s). Also, the cause that the symptom need not be medically unexplainable was removed. A new rubric, illness anxiety disorder, was added to include patients with hypochondriasis with anxiety who do not have any symptom focus (Association and Association 2013).

The etiology for the origin of these symptoms includes both biological and psychosocial factors. These include genetic and environmental factors as observed to contribute to the increased incidence of the cases in females with relatives diagnosed with somatization disorder, the presence of neuroticism personality traits, and childhood adversity. Other factors include increased physiological activation probably resulting in increased chance of perception and subsequent misattribution of bodily sensations, altered cortisol levels, impairment in attention, and filtering process studied using EEG evoked potentials, individual differences in the attentional biases, making illness attributions, and lack of normalizing attributions (Rief and Barsky 2005; Deary et al. 2007).

Dissociation

Dissociative phenomena have been recognized in the history for a very long time. Till the end of the eighteenth century, spirit possession remained a dominant explanation for experiences of altered states of identity. The practice of exorcism of demons and evil spirits was the preferred treatment for such problems around this time. The term hysteria, derived from the Greek word hystera (signifying the uterus), dates back to the time of Hippocrates, when it was thought that the uterus became physically displaced from its normal position in the pelvis, wandering throughout the body to create symptoms in the various places that it inhabited. In 1697, the English physician Thomas Sydenham described hysteria as an emotional condition rather than as a physical disorder, attributing the source of the disorder to the central nervous system. In the nineteenth century, the French physician Paul Briquet (who first coined the term dissociation) used a syndromic approach defining hysteria operationally as a chronic disorder characterized by the presentation of many medically unexplained symptoms in the body's multiple organ systems (Allin et al. 2005).

Nosology

The dissociative disorders first appeared in the classificatory system in the first edition of the *Diagnostic and Statistical Manual of Mental Disorders* (DSM) of the American Psychiatric Association as "dissociative reaction" together with "conversion reaction" in a section for "psychoneurotic disorders" that also included anxiety (e.g., "anxiety hysteria") and depressive "reactions." Significant changes have been made in the subsequent versions of the DSM. In DSM-V, the diagnosis of depersonalization disorder was changed to "depersonalization/derealization disorder," and dissociative fugue was added as a specifier for the diagnosis of dissociative amnesia (North 2015).

Somatoform Disorders in the Rural Settings

Somatic symptoms are the most common expression of emotional distress across the world. Although considered to be more common in the non-Western population, there is no definitive evidence for the difference of prevalence between cultures due to lack of studies using comparable methodologies and due to lack of comparison groups in the studies. In the ECA study, the prevalence of somatization was found to be more in the African-American men and women compared to the overall population which may be due to difference in the educational status (Robins and Regier 1991). However, high rates of somatization and related disorders have been found in a study in a North American study across all ethnical and cultural groups which challenges the view that it is more common in non-Western or Asian population (Laurence J Kirmayer and Young 1998). In addition, a WHO cross-national study on mental disorders in primary care across 14 countries found no economical, geographic, and cultural factors attributing to increased rates of somatization in the two South American cities, namely, Santiago and Rio de Janeiro (Gureje et al. 1997).

In order to decrease the influence of socioeconomic factors while studying the cultural factors influencing somatization across various ethnocultural groups, a community survey was conducted in the multiethnic neighborhood of Montreal by providing equal access to health care to everyone irrespective of the socioeconomic strata. No significant differences in the rates of reporting somatic symptom between the immigrant groups and the Canadian-born groups were found after controlling for age, gender, employment, and education. However, between the five ethnocultural groups identified in the study, Vietnamese men reported somatic symptoms more than men and women in all other groups and more than Vietnamese women. Increasing age, female gender, unemployment, and certain ethnic groups, Vietnamese in this study, were associated with higher rates of somatic symptoms in the last 1 year (L J Kirmayer and Young 1996).

Although the classificatory systems emphasize distinction between the psychological and somatic systems, this distinction is blurred in certain ethnic and cultural groups. Hence, diagnostic syndromes, namely, culture-bound syndromes which include both groups of symptoms, have been popular and may not be recognized within the culture as an illness. These are mostly characterized by predominant somatic symptoms and emotional distress which are usually associated with psychosocial stressors. The various presentations in these syndromes are influenced by the local beliefs about the body. Common syndromes include the brain fag in Nigeria which is characterized by the sense of heaviness or heat in the head while studying; Dhat syndrome in India in which patients present with a range of somatic and psychological symptoms secondary to the loss of vital essence semen in the urine and the *hwa-byung* syndrome was found in the married past middle-age Korean women living in the United States. The symptoms are secondary to the suppression of feelings of anger and resentment which forms a mass in the chest as per the sufferers leading to constellation of numerous somatic symptoms and depressive, anxiety symptoms as well as irritability.

These syndromes are hypothesized to arise from somatic amplification which is secondary to interaction between emotional arousal, bodily focused attention, symptom attribution, and cognitive appraisal leading to increased distress in various functional systems. The above mentioned psychological processes are influenced by social, cultural, and interpersonal factors that reinforce and contribute to the somatic expression of distress and pathological behavior (Kirmayer and Young 1998).

When it comes to the psychiatric presentations in Indian culture, Gautam and Kapur in a study of psychiatric patients presenting with somatic complaints reported that more patients from Muslim ethnic group presented with somatic symptoms in South Indian population. Gautam et al. repeated the study in North Indian population and found that the predominant somatic complaint was constipation and feeling of gas in the abdomen (Gautam and Jain 2010). Chaturvedi SK et al. in their work "Dissociative disorders in a psychiatry institute in India – A selected review and patterns over a decade" emphasized that unlike in the West, dissociative identity disorders were rarely diagnosed; instead, possession states were commonly seen in the Indian population, indicating cross-cultural disparity (Chaturvedi et al. 2010).

Patients with somatoform disorders frequently visit primary care physicians for treatment and most often the diagnosis is missed. Although the disorder is considered to be more common in rural areas, recent studies have found that the prevalence is fairly equivalent in both urban and rural settings (Ng et al. 2011). There exists wide heterogenicity in the prevalence rates of somatoform disorder in the primary care depending on the diagnostic criteria used (ICD 10 or DSM-IV), type of study-survey, or clinical interview. Overall, as per a recent meta-analysis, the point prevalence of somatoform disorder as per DSM or ICD was around 26% and nearly 35% when only high-quality studies were included. The lifetime prevalence rates were around 41%. Lifetime prevalence rates for different subcategories of somatoform disorders were 20.5% for unspecified somatoform disorder, 13.4% for undifferentiated somatoform disorder, 9.2% for chronic pain disorder, and 5.9% for somatization disorder (Haller et al. 2015). In a retrospective study estimating the prevalence and comorbidities of somatoform disorder in a rural clinic, 5% of the patients had somatoform disorder. Patients with somatoform disorders had increased comorbidity of psychiatric and medical disorders. In addition, these patients were observed to be more sensitive to side effects of medications and hence discontinue them (Ng et al. 2011). In a study in rural setting conducted in India alexithymia, life events and coping skills in women with functional somatic symptoms (FSS) were compared with women without FSS. It was found that women with FSS had greater number of overall life events especially in the areas of family, finance, and marital issues (Geetha and Sekar 1995).

In a study conducted in a rural primary care setting investigating the relationship between patient reported physical symptoms as a predictor of a psychiatric disorders, it was found that somatoform physical symptoms were highly predictive and the odds of psychiatric disorder increased as the number of these symptoms increased (Rasmussen et al. 2008).

Dissociative Disorders in Rural Setting

There are few studies done in India on dissociative disorder, and most of the studies were on children and adolescents. In a study done in tertiary psychiatry hospital, among children and adolescents found that most of the subjects were prepubertal, had equal gender distribution and pseudo-seizures was the most common presentation (Srinath et al. 1993). In a follow-up study, 38 women with a diagnosis of hysterical neurosis were evaluated after a period of 5 years. They found that 63% of the patients remained totally asymptomatic and premorbid hysterical personality showed significant relationship with outcome (Chandrasekaran et al. 1994). In another prevalence study from a tertiary center among the patient visited hospital from 1999 to 2008, the prevalence ranged between 1.5 and 15.0 per 1,000 for outpatients and between 1.5 and 11.6 per 1,000 for inpatients. There was female predominance among the patients, and dissociative motor disorder was the commonest type (Chaturvedi et al. 2010). A study from Assam reported that conversion disorder is more common in young adults (57.5%), females (92.5%), and among students belonging to nuclear family of lower socioeconomic status. The common precipitating factors were family-related (40%) and school-related (30%) with motor symptoms as the predominant presentation (Deka et al. 2007).

Management

Understanding the symptoms is essential in providing treatment. A person presenting with physical symptoms may hold a varied explanatory model from that of the health providers. There may be cultural variations in understanding, interpretation, and experiencing of symptoms. Faith healers, religious treatments, and indigenous treatments might be the first contact rather than allopathic services. Tricyclic antidepressants are often used in the treatment of somatoform disorders and have been found to be effective (Kroenke 2007). Psychosocial treatments have been demonstrated to have some efficacy in treatment of somatoform disorders (Van Dessel et al. 2014). However, most of these interventions are in the Western urban settings. Single-session interventions might have role in the rural settings which have been described in a similar setting (Hoch 1977).

Conclusions

Challenges of treatment of somatoform disorders and dissociative disorders in rural settings include the cultural explanatory models for dissociations, acceptability of medical models of mental illness, lack of services for psychosocial interventions, and stigma associated with seeking treatment for mental illnesses. Integrated care is a possible answer for the above difficulties in management.

In conclusion, somatoform and dissociative disorders are common in rural setting though not limited to the same. It is essential to provide care by understanding the sociocultural context.

References

Allin M, Streeruwitz A, Curtis V (2005) Progress in understanding conversion disorder. Neuropsychiatr Dis Treat 1(3):205–209

Association, AP, Association, AP (2013) Diagnostic and statistical manual of mental disorders: DSM-5. American Psychiatric Association, Washington, DC

Chandrasekaran R, Goswami U, Sivakumar V, Chitralekha n (1994) Hysterical neurosis – a follow-up study. Acta Psychiatr Scand 89(1):78–80

Chaturvedi SK, Desai G, Shaligram D (2010) Dissociative disorders in a psychiatric institute in India – a selected review and patterns over a decade. Int J Soc Psychiatry 56(5):533–539. https://doi.org/10.1177/0020764009347335

Deary V, Chalder T, Sharpe M (2007) The cognitive behavioural model of medically unexplained symptoms: a theoretical and empirical review. Clin Psychol Rev 27(7):781–797

Deka K, Chaudhury PK, Bora K, Kalita P (2007) A study of clinical correlates and sociodemographic profile in conversion disorder. Indian J Psychiatry 49(3):205–207. https://doi.org/10.4103/0019-5545.37323

Gautam S, Jain N (2010) Indian culture and psychiatry. Indian J Psychiatry 52(Suppl1):S309–S313. https://doi.org/10.4103/0019-5545.69259

Geetha PR, Sekar K (1995) Alexithymia in rural health care. NIMHANS J 13:53

Gureje O, Simon GE, Ustun TB, Goldberg DP (1997) Somatization in cross-cultural perspective: a World Health Organization study in primary care. Am J Psychiatr 154(7):989–995

Haller H, Cramer H, Lauche R, Dobos G (2015) Somatoform disorders and medically unexplained symptoms in primary care. Dtsch Arztebl Int 112:279–287

Hoch ME (1977) Psychotherapy for the illiterate. In: Arieti S, Chrzanowski G (eds) A new dimension in psychiatry, a world view. Wiley, New York, pp 75–92

Kirmayer, L. J., & Young, A. (1996). Gaulbaud du Fort G, et al: Pathways and barriers to Mental Health Care: a community survey and ethnographic study. Montreal, Culture & Mental Health Research Unit, Institute of Community & Family Psychiatry, Sir Mortimer B. Davis-Jewish General Hospital

Kirmayer LJ, Young A (1998) Culture and somatization: clinical, epidemiological, and ethnographic perspectives. Psychosom Med 60(4):420–430

Kroenke K (2007) Efficacy of treatment for somatoform disorders: a review of randomized controlled trials. Psychosom Med 69(9):881–888

Ng B, Tomfohr LM, Camacho A, Dimsdale JE (2011) Prevalence and comorbidities of somatoform disorders in a rural California outpatient psychiatric clinic. J Prim Care Community Health 2(1):54–59

North CS (2015) The classification of hysteria and related disorders: historical and phenomenological considerations. Behav Sci 5(4):496–517. https://doi.org/10.3390/bs5040496

Pichot P (1986) DSM-III: the 3d edition of the diagnostic and statistical manual of mental disorders from the American Psychiatric Association. Rev Neurol 142(5):489–499

Rasmussen NH, Bernard ME, Harmsen WS (2008) Physical symptoms that predict psychiatric disorders in rural primary care adults. J Eval Clin Pract 14(3):399–406

Rief W, Barsky AJ (2005) Psychobiological perspectives on somatoform disorders. Psychoneuroendocrinology 30(10):996–1002

Robins LN, Regier DA (1991) Psychiatric disorders in America: the epidemiologic catchment area study. Free Press, New York

Srinath S, Bharat S, Girimaji S, Seshadri S (1993) Characteristics of a child inpatient population with hysteria in India. J Am Acad Child Adolesc Psychiatry 32(4):822–825. https://doi.org/10.1097/00004583-199307000-00017

Trujillo M (2001) Culture and the organization of psychiatric care. Psychiatr Clin North Am 24(3):539–552

Van Dessel N, Den Boeft M, van der Wouden JC, Kleinstäuber M, Leone SS, Terluin B, ... & van Marwijk H (2014) Non-pharmacological interventions for somatoform disorders and medically unexplained physical symptoms (MUPS) in adults. Cochrane Database of Systematic Reviews (11)

Alcoholism, Substance Use, and Other Addictive Disorders

5

Venkata Lakshmi Narasimha and Arun Kandasamy

Contents

V. L. Narasimha · A. Kandasamy (✉)
Centre for Addiction Medicine, Department of Psychiatry, National Institute of Mental Health and
Neurosciences, Bangalore, Karnataka, India
e-mail: arunnimhans05@gmail.com

© Springer Nature Singapore Pte Ltd. 2020
S. Chaturvedi (ed.), *Mental Health and Illness in the Rural World*, Mental Health and
Illness Worldwide, https://doi.org/10.1007/978-981-10-2345-3_13

Abstract

Urban-rural divide across the world plays an important role from economy to health. Health-related gap in general and addiction services in specific take a toll on rural population. Variations round the globe in terms of defining these population groups, problems, and barriers faced by rural areas exist. Dynamic changes in prevalence of substance use disorders (SUDs) and epidemiological differences across the countries make it difficult to study and understand these differences. The differences across the countries based on the economies have been described in the chapter. Special population like pregnant women, prisoners, and migrant laborers have been observed to get effected by the urban-rural divide. Factors like availability of treatment centers, use of evidence-based services, perceived ease of access, screening and follow-up services, financial status, perception of substance use as problem, and gender inequality have been shown to contribute for urban-rural differences. These inequalities also play an important role in preventive dimension, i.e., in initiation of substance use. Urban-rural differences play an important role in planning, organizing, and implementing preventive programs. Understanding/studying urban-rural differences is important in allocation of resources, policy making, and planning interventions for the management of SUDs. Further research is required in this area.

Keywords

Substance use disorders · Rural · Urban · Alcohol

Introduction

Globally, approximately one half of the population lives in rural areas, but less than 38% of the nurses and less than 25% of the physicians work there (WHO 2010). While appointing and sustaining health workers in rural and remote areas is a challenge for every country, the situation is worse in the 57 countries that have an absolute shortage of health workers. When it comes to specialty services like deaddiction, these numbers fall further and implementation of these services by primary health care physicians is indeed a difficult task. Starting from population dynamics to implementation of health programs, there is a lot of difference between urban and rural populations. In this chapter we tried to discuss urban-rural differences across the globe and how it becomes an important in substance use disorders.

Defining Urban and Rural Areas

The challenge starts with the difficulty to define what is a rural or an urban area. The definition varies from countries to countries. For example, in highly developed countries like United States, the first definition developed by the census bureau identifies two types of urban areas: Urbanized Areas (UAs) of 50,000 or more people; Urban Clusters (UCs) of at least 2500 and less than 50,000 people. The Census does not actually define "rural." "Rural" encompasses all population, housing, and territory not included within an urban area (HRSA 2017). According to the United Nations Office of Economic and Social Affairs system, counties have been categorized into three types: "rural" (nonmetropolitan counties with urban population less than 20,000), "urbanized nonmetropolitan" (nonmetropolitan counties with urban population 20,000 or higher), and "metropolitan" (counties in metropolitan areas with 500,000 or more inhabitants) (Gfroerer et al. 2007; United Nations Department of Economic and Social Affairs). This plays a major role not only in understanding the dynamics of population but also for planning the interventions.

In most of the countries, the rural economy plays a major role in the economy. For example, India has about 650,000 villages. These villages are inhabited by about 850 million consumers making up for about 70% of population and contributing around half of the country's Gross Domestic Product (GDP) (Chandrasekhar and Murali 2016). The picture is same across many countries in world. And the differences between the two are not restricted to the population number alone. It is well known from the population based studies that there are fairly significant amount of differences between rural and urban area with regards to their health index and the economy.

Urban-Rural Differences in Terms of Health and Economy

Rural population is different from urban in terms of health and economy (https://www.unodc.org/documents/drug-prevention-and-treatment/16-10463_Rural_treatment_ebook.pdf) in various parameters, which includes,

- Higher rates of poverty.
- Child marriages are more common.
- Rural youth are less likely to stay in school, with young men having higher educational advantages and higher completion rates in both settings than young women (with the United States being one exception).
- Higher infant mortality rates and a lower likelihood of receiving antenatal care and skilled care at delivery.
- Rural women have more children than urban women.
- A greater percentage of children who are underweight, a greater incidence of food insecurity, and lower access to safe drinking water and sanitation.
- Higher rates of maternal mortality among women living in rural areas and poorer communities, with 99% of all maternal deaths occurring in developing countries (Alkema et al. 2016).
- While 56% of the global rural population lacks health coverage, only 22% of the urban population is not covered (Scheil-Adlung 2015).
- The situation is aggravated by extreme health workforce shortages in rural areas impacting on the delivery of quality services: in rural areas a global shortfall of about seven million health workers to deliver services is observed, compared to a

lack of three million staff in urban areas. Due to these rural health workforce shortages, half the global rural population lacks access to urgently needed care.
- Deficits in per capita health spending are twice as large in rural areas as in urban areas.
- These deficits result in unnecessary suffering and death, as reflected, for example, in rural maternal mortality rates that are 2.5 times higher than urban rates.

Problems and Barriers to Treatment in Rural Areas

Barriers to effective substance use treatment experienced by rural residents (https://www.unodc.org/documents/drug-prevention-and-treatment/16-10463_Rural_treatment_ebook.pdf).

- Fewer treatment options for rural clients
- Lack of service providers delivering services (Mpanza and Govender 2017)
- Challenges in getting to treatment facilities, including the lack and cost of public transportation, long travel distances, geographic isolation
- Reliance on friends and family for transportation
- Lack of good facilities, inadequate infrastructure (e.g., building resources)
- Challenges in meeting housing and other support needs of people in treatment
- Lack of educational resources for clients and early education about substance use risks
- Limited continuing education opportunities for counselors
- Higher rates of unemployment, financing issues, and poverty
- Poor prioritization and lack of monitoring
- Health gap
- Cultural differences

To explain the importance of each factor, we have taken an example of cultural differences. Indigenous people form a part of rural population. Around 370 million indigenous people are there worldwide who are culturally different from the rest of the population. Prevention and treatment programs are more effective when they recognize and understand these contextual issues. Culture and ethnicity plays an important role. For example, in 1992–1997, National Treatment Improvement Evaluation Study, a prospective cohort study of substance abuse treatment programs and their clients found that racial/ethnic minorities are underserved compared to whites in the substance abuse service system. Different racial/ethnic groups come into treatment with distinct needs and receive distinct services. Although groups respond differentially to service types, substance abuse counseling and matching services to the cultural needs is an effective strategy both for retaining clients in treatment and for reducing posttreatment relapse for African Americans and Whites (Marsh et al. 2009).

Even though we tried to highlight problems faced by rural areas, there are problems which the urban areas do face. In a study (Pullen and Oser 2014), counselors' perspective revealed that there is predominantly lack of funding and bureaucratic challenges which includes heavy caseloads with understaffing, lack of technological support, language barriers, cultural differences, lack of case management. Apart from these, there are common challenges between these two populations for example lack of interagency cooperation.

Constantly Changing Pace of Prevalence of SUDs

Research demonstrated that there are continued shifts in trends in illicit drug use in the United States and called attention to rising rates of prescription drug misuse and abuse. This is also true with other parts of the world. Findings have also continued to highlight the substantial co morbidity of SUDs with other psychiatric disorders and with the ongoing HIV epidemic (Schulden et al. 2009).

Epidemiology of Substance Use Disorders and Its Importance

Studying epidemiology of substance use disorders and trying to understand urban-rural differences in substance use disorders help in planning the prevention programs to providing the rehabilitation services (Schulden et al. 2009).

The key to developing effective policies, practices, and interventions related to the substance use disorders is to select the spatial units and characteristics of rurality that are most important and relevant to stakeholders and capture demographic and population changes as they occur.

An enhanced understanding of these types of differences may enable policymakers and treatment providers to direct limited resources more effectively and increase the quality of care received in different geographic contexts and suggest a diverse set of needs.

However, studies also suggest that there are methodological issues in studying epidemiology. Building on these foundations, future challenges for research in substance abuse epidemiology will include using novel methodological approaches to further unravel the complex interrelationships that link individual vulnerabilities for SUDs, including genetic factors, with social and environmental risk factors.

Early prevention and intervention efforts in rural areas may help to mitigate future substance dependence and abuse considering the high rate of adolescent initiation among rural admissions. It can alleviate the strain which the substance abuse and its associated problems (e.g., negative health outcomes, crime) levy on rural substance abuse treatment, health care, and law enforcement systems.

Urban and Rural Differences Across World

Despite the lack of global prevalence data on drug use, available data from different countries indicate that rural areas suffer from drug use. Many studies dispel the notion that substance abuse is only an urban problem and provides information useful in developing and implementing interventions that consider the unique characteristics of rural residents.

In this section we tried to look at urban-rural differences based on their economies as defined by World Bank (2016).

High Income Economies: USA and Australia

Most of the studies on urban rural differences come from high income countries like United States of America.

Differences in adolescent substance use: In 2008, Community Youth Development Study (CYDS) done in USA (Rhew et al. 2011), stated that current alcohol use, smokeless tobacco use, inhalant use, and other illicit drug use were more prevalent among high school-aged youths living on farms than among those living in towns. Prevalence of drug use did not significantly vary across youths living in different residential contexts among middle school youths. The findings suggest that outreach activities to farm-dwelling youths may be particularly important to prevent adolescent drug use in rural settings.

Differences in inpatient admissions: In 2012 SAMSHA (2012), published a report on differences in admissions among the urban and rural areas. Rural admissions were younger and less racially and ethnically diverse than urban admissions. Rural admissions were more likely than urban admissions to report primary abuse of alcohol (49.5 vs. 36.1%) or nonheroin opiates (10.6 vs. 4.0%); urban admissions were more likely than rural admissions to report primary abuse of heroin (21.8 vs. 3.1%) or cocaine (11.9 vs. 5.6%). Rural admissions were more likely than urban admissions to be referred by the criminal justice system (51.6 vs. 28.4%) and less likely to be self- or individually referred (22.8 vs. 38.7%).

Differences in opioid misuse: In 2012, National Survey on Drug Use and Health (Monnat and Rigg 2016) observed that the prescription opioid misuse is more common in rural and small urban adolescents. Some of the important findings from this survey are that criminal activity, lower perceived substance use risk, and greater use of emergency medical treatment partially contribute to higher odds among rural adolescents, but they are also partially buffered by less peer substance use, lesser access to illicit drugs, and stronger religious beliefs. However, urban adults have more misuse compared to the rural adults (Rigg and Monnat 2015). This brings the need for early interventions in rural population and tailoring of interventions accordingly in urban populations.

Reasons for differences in Opioid misuse: Data from the 2011 and 2012 National Survey on Drug Use and Health, urban adults were more likely to engage in prescription Opioid medications compared to rural adults because of their higher use of other substances, including alcohol, marijuana, and other illicit and prescription drugs, and because of their greater use of these substances as children (Rigg and Monnat 2015).

Differences in implementation of Opioid treatment program: In 2004, National Survey of substance abuse treatment services looked at urban-rural differences.

Substance abuse treatment overall and intensive services in particular is limited in rural areas, especially among counties not adjacent to metro areas. Less populated areas with greater commuting distances contain a small proportion of facilities offering a range of core services and varying levels of outpatient care. This situation is particularly striking for opioid treatment programs, which are nearly absent in rural areas. The greater proportion of rural-based facilities accepting public payers and providing discounted care may indicate greater challenges to financing treatment in rural areas (Gale and Lenardson 2007).

Differences in stimulant use: In 2002, National survey of drug use and health reported some of the important differences in the use of stimulants. The use of Ecstasy is higher among youth in metropolitan and urbanized nonmetropolitan counties than rural counties, while rural youth have a higher prevalence of stimulant and methamphetamine use than metropolitan youth. Rural adults had generally lower rates of illicit drug use than metropolitan adults, but adults in rural and urbanized nonmetropolitan areas had higher rates of methamphetamine use than those in metropolitan areas. They also found that rural youth had a higher prevalence of past month use of tobacco and alcohol. Rural adults had higher rates of tobacco use but lower rates of alcohol use (Gfroerer et al. 2007).

Australia: Prevalence and patterns of illicit drug use vary between rural and metropolitan residents. People living in remote and very remote areas were twice as likely as people in major cities to have recently used meth/amphetamines, but less likely to have used ecstasy compared with those from major cities. Cannabis use and the use of pharmaceuticals not for medical purposes are higher in remote/very remote areas than in major cities.

Aboriginal and Torres Strait Islander people, of whom 70% live in rural Australia, were 1.7 times more likely to have used illicit drugs recently compared to the general population. Social disadvantage is a contributing factor to illicit drug use and strategies to combat illicit drug use should address its social determinants.

Rural residents face barriers to accessing drug treatment services, including limited access to health services in general and drug treatment options in particular, greater distance from services and a lack of transport. Drug services that are particularly limited in rural areas include methadone programs, withdrawal and detoxification services, as well as needle and syringe programs.

Other barriers include lack of motivation to seek treatment, unfavorable attitudes, such as resistance to treatment and fear of what may be involved, and concern about confidentiality. Rural residents may also be more reluctant to disclose their drug use to healthcare professionals who are more likely to be personally known to them. Programs targeted at rural communities must consider their diversity and unique perspectives and be locally tailored to maximize their chances of success. These strategies should be aimed at promoting social inclusion, building individual and community resilience, enhancing protective factors, reducing risk factors, and providing support to families affected by illicit drug use. Interventions designed to target illicit drug use among rural residents will require strong community consultation so as to engage and empower rural communities.

Higher Middle-Income Economies

Middle East-Iran Rural Prevalence

Rural household survey looked at substance abuse in one of the rural areas of southeast Iran, in a 12-year period (2000 and 2012 (Ziaaddini et al. 2013)). Demographic characteristics, frequency of substance abuse, and ease of access to various drugs were studied. Majority of the participants (61.8%) were below 30 years of age and among them 54.4% were male. Cigarette (17.0%), opium (15.7%) and opium residue (9.0%) were the most frequent substances abused on a daily basis. Based on the participant's opinion, they concluded that the ease of access to cigarette, water pipe, and opium contributed to their increase in consumption compared with earlier years. The steady rise in substance abuse in rural communities demands immediate attention and emergency preventive measures from policy makers.

Lower-Middle Income Economies

Urban-Rural Differences in India

Earlier studies identified urban-rural differences in prevalence of mental disorders (Ganguli 2000; Reddy and Chandrashekar 1998). Chandrasekhar (1998) mentioned that the urban rates were twice as much as the rural prevalence rates, whereas quite contrastingly, Ganguli (2000) showed that for every 100 rural persons afflicted with a mental disorder, there existed about 157 urban people with a mental disorder double in urban when compared to rural.

In the latest National mental health survey 2015 (National Mental Health Survey of India 2015), the prevalence of substance use disorders was more in rural areas (24.1%) as compared to urban non metro (20.3%) and urban metro areas (18.3%). The burden of use of tobacco was relatively more in rural areas (22.7%). The prevalence of other substance use disorders excluding tobacco and alcohol in the urban metro areas (1.0%) was twice as much as in the urban nonmetro or rural areas.

In a country with diverse cultures like India, society is very much concerned for substance related issues, but sometimes promotes (cultural sanctioned use of cannabis and alcohol) it; "especially in rural places, alcohol use is a societal norm during celebrations and there are many events round the year to celebrate" in states like Jharkhand; "Bhukki (opium) and alcohol use are socially sanctioned and so less stigmatizing" in states like Punjab.

While the causes, risk factors, and protective factors vary in urban and rural populations, availability, accessibility, and affordability of care are different in both areas; awareness is still limited. Thus, the need for coverage of mental health services across India on an equal basis merits importance. Factors ranging from awareness to affordability, varying between rural and urban areas, need to be critically delineated to address specific issues in bridging treatment gap.

The National Household Survey 2004 revealed that rural individuals were 1.5 times more likely to use alcohol compared with urban users. This would probably be attributed to education, income, occupation, and other social factors (Ray 2004).

Subramanian et al. (2005) reanalyzed the data from the NFHS – 2 and observed that the prevalence of alcohol use among both men and women was significantly higher in towns and villages as compared to large and small cities (Subramanian et al. 2005). In the same way despite lack of good studies, earlier studies done during the last five decades reported that it is clear that the problem of alcohol use is significantly higher in rural areas, urban slums, transitional towns, and tribal areas (Neufield et al. 2005; Isaac 1998; Benegal et al. 2003; Ray and Sharma 1994; Thimmaiah 1979; Gururaj et al. 2004, 2006, 2011; Anand et al. 2007).

In the GENACIS study undertaken in the state of Karnataka, the prevalence of drinking among men was 23% in rural areas and 41% in urban areas among men, while similar rates among women was 4.4% and 7%, respectively (Benegal et al. 2005). However, NFHS-3 had opposite results with an increased prevalence in rural women compared to urban women. Among females the ratio between urban to rural was 1:5 (0.6%: 3.0%) (2007).

In India, Tobacco use is more in rural population when compared to urban slums which is significantly higher than urban areas and also reportedly nontaxed forms of nicotine product consumption is found to be higher in these areas (Gupta et al. 2010). The antitobacco policies of India need to focus on bidis in antitobacco campaigns. The program activities must find ways to reach the rural and urban-slum populations.

Low Income Economies

In **Afghanistan**, INL survey of drug use (which included toxicology testing) found that 31% of households and 11% of the population tested positive for one or more drugs and drug use was found to be three times greater in rural areas than in urban ones (Bureau for International Narcotics and Law Enforcement Affairs 2016).

In **rural African** areas, high rates of fetal alcohol syndrome and partial fetal alcohol syndromes were present especially in isolated communities (Olivier et al. 2013). The negative consequences of substance use disorders (SUDs) in rural settings are very serious and require immediate responses. However, another isolated region of rural African region has a very low prevalence of substance use disorders (Tshitangano and Tosin 2016).

Urban-Rural Differences in the Special Population

Pregnant Women

A 3 years' study (Shaw et al. 2015), which looked at the urban-rural differences in pregnant women using alcohol from Washington state of United States of America, found results to be troubling.

1. Rural participants were more likely to report alcohol use and binge drinking at program intake and at the 3-year program exit.
2. Throughout the program, rural women were less likely to complete outpatient substance abuse treatment compared to urban participants.
3. Rural women also used less services during the last year including alcohol/drug support and mental health provider services.
4. At program exit, rural participants also reported higher use of alcohol and more suicidal thoughts than those residing in urban areas.
5. Implication of the study was identifying community-specific needs of substance abusing pregnant or parenting women in both rural and urban settings is crucial for the successful development and improvement of treatment and intervention programs for this vulnerable population of women.

Another study (Shannon et al. 2010) examined differences in substance use (predominantly opioids) among pregnant women from rural and urban areas. Rural pregnant women had higher rates of illicit opiate use, illicit sedative/benzodiazepine use, and injection drug use (IDU) in the 30 days prior to admission. Additionally, a greater proportion of rural pregnant women reported the use of multiple illegal/illicit substances in the 30 days prior to entering detoxification. The increased rates of prescription opiate and benzodiazepine use as well as IDU among rural pregnant women are concerning. In order to begin to understand the elevated rates of substance abuse among rural pregnant women, substance use must be considered within the context of demographic, geographic, social, and economic conditions of the region.

Prisoners

1. In study (Warner and Leukefeld 2001) that examined differences in drug use and treatment utilization of urban and rural offenders, chronic drug abusers from rural and very rural areas have significantly higher rates of lifetime drug use, as well as higher rates of drug use in the 30 days prior to their current incarceration than chronic drug abusers from urban areas. Nonetheless, being from a very rural area decreased the likelihood of having ever been in treatment after controlling for the number of years using and race. While problem recognition appears to explain much of the effect of very rural residence on treatment utilization for alcohol abuse, the effects of being from a very rural area on seeking treatment for drug abuse remain statistically significant even after controlling for several other variables. The findings point to the importance of providing culturally appropriate education to very rural communities on the benefits of substance abuse treatment and of providing substance abuse treatment within the criminal justice system.
2. The results of this study indicate that rural DUI (Malek-Ahmadi and Degiorgio 2015) (driving under influence) offenders have a significantly greater risk of heavy alcohol use when compared to urban DUI offenders.
3. In a study (Elgar et al. 2003) examining the role of temperament using clinical cutoffs to assess the outcome behaviors between urban and rural populations,

significant differences have been noted. Urban delinquent youths showed higher rates of attention problems, delinquent behaviors, and externalizing behaviors than those in rural communities. Incarcerated young offenders show elevated rates of psychological problems that require treatment. Rural and urban differences in the rates of these problems may reflect differences in community service availability in these areas or in environmental influences on the development of child behavioral problems.

Migration

In a study Chen et al. (2008) to assess the effect on substance use on urban rural migration found that substance use is prevalent among rural-to-urban migrants, especially among female migrants. Workplace, income, and depression are associated with substance use interactively. Tailored substance use prevention is needed to target high-risk workplaces with specific efforts devoted to female migrants.

Factors Involved in Urban-Rural Differences

Treatment Centers

Bond Edmond et al. (2015) looked at the structural and quality differences between rural and urban treatment centers and found that the rural centers had reduced access to professionally trained counselors, were more likely to be nonprofit organizations and dependent on public funding, offered fewer wrap around services, and had less diverse specialized treatment options. This author also indicated that rural centers were less likely to prescribe buprenorphine (Opioid Substitution) as part of their treatment but were more likely to employ nursing staff and offer specialized treatment for adolescents. Increasing the resources, provision of funding, change in the policy with a continuous monitoring would be some of the strategies in improving the quality of the centers.

Using Evidence Based Services

A study (Dotson et al. 2014), which looked at delivery of evidence based practice (EBP) between urban and rural population, found that most mental health and substance abuse treatment agencies used more than 1 EBP, although rural substance abuse agencies were less likely to do so than urban agencies. Rural substance abuse agencies were more likely to be solo than group practices. Urban agencies reported significantly more collaboration with universities for EBP training, although training by internal staff was the most commonly reported training mechanism regardless of agency focus or location. Over half of agencies reported conducting no systematic assessment of EBPs, and of those who did report systematic assessment, most used outcome monitoring more than program evaluation or benchmarking. Urban and rural mental health and substance abuse prevention providers reported shortages of appropriately trained workforce and financing issues available to pay for EBPs as the greatest barriers to utilization.

Perceived Ease of Access

Ease of access to substance has been shown to have a direct and significant relationship with substance use for school-aged children. It has been shown that ease of access is an important predictor of recent drug use among the rural adolescents (Warren et al. 2015). It appeared the rural-urban differences fell along legal/illicit lines (Warren et al. 2015). Rural students reported higher level of access to legal substances and urban students reported higher level of access to illicit substances. Studies focusing on limiting the ease of access would make an impact on policy making. For middle school students, a significant difference in perceived ease of access was found for each substance, with rural students reporting greater access to smoking tobacco, chewing tobacco, and steroids and urban students reporting greater access to alcohol, marijuana, cocaine, inhalants, ecstasy, methamphetamine, hallucinogens, and prescription drugs. Rural high school students reported higher access to alcohol, smoking tobacco, chewing tobacco, and steroids, but urban students reporting higher access to marijuana, cocaine, inhalants, ecstasy, and hallucinogens. Perceptions of ease of access more than doubled for each substance in both geographies between middle and high school. More than 60% of both rural and urban high school students reported easy access to alcohol. Future research should investigate ways to decrease the perceptions of access to substances in order to prevent use and abuse.

Screening and Follow-Up Services

Screening for substance use disorders in population with psychiatric illness seems to be high when compared to general population. However, follow-up after the initial screening is less in rural population when compared to urban population (Chan et al. 2016). There is a need to think about generalizability of this finding, as availability of screening and follow-up for the substance use disorders at primary care may not be available at many centers all over the world.

Availability of Money

A clear association was made by respondents between greater availability of money and increasing alcohol use. In the localities where availability of money had increased due to work for pay schemes in rural areas, alcohol use had increased because a large proportion of these money were spent on alcohol (National Institute of Mental Health and Neuro Sciences 2012).

Perception of Substance Use as a Problem

In a study which looked at intimate partner violence and violence towards children between urban and rural areas from India, there appears to be a greater normalization and acceptance of alcohol use from rural respondents and therefore less of causal attribution of alcohol as a factor in violence and other harm. Urban respondents appear to attribute a greater proportion of harm to alcohol misuse (National Institute of Mental Health and Neuro Sciences 2012).

Gender

Gender differences in substance usage have been evaluated and understood. When it comes to urban rural differences, culture, specific occupations (toddy tapping), or specific communities seems to play an important factor in determining the substance usage in rural population. For example, a study done from the rural part of an Indian state (Surat in Gujarat) where alcohol is in prohibition for long period of time, historically some communities have tolerated alcohol use by women. However, alcohol use among women was generally still looked down upon with a perception: "(women) drinking will adversely affect child rearing and ruin the family. Such girls find difficulty in getting married. It is not good for her safety and culture. Her drinking is a big societal loss (National Institute of Mental Health and Neuro Sciences 2012)." Differences in mental health, substance use, and sexual behavior have been noted in transgender population (Horvath et al. 2014). Significant higher amount of cannabis use along with unprotected sexual behavior has been observed in this population.

Barriers and Facilitators

A study (Browne et al. 2016) which did a qualitative thematic analysis of stake holders and patients attending service agencies in a particular region found that four predominant themes noted as barriers and facilitators: availability of services for individuals with substance use disorders; access to the current technology for client services and agency functioning; cost of services; and stigma.

Prevention

Rural populations are often composed of numerous vulnerable subgroups with different cultural, ethnic, and/or religious belief structures or with differing levels of marginalization. Rural communities have diverse characteristics and interventions will need to be localized rather than follow a one-size-fits-all approach.

The primary goal of substance use prevention is to help nonsubstance users avoid or delay the initiation of substance use. For those who are vulnerable to be dependent on substances, prevention seeks to minimize their likelihood of developing a substance use disorder (e.g., dependence). Researchers, policy makers, and treatment providers must consider the complex array of individual, social, and community risk and protective factors to understand rural/urban differences in prevention of substance use at an early stage.

In a study that looked at prevention of adolescent prescription opioid misuse (POM), potential points of intervention to prevent POM in general and reduce rural disparities include early education about addiction risks, use of family drug courts to link criminal offenders to treatment, and access to nonemergency medical services to reduce rural residents' reliance on emergency departments where there is a higher likelihood of prescription of opioids (Monnat and Rigg 2016).

Conclusion

Urban-rural differences play an important role in planning, organizing, and implementation of preventive programs. Understanding/studying urban-rural differences is important in allocation of resources, policy making, planning interventions for the management of SUDs. It is clear that the pattern and prevalence of SUD differ between these two parts with respect to the nature of substances, gender differences, cultural factors, accessibility to services, and other factors. Rural areas are mostly deprived of mental health care and addiction services and also suffer significant problems in terms of barriers to treatments compared to the urban areas. Studies which have been conducted to understand the urban rural differences are predominantly restricted to developed countries. More countries with predominantly rural economies need to be included in this research.

References

Alkema L, Chou D, Hogan D, Zhang S, Moller A-B, Gemmill A et al (2016) Global, regional, and national levels and trends in maternal mortality between 1990 and 2015, with scenario-based projections to 2030: a systematic analysis by the UN Maternal Mortality Estimation Inter-Agency Group. Lancet Lond Engl 387(10017):462–474

Anand K, Shah B, Yadav K, Singh R, Mather P, Paul E et al (2007) Are the urban poor vulnerable to non communicable diseases? A survey of risk factors for non-communicable diseases in urban slums of Faridabad. Natl Med J India 20(3):115–120

Benegal V, Gururaj G, Murthy P (2003) Report on a WHO Collaborative Project on unrecorded consumption of alcohol in Karnataka, India. Available from: http://www.nimhans.kar.nic.in/Deaddiction/lit/UNDOC_Review.pdf. Accessed 6 Sept 2005

Benegal V, Nayak M, Murthy P, Chandra P, Gururaj G (2005) Women and alcohol in India. In: Obot IS, Room R (eds) Alcohol, gender and drinking problems. Perspectives from low and middle income countries. World Health Organization, Geneva

Bond Edmond M, Aletraris L, Roman PM (2015) Rural substance use treatment centers in the United States: an assessment of treatment quality by location. Am J Drug Alcohol Abuse 41 (5):449–457

Browne T, Priester MA, Clone S, Iachini A, DeHart D, Hock R (2016) Barriers and facilitators to substance use treatment in the rural south: a qualitative study. J Rural Health 32(1):92–101

Bureau for International Narcotics and Law Enforcement Affairs. INCSR 2016 [Internet]. [cited 7 Jul 2017]. Available from: https://www.state.gov/documents/organization/253983.pdf

Chan Y-F, Lu S-E, Howe B, Tieben H, Hoeft T, Unützer J (2016) Screening and follow-up monitoring for substance use in primary care: an exploration of rural-urban variations. J Gen Intern Med 31(2):215–222

Chandrasekhar BVNG, Murali KB (2016) Rural marketing in India: prospects and challenges. ITIHAS J Indian Manag 6(1):73–85. Print ISSN: 2249–7803

Chen X, Stanton B, Li X, Fang X, Lin D (2008) Substance use among rural-to-urban migrants in China: a moderation effect model analysis. Subst Use Misuse 43(1):105–124

Dotson JAW, Roll JM, Packer RR, Lewis JM, McPherson S, Howell D (2014) Urban and rural utilization of evidence-based practices for substance use and mental health disorders. J Rural Health 30(3):292–299

Elgar FJ, Knight J, Worrall GJ, Sherman G (2003) Behavioural and substance use problems in rural and urban delinquent youths. Can J Psychiatr Rev Can Psychiatr 48(9):633–636

Gale J, Lenardson J (2007) Distribution of substance abuse treatment facilities across the rural-urban continuum. Popul Health Health Policy [Internet], 5 Oct 2007. Available from: http://digitalcommons.usm.maine.edu/healthpolicy/26

Ganguli HC (2000) Epidemiological findings on prevalence of mental disorders in India. Indian J Psychiatry 42(1):14–20

Gfroerer JC, Larson SL, Colliver JD (2007) Drug use patterns and trends in rural communities. J Rural Health 23(Suppl):10–15

Gupta V, Yadav K, Anand K (2010) Patterns of tobacco use across rural, urban, and urban-slum populations in a North Indian community. Indian J Community Med 35(2):245

Gururaj G, Isaac M, Girish N, Subbakrishna DK (2004) Final report of the study health behaviour surveillance with respect of mental health submitted to the Ministry of Health and Family Welfare. Government of India, New Delhi

Gururaj G, Girish N, Benegal V (2006) Alcohol control series 1: Burden and socio-economic impact of alcohol – the Bangalore study. World Health Organisation, Regional Office for South East Asia, New Delhi

Gururaj G, Murthy P, Girish N, Benegal V (2011) Alcohol related harm: implications for public health and policy in India, Publication no. 73. NIMHANS, Bangalore

Horvath KJ, Iantaffi A, Swinburne-Romine R, Bockting W (2014) A comparison of mental health, substance use, and sexual risk behaviors between rural and non-rural transgender persons. J Homosex 61(8):1117–1130

HRSA. Defining rural population [Internet]. [cited 10 Aug 2017]. Available from: https://www.hrsa.gov/ruralhealth/aboutus/definition.html

https://www.unodc.org/documents/drug-prevention-and-treatment/16-10463_Rural_treatment_ebook.pdf [Internet]. [cited 7 Jul 2017]

International Institute for Population Sciences (IIPS) and Macro International. 2007. National Family Health Survey (NFHS-3), 2005–06: India: Volume I. Mumbai: IIPS

Isaac M (1998) Contemporary trends: India. In: Grant M (ed) Alcohol and emerging markets: patterns, problems and responses. Taylor and Francis, Baltimore, pp 145–176

Malek-Ahmadi M, Degiorgio L (2015) Risk of alcohol abuse in urban versus rural DUI offenders. Am J Drug Alcohol Abuse 41(4):353–357

Marsh JC, Cao D, Guerrero E, Shin H-C (2009) Need-service matching in substance abuse treatment: racial/ethnic differences. Eval Program Plann 32(1):43–51

Monnat SM, Rigg KK (2016) Examining rural/urban differences in prescription opioid misuse among US adolescents. J Rural Health 32(2):204–218

Mpanza DM, Govender P (2017) Rural realities in service provision for substance abuse: a qualitative study in uMkhanyakude district, KwaZulu-Natal, South Africa. South Afr Fam Pract 59(3):110–115

National Institute of Mental Health and Neuro Sciences (2012) Patterns & consequences of Alcohol Misuse in India – an epidemiological survey. Bangalore: National Institute of Mental Health and Neuro Sciences. [cited 9 Aug 2017]. Available from: http://nimhans.ac.in/cam/sites/default/files/Publications/WHO_ALCOHOL%20IMPACT_REPORT-FINAL21082012.pdf

National Mental Health Survey of India, 2015–2016. Prevalence, patterns and outcomes. Supported by Ministry of Health and Family Welfare, Government of India, and implemented by National institute of Mental Health and Neurosciences (NIMHANS), Bengaluru, in collaboration with partner institutions

Neufield KJ, Peters DH, Rani M, Bonu S, Brooner RK (2005) Regular use of alcohol and tobacco in India and its association with age, gender, and poverty. Drug Alcohol Depend 77(3):283–291

Olivier L, Urban M, Chersich M, Temmerman M, Viljoen D (2013) Burden of fetal alcohol syndrome in a rural West Coast area of South Africa. S Afr Med J 103(6):402–405

Pullen E, Oser C (2014) Barriers to substance abuse treatment in rural and urban communities: a counselor perspective. Subst Use Misuse 49(7):891–901

Ray R (2004) The extent, pattern & trends of drug abuse in India, National Survey. Ministry of Social Justice & Empowerment, Government of India & United Nations Office on Drugs & Crime, Regional Office for South Asia, New Delhi

Ray R, Sharma HK (1994) Drug addiction – an Indian perspective. In: Bashyam VP (ed) Souvenir of ANCIPS 1994. Indian Psychiatric Society, Madras, pp 106–109

Reddy VM, Chandrashekar CR (1998) Prevalence of mental and behavioural disorders in India: a meta-analysis. Indian J Psychiatry 40(2):149–157

Rhew IC, Hawkins JD, Oesterle S (2011) Drug use and risk among youth in different rural contexts. Health Place 17(3):775–783

Rigg KK, Monnat SM (2015) Urban vs. rural differences in prescription opioid misuse among adults in the United States: informing region specific drug policies and interventions. Int J Drug Policy 26(5):484–491

Scheil-Adlung X (2015) Global evidence on inequities in rural health protection: new data on rural deficits in health coverage for 174 countries, Extension of Social Security series, no. 47. International Labour Organization, Geneva [Internet]. [cited 7 Jul 2017]. Available from: http://www.ilo.org/wcmsp5/groups/public/%2D%2D-ed_protect/%2D%2D-soc_sec/documents/publication/wcms_383890.pdf

Schulden JD, Thomas YF, Compton WM (2009) Substance abuse in the United States: findings from recent epidemiologic studies. Curr Psychiatry Rep 11(5):353–359

Shannon LM, Havens JR, Hays L (2010) Examining differences in substance use among rural and urban pregnant women. Am J Addict 19(6):467–473

Shaw MR, Grant T, Barbosa-Leiker C, Fleming SE, Henley S, Graham JC (2015) Intervention with substance-abusing mothers: are there rural-urban differences? Am J Addict 24(2):144–152

Subramanian SV, Nandy S, Irving M, Gordon D, Smith GD (2005) Role of socio economic markers and state prohibition policy in predicting alcohol consumption among men and women in India: a multilevel statistical analysis. Bull World Health Organ 83(11):829–836

Substance Abuse and Mental Health Services Administration, Center for Behavioral Health Statistics and Quality (2012) The TEDS report: a comparison of rural and urban substance abuse treatment admissions, Rockville [Internet]. [cited 7 Jul 2017]. Available from: https://www.samhsa.gov/sites/default/files/teds-short-report043-urban-rural-admissions-2012.pdf

Thimmaiah G (1979) Socio-economic impact of drinking, state lottery and horse-racing in Karnataka. Sterling, New Delhi, p 43, 120

Tshitangano TG, Tosin OH (2016) Substance use amongst secondary school students in a rural setting in South Africa: prevalence and possible contributing factors. Afr J Prim Health Care Fam Med 8(2):934. Available from: http://www.ncbi.nlm.nih.gov/pmc/articles/PMC4845911/

United Nations Department of Economic and Social Affairs. Population density and urbanization. Available from: https://unstats.un.org/unsd/Demographic/sconcerns/densurb/default.htm

Warner BD, Leukefeld CG (2001) Rural-urban differences in substance use and treatment utilization among prisoners. Am J Drug Alcohol Abuse 27(2):265–280

Warren J, Smalley K, Barefoot K (2015) Perceived ease of access to alcohol, tobacco and other substances in rural and urban US students. Rural Remote Health [Internet] 15:3397. [cited 7 Jul 2017]. Available from: http://www.rrh.org.au/articles/subviewnew.asp?ArticleID=3397

WHO (2010) Increasing access to health workers in remote and rural areas through improved retention: global policy recommendations [Internet]. World Health Organization (WHO Guidelines Approved by the Guidelines Review Committee), Geneva. Available from: http://www.ncbi.nlm.nih.gov/books/NBK138618/

World Bank Country and Lending Groups – World Bank Data Help Desk. (n.d.). Retrieved May 29, 2019, from https://datahelpdesk.worldbank.org/knowledgebase/articles/906519-world-bank-country-and-lending-groups

Ziaaddini H, Ziaaddini T, Nakhaee N (2013) Pattern and trend of substance abuse in eastern rural Iran: a household survey in a rural community. J Addict [Internet]. [cited 2 Jul 2017]. Available from: https://www.hindawi.com/journals/jad/2013/297378/

Stress and Rural Mental Health

6

Bettahalasoor Somashekar, P. S. Reddy, and Balaji Wuntakal

Contents

B. Somashekar (✉)
Adult Community Mental Health Services, Swanswell Point, Coventry and
Warwickshire Partnership NHS Trust, Coventry, UK
e-mail: bettahalasoor.somashekar@covwarkpt.nhs.uk; bsomashekar@hotmail.com

P. S. Reddy
Department of Psychiatry, Queens University, Barrie, ON, Canada
e-mail: Psreddy50@hotmail.com

B. Wuntakal
Solent NHS Trust, Langston Centre, St James Hospital, Portsmouth, UK
e-mail: balaji.wuntakal@solent.nhs.uk

© Springer Nature Singapore Pte Ltd. 2020
S. Chaturvedi (ed.), *Mental Health and Illness in the Rural World*, Mental Health and
Illness Worldwide, https://doi.org/10.1007/978-981-10-2345-3_27

Abstract

Stress and rural are difficult words to define conceptually, and this chapter gives
some ideas about operational definitions to use in research framework. A brief
overview on evolutions of stress concept has been provided. Rural stress is
conceptualized as stress unique to rural area and stress in the rural context
but common to all. The very factors, which define rural areas, may sometimes
work as perpetuating factors for stress in rural population. The manifestations of
stress can be both physical and psychological and depend largely on person's
coping abilities. As manifestations are seen as "subthreshold" for categorical
classificatory systems, research is focused on "disorders" rather than manifestations of stress, sometimes giving an impression that stress may not be important
despite significant burden to the individual from the symptoms. Review of
the literature on "stress" reveals that it has been focused mainly in urban
populations when in reality the majority of population is scattered around in
rural areas. Farming is exclusively a rural activity and hence farming stress is
discussed in a separate section. It is important to recognize that seasonal variations in farming can affect the prevalence of stress manifestations throughout
the year. Most interventions for rural stress are extrapolation from general
stress research and prescriptive with limited empirical evidence. Therefore any
suggested interventions need to be adapted carefully to rural settings.

Keywords

Rural · Stress · Stressors · Coping · Mental illness · Farming · Agriculture

Introduction

Circumstances rule men; men do not rule circumstances – Herodotus

The terms rural and stress both are fluid and not precisely definable. Although there
is a notional understanding and acceptance of what stress and rural means, there is no
standard definition yet acceptable to all. The meaning with which these words are
used depends on the person's profession using them and also for the purpose they are
used. For example, a sociologist considers life events as stress, while a psychiatrist
considers symptoms/reaction to life events as stress, and a psychologist may see
faulty coping pattern as stress. Similarly, the word rural can be defined in several
ways using parameters such as geographical, population density, and perception of

individual, as a result the definition differs depending on the purpose for which it is needed. This makes it difficult to conduct research and generalize the findings. Despite our difficulties in defining these terms, an operational definition is still required to understand and conduct research and make policies.

There is overwhelming evidence that stress plays an important role in the development and maintaining several physical and mental illnesses. This chapter is focused on stress-related mental health problems from rural context and effect of stress on physical health are not covered here.

This chapter has attempted to address conceptual issues around the use of terms "rural" and "stress" followed by a review on "rural stress." There has been limited research focused exclusively on rural mental health, and often the main focus has been on farming and therefore a separate section is devoted to stress of farming.

Conceptual Issues Related to Rural

A universally acceptable definition of rural area is logically not possible as the idea of rurality is relative and changing encompassing several aspects such as population density, land use, and social and physical environment. The definition of rural area varies with countries and within individual countries. In most circumstances, the qualifiers used to define rurality are population based, geography based, or perception of individual. Population-based factors include population size, population density, and settlement factor, and geography-based factors include distance from urban center and postcode. Therefore, the percentage of population living in rural area depends on the criteria used to define rural and varies with countries. The definition of "rural" for policymaker is different to what is "rural" for health professionals. Similarly what is rural in India is different to what is rural in Canada or the United Kingdom. In sociological and health research, there is no single accepted definition used, and in some studies it is arbitrary based on the perception of the subject or the investigator.

Although there are several definitions for rural, the key characteristics are common across all definitions. They include less density of population, lesser amenities, less ability or access to various modes of transport, and lesser density of communal buildings. It is worth recognizing that some of the so-called urban areas/semi-urban areas adjacent to larger cities have characteristics not too distinct or dissimilar to rural areas.

Stress: Conceptual Issues

Stress is a universal phenomenon seen in inanimate objects, plants, animals, and humans. In living organisms, the common denominator of stress may be simple activation of survival process involving physiological arousal reactions to threatening or unpleasant stimuli. This process has become complex with evolution and development of specialized organs with specific functions. In higher species with a well-developed nervous system, the responses to stressor (threat) are both physical and emotional and are directed at survival. In humans with highly specialized complex organ systems, the stress response is dynamic and multidimensional involving various systems (neuronal, hormonal, immune, etc.).

The concept of stress in humans has evolved over several hundred years but with use of different terms (Cannon 1963; Selye 1956, 1976). The notion of stress affecting health, particularly cardiovascular health, has a long history. In 1628, William Harvey has made reference to the relation between the mind and heart stating that activities that afflict the mind in the form of pain or pleasure and fear or hope extend their influence on the heart.

In the seventeenth century, the terms such as hardship, strain, adversity, and afflictions were used to denote stress. In the eighteenth and nineteenth centuries, the terms such as force, pressure, strain, and strong effort excreted upon a person or object were used to denote stress.

The importance of stress on health was widely recognized in the late nineteenth and early twentieth century. The initial studies on stress were carried out by physiologists. In second half of the nineteenth century, French physiologist Claude Bernard described stress without using the word stress. He stated that the internal environment (mallei interna) must remain constant despite changes in external environment, and in his opinion the fixity of the internal environment is the condition of free and independent life. In 1842, Thomas B. Curling noted the development of gastrointestinal ulcers in patient with severe skin burns, which are called stress ulcers. In 1867 Christian Albert Theodor Billroth noted similar finding after major surgical intervention complicated by infections.

About 50 years later (1914), an American physiologist Walter B. Cannon developed the concept of theory of flight and fight syndrome. He explained that the coordinated physiological process maintains the steady state in an organism referred to as homeostasis. Cannon conceived stress as a disturbing homeostasis causing the movement away from physiological equilibrium which results from exposure to physiological and psychological stimuli. Bodily responses to physiological stimuli have a survival advantage and form the basis of evolutionary perspective of stress.

Hans Selye, very early in his career (as a medical student), noted a nonspecific response from many diseases that threaten homeostasis, which have many common symptoms. However, after several years later in his animal experiments, he confirmed the nonspecific response to stressful stimuli and introduced the term general adaptation syndrome (GAS) or biological stress syndrome. He elaborated this further and divided the process into three stages with introduction of key concepts. Firstly, adaptation energy is finite, for example, an animal can cope with extreme cold (stress) up to certain extent. Secondly, depending on the innate capacity, the organism eventually begins to resist. He explained this by comparing to a machine, machines gradually wear out even if the machine has enough oil to function, and similarly man becomes a victim of constant wear and tear. Finally, the organism gives up with exhaustion.

Although Hans Selye worked on stress from 1936, he first used the word stress in 1946. Walter Cannon developed the stress response concept well before Hans Selye in his work on fight or flight response in 1914. Both Cannon's theory and Selye's theory focused on the physiological aspect of stress but not on the psychological aspect except recognition that psychological stimuli can be a possible stressor. These two theories influenced stress research and formed the basis of neuroendocrine research into stress, but they did not clarify the process of psychological or social aspect of stress.

Many other prominent people like Freud and Pavlov worked on issues related to stress during the contemporary period; however, it was the introduction of GAS by Hans Selye that made a profound impact. Over the years, the stress-related research expanded with the involvement of professionals from medicine, psychology, and sociology and new concepts/explanations developed (e.g., effect of physical environment on occupation (studies of performance and noise), coping, effect of war on soldiers, public, etc.).

Around Selye's period, several other developments also influenced the course of stress research. Harold Wolff (1953), a physician, floated an idea that life stress played a role in etiology of disease. Around the same time, the psychoanalytical theories were taken over by the behavior approach in psychosomatic medicine with animal experiments. In 1945, the Second World War provided another impetus to stress research with development of concept such as combat stress. The interest in combating stress further expanded stress research on individual variability and vulnerability to stress. The developments in several scientific disciplines during this period helped further expand research stress. For example, biochemistry helped in extraction and reliable and objective measurement of stress hormones such as adrenaline and cortisol. Similarly the developments in psychophysiological technology allowed for precise measurement of variables such as heart rate, galvanic skin response, and other indices of autonomic arousal. Stress research between 1954 and 1970, starting from military work, expanded into other areas such as travel in space and effect of physical environment on performance (e.g., noise on performance) with development of concept of occupational stress (Cox 1993).

From 1960 psychological research attempted to provide theoretical account to explain individual differences to stress. In 1966 Richard Lazarus studied performance under stress and developed the concept of cognitive appraisal to explain the individual differences stating that stress is stressful when it is perceived that way. This not only helped in explaining individual differences but also helped in the development of coping strategies.

In the same period, sociological research looked at effects of life changes (positive and negative) on health. This has led to the life events research and development of an assessment tool called Social Readjustment Rating Scale (SRRS) by Holmes and Rahe (1967) to measure the stressfulness of the life events.

The initial focus on physiological response moved toward studying individual difference to a stressful situation (1960–1970). This has paved way to study environmental adversities on health and illness, and life events research became prominent. Several studies showed that negative or stressful life events have negative effect on health. This has led to examination of social events such as riots, trends in suicide rates, and rates of mental illness and paved the way for the development of a new concept of social support. Although there is conflicting evidence to support this idea, the evidence suggests that social support in the form of emotional and instrumental support helps such as material and practical support which reduce the negative effects of stress on health.

Stress is either positive (pleasant) or negative (unpleasant) . It is well accepted that some degree of stress is necessary for organisms to function effectively. It acts as a stimulus to take positive action regarding survival and as a source of motivation. This kind of stress is referred as "eustress." However, if the stress exceeds one's ability to adapt or is beyond one's resources to cope or persists beyond the point of relevance, it becomes negative or unpleasant and commonly called as "distress."

Although there is no standard definition of stress acceptable to all, a single definition of stress is difficult because it has several dimensions such as well-being, health, illness, social, psychological, etc. In addition, like happiness, success, and failure, stress is subjective and makes it even difficult to define precisely. Everyone knows what stress is and experiences it but cannot easily translate it into words. For these reasons, understandably in most stress-related discourses (and research), the meaning of stress used is often "tangential." The discourse begins by giving an oblique reference to stress and take off into areas which are either causes (stressors) or the consequences (e.g., health problems). Furthermore, the causes and consequences are often used interchangeably, for example, isolation can be a cause or consequence of an illness. Until a universally agreed or at least operational definition is found, stress research will continue to be imprecise with reduced clarity and hence reduced clinical utility.

The definitions of stress used by various professionals fall into three areas such as stress as stimulus, stress as a response, and stress as interaction between person and environment. However, in reality stress encompasses all three aspects and often all these events happen concurrently. Therefore, it is best to consider stress as a process consisting of three phases such as cause (environment), interaction with individual (coping and adaption), and the consequences (effect of stress). Thus, stress can be conceptualized as a process in which stressful (environmental) events cause symptoms/suffering mediated by individual perception of the event, coping pattern, and availability of resources. The perception and coping are determined by factors such as personality, coping style (trait), and resources (example social support). The end point of stress process such as physical and psychological symptoms is usually the consequences of perception of events and the behavioral and psychological attempts to cope with the situation rather than from the event itself.

Thus a comprehensive understanding of stress is possible by studying all three aspects of stress in a single individual at given time as these factors do not operate in isolation.

The three components of stress process are:

1. Stressors
2. Coping and adaptation
3. Consequences

Stressors (stimulus) can be classified in several ways. They can be internal or external, physiological or psychological, positive or negative, and exit or entry as commonly used in life events research. They may be acute or chronic.

The consequences of stress are many and one of which is impact on health, both physical and mental. The health manifestation of stress response has several components such as physical, psychological, behavioral, cognitive, etc. The mental health manifestations of stress are a spectrum from a minor annoyance at one end to onset of serious mental illness (reactive psychosis) or suicide at the other extreme.

The common denominator of all definitions of stress is imbalance, mismatch, or lack of equilibrium between demand and individual ability to meet that demand. The ability to meet the demand is called coping.

Coping is defined as the process of managing external and/or internal demands that are appraised as taxing or exceeding the resources of the person (Folkman 2013). Understanding different ways people cope with stress is helpful in developing strategies to both prevent and manage stress. The coping may actually begin even before the actual stress happens (anticipatory coping).

Coping is broadly divided into two types: (1) problem-focused coping in which the focus is to solve the problem and (2) emotion-focused coping which deals with the emotions that arise as a result of the stressful situations. Although there are several ways people cope, Lazarus and his colleagues (Lazarus and Folkman 1984) identified the following eight ways of coping:

1. Standing up for one's rights and fighting for what he wants
2. Active problem-solving – involving defining the problem, exploring options, and finding solutions for problems
3. Distancing oneself psychologically from the situation
4. Self-control – exercising self-control over expression of emotions
5. Accepting responsibility – for the events that has led to stress
6. Escape avoidance – getting away from the problem
7. Positive reappraisal – trying to see situation from a positive point way
8. Seeking social support – either to get practical help to solve the problem or for emotional support

The first two are problem-focused coping, also referred to as confrontive coping, aiming to reduce the impact of loss or suffering, the next five are emotion-focused coping, and the last a mixture of both.

In addition, people cope by anticipating and preparing for stress which is termed prospective or prophylactic coping. Stress inoculation, a concept derived from prospective coping, is another problem-focused method.

Although there are several methods of coping and different methods are effective in different situations, people usually have a habitual way of coping, which is referred to as coping trait or coping style. It is likely that people use "problem-focused coping" strategies if they appraise a situation in such a way that they believe they have control on the situation and they drift into "emotion-focused coping" strategies if they believe they have no control. Some may use prospective coping if they anticipate losing control on the situation, for example, preparing well or taking insurance for loss of crop due to unpredictable climatic changes or pest infestation.

There has been criticism for almost every concept related to stress such as definition of stress, methods of coping, social support, and life events research. We have therefore deliberately avoided going into details of this, as there was no consensus. For more in-depth understanding of stress-related issues, readers are encouraged to consult Selye (1956, 1974, 1976), Lazarus (1966), Lazarus and Folkman (1984), Bartlett (1998), Cox (1993), Carver and Connor-Smith (2010), and Sarafino and Smith (2014).

Rural Stress

In a broad sense, rural stress encompasses two aspects such as stress which is unique to rural areas (e.g., farming) and stress experienced in rural context but common to all (financial difficulties) (Lobley 2005). Although a wide range of stress is common to both urban and rural population, stress experienced in rural area is magnified. For example, lack of service in rural area makes a person to travel for long distances; additional time and travel expense make the same stress experience more severe and unpleasant.

Defining rural stress is even more complicated, and most of the studies that looked at rural stress have not defined it. To identify the stressors and their consequences, studies have used demographic data, items in life events scales, and components of popular assessment tools such as PHQ (Patient Health Questionnaire), HDRS (Hamilton Depression Rating Scale), and GHQ (General Health Questionnaire). In addition the voluntary sector organizations in some developed countries like the United Kingdom (MIND, Rethink, Samaritans, etc.) that work on rural stress have not given any acceptable definition; instead they list the causes or consequences of stress and offer pragmatic solutions.

For the purpose of simplicity and better understanding, research into rural stress can be categorized into three components; (1) stressors, (2) coping, and (3) adaption and response or outcome of stress. Let us examine each in more detail.

Stressors

Life events research and demographic characteristics are typical examples of studies looked into this area. Several factors considered as stressors by the rural population are also considered as stressors in urban setting. However some stressors are relatively unique to rural setting such as geographical isolation and hazards associated with certain occupations (e.g., farming).

Although many factors, which have been listed as stressors, are common for both urban and rural settings, the events and the manifestations of stress in rural areas are hidden. The "iceberg stress model" has been adopted to explain this phenomenon suggesting that stress among rural worker is revealed by high suicide rates (Lobley et al. 2004). Stress is prevalent in all groups of rural population;

however, certain groups are considered "at risk." They include the elderly, unemployed, families or mothers with young children, homeless, and farmers and their families (Jones et al. 1994; O'Brien et al. 1994; Monk and Thorogood 1997; Paykel et al. 2003).

Several studies have identified different forms of isolation such as physical, social, cultural, and psychological isolation as an important stressor. The consequences of isolation are decreased economic activity, limited opportunities, and poor service infrastructure. Self-imposed psychological isolation stemming from social attitudes reflecting a strong sense of self-reliance was identified (Hoyt et al. 1995, 1997; O'Brien et al. 1994).

Coping, Adaptation, and Social Support

A few studies looking at coping and adaptation used by farmers in rural area reveal that men and women adapt differently to stress. Men seem to use problem-focused coping more often, while women use emotion-focused coping. This is perhaps reflective of traditional roles such as need for achievement for men and caring role for women. Women's adaptive strategies center around consumption, making sacrifices in spending on the household budget, while men worry more about the stigma associated with their perceived failure. The long-held belief that men should have control on the situation makes the rural men more vulnerable to financial and job-related stress.

Social support is known to have a direct relationship on health and well-being. The rural population holds strong relationships within family and friends. Social support allowed people to live alone and maintain independence. Perceived quality of social support is a major indicator of life satisfaction. People with high level of social support experience less stress and are able to cope more successfully when faced with stressful situation. Social support provides individual with security, worthiness, and a sense of identity during the time of crisis and decreases feeling of hopelessness (Letvak 2002; Hiott et al. 2008).

Response or Outcome of the Stress

In contemporary classificatory systems such as the International Classification of Diseases (ICD 10) (WHO 2004) and Diagnostic and Statistical Manual of Mental Disorders (DSM-5R) (American Psychiatric Association 2013), stress-related disorders fall into a spectrum consisting of acute stress reaction, adjustment disorders, post-traumatic stress disorder, and psychosis (reactive psychosis, an older term but clinically more useful). Unfortunately stress research in general has not been designed to identify these specific disorders. This is partly due to a wide spectrum and subthreshold nature of symptomatology, and hence it is often difficult to generalize the findings of stress research across countries and populations.

A few studies which have looked into relations of stress and symptomatology have documented diagnosis, for example, studies on isolation revealed that isolated people are more likely to be depressed. Similarly people who have financial difficulties are likely to experience increased level of anxiety and stress (Hoyt et al. 1997; Jones et al. 1994).

Studies did not look into diagnostic categories in a systematic way; therefore, it is difficult to draw a conclusion about the nature of disorders caused by stress.

As highlighted above (under conceptualization of stress), the consequences of stress manifest in several domains such as physical, emotional, cognitive, and behavioral symptoms. Several studies have identified them and they are summarized in the table (Jones et al. 1994; Simkin et al. 1998; Hiott et al. 2008; Paykel et al. 2003; Lobley et al. 2004; Monk and Thorogood 1997) (Table 1).

Table 1 Manifestation of stress: common symptoms and signs of stress

Physical symptoms	Emotional symptoms	Cognitive symptoms	Behavioral symptoms
Nausea	Apathy	Preoccupation	Withdrawal
Shortness of breath	Anxiety	Confusion	Argumentative
Headache	Despair	Forgetfulness	Aggressive
Backache	Fear	Poor judgment	Impulsive behavior
Dizziness	Depression	Self-attribution	Critical of others
Exhaustion Palpitation	Ideas of	of blame	Doing risky or careless
Indigestion or stomach	worthlessness	Persistent	things such as excessive
upset	Excitement	worrying about	drinking, gambling, or drug
Skin rashes	Feeling	little things	use
Muscle tension	irritable	Memory	Isolation by avoiding
Pain	Feeling of	impairment	people, places, and events
Clenched teeth	tension	Poor	Working longer hours but
Rapid breathing heart rate	Impatient or	concentration	with diminished
Raised blood pressure	emotional for	or decision-	effectiveness
Loss of enjoyment and	no apparent	making	Poor work performance
interest in activities usually	reason	problem	Loss of output
enjoyed	Frustration	Procrastination	Low quality
Loss of energy and constant	Anger		Deteriorating relationship
tiredness	outbursts		with colleagues
Changes in sleep pattern	Difficulty		Communication problems
Sleeping difficulties despite	controlling		Verbal and physical abuse
physical exhaustion or	emotions		Sarcastic arguments
sometimes sleeping too	Loss of		Poor time keeping
much	confidence		Difficulty retaining
Loss of appetite or	Low self-		information
overeating	esteem		Trouble adapting to
	Continuous		changes or circumstances
	feeling of		
	anxiety and		
	tension		
	Loss of		
	motivation or		
	commitment		
	A sense of		
	being out of		
	control		

From the available literature, it is difficult to cluster symptoms to arrive at definitive diagnosis of "stress-related disorders" as defined by classificatory systems because the symptoms are widespread and do not fit in the existing classificatory system. However if a strict criterion is used to make a diagnosis, most patients are likely to end up receiving a diagnosis of adjustment disorder according to ICD 10.

Contradictions About Rural Mental Health

There is a popular perception that rural life is calmer, slower, happier, and healthier than busy urban life. If this were to be true, it is difficult to explain the outward migration of rural population to urban areas and general decline in rural population. The reason for this perception is unclear and has not been systematically examined. It is possible that the natural beauty of the scenery and slow pace of life may make the occasional visitor to the countryside to perceive that living in rural area is better than urban area. However people living in rural areas have a different set of problems which can be stressful.

The findings of rural stress research are contradictory. Review of the literature about urban and rural differences suggests that mental health is probably better in rural areas than in urban areas with exception of suicide (Paykel et al. 2003; Nicholson 2008; Gregoire and Thornicroft 1998; Gregoire 2002). Suicide rate is higher in both male and female in rural areas compared to urban areas in most countries (Jones et al. 1994). Although there is problem in defining rural and urban, evidence suggests that major mental illness is less common in rural than urban areas, especially strongest for schizophrenia. Similarly, the incidence of nonpsychotic or common mental disorder such as depression and anxiety is either the same or less common in rural area.

However, on the other hand, the indirect measures of mental health and stress such as suicide rates, unemployment, poverty, intellectual disability, and deprivation are all common in rural population (Wilkinson and Marmo 2003; WHO and Calouste Gulbenkian Foundation 2014; Allen et al. 2014).

It is well known that health outcomes of people vary depending on the location where they live. Rural community across countries have reduced access to medical services and health insurance. The measures of health and well-being indicate that rural populations have worse health outcomes. They have premature mortality across all age group due to unintentional injuries, suicide, and chronic obstructive pulmonary disease. Life expectancy is lower in rural area because it is directly proportional to economic condition and provision of heath (Wilkinson and Marmot 2003). Variables such as climate, soil, rainfall, temperature, altitude, and seasonality have great influence on the food crop production and economic situation. The availability and quality of water and vector-borne disease affect health and contribute to lower life expectancy.

In 2003, the WHO Europe suggested that the social determinants of health included stress as one of the determinants in their list because of its influence on health (Wilkinson and Marmot 2003). The relationship between chronic stress and negative health outcomes are well established. The negative health outcome was brought about by both direct and indirect effects of stress. The direct effect of stress on the physiology of a person is through organ systems such as adrenocortical, neuroendocrine, and immune system. The indirect effect is through strain on the psychological resources on the person who is stressed. The chronic stress is common among poorer patients due to excessive worry about their financial situation, housing, and employment which are often seen in rural population.

Poverty predisposes people to unemployment, financial hardships, frustrated aspirations, crime, victimization, and illnesses. These factors indirectly affect people's self-confidence, self-esteem, and sense of no control on life seeding to family and marital conflicts and overall poor quality of life. The indirect measures of mental health such as suicide rates, employment status, isolation, and homelessness are worse in rural areas than urban areas (WHO and Calouste Gulbenkian Foundation 2014; Reddy and Mishra 2008; ILO 2017).

The apparent contradictory findings may indicate different problems such lack of availability of treatment facilities for serious mental illness forcing people to move nearer to places where services are available (urban areas). It is possible that some of the urban problems are reflection of displaced rural issues (Breslow et al. 1998) because people drift to places where services exist. Rural patients with mental illness may eventually become urban service users due to lack of specialist services locally. Therefore, migration of population is not always necessarily due to better employment opportunities but due to availability of other services including access to mental health services. The other possibility for the contradiction could be high tolerance and belongingness in the rural community.

The issues contributing to contradictory findings are discussed in more details below under management section.

Farming Stress

A separate section is devoted to farming stress in the chapter because it is the largest occupation in the world, and farming is exclusively a rural activity. According to the International Labour Organization, it is estimated that 1.1 billion or over 31% of global workforce were engaged in agriculture in 2013, which has reduced from 45% in 1991 (ILO 2017). There has been progressive decline in people working in agriculture due to mechanization, improved method of cultivation, and use of advanced technology. The percentage of people engaged in farming varies across countries from 2% in developed countries like the United States and Canada to 80% in underdeveloped countries like African countries.

A combination of high suicide rate in farming community compared to other occupations across all cultures and media attention to farmer's suicide has contributed to more systematic studies. The focus of most studies includes identifying stressors and outcome of stress, and only a few studies focused on coping and adaptation.

Several stressors identified in farming are common across cultures (Simkin et al. 1998; Allen 2000; Lobley 2004; Reddy and Mishra 2008; Gray 2011; Leonard 2015; Goffin 2014). They range from global issues shared by all farmers to individual factors and are studied under separate categories (see below). The following are general summary of the findings.

The stress of farming varies with type of farming. Subsistence farming where the farmer's aim is to satisfy the family needs and very little for selling has different problems, while intensive farming cultivated for commercial purposes has its own issues. Over the years, agriculture farming has progressed from subsistence farming to intensive farming, particularly in developed countries. The use of synthetic fertilizers and pesticides has resulted in increased production, but pollution, occupational hazards, soil erosion, and degradation are the major problems/concerns for agriculture today.

The Most common identified stressors in farming include machinery breakdown, uncertainty over harvest (weather, prices, and market), finances, isolation, and hazardous working conditions. The other stressors include government policy, time pressure, and future of farming family. It appears that the level of stress varies depending on the prevailing health and social situation of the countries.

Although most farmers feel farming is stressful, livestock farmers feel more stressed than agricultural farmers.

Seasons such as lambing, harvesting, and planting are considered as more stressful than other seasons. The feeling of isolation is associated with high level of anxiety and depression. Geographical isolation is one of the common stressful factors because working alone and making daily independent decisions contribute to elevated level of personal distress. The stress levels are further worsened by a feeling that farming has a bleak future. Understandably, the lack of control on factors such as fluctuations in prices, unpredictable weather, and administrative paperwork adds to stress.

Farming men and women react differently to stress. Farm women report more stress than men. Men usually consider personal failure due to traditional role for achievement. Farmers who perceive high level of psychological distress score high on neuroticism personality trait. They are likely to use emotion-focused coping than problem-focused coping, blame self for the origin of problems, and think that they cannot cope.

Global Factors

Global stressors are shared by all farmers across cultures although some experience them more intensely than others. They include climate change and its consequences, commodity markets, high interest rates and decline in rural infrastructure, and limited opportunities for social connection. Most of these stressors are usually beyond the control of farmers and have little influence. A sense of loss of control makes it difficult for them to cope and usually resort to emotion-focused coping by distancing or giving up. Globally, rural population has a higher rate of premature deaths due to suicide, poisoning, and accidents across all culture.

Change in demography of rural setting over the last three decades has contributed additional stress to families who continued to choose farming, particularly agriculture. It has become difficult to recruit skilled farmworkers due to general decline in rural population, outmigration, and people giving up farming for other occupations. Increased mechanization has also contributed to decline in employment opportunities with a consequent decline in intergenerational farm families, and also in a skilled labor force and increased demand for existing skilled workers. Increasing rural unemployment is due to unskilled people not getting jobs and well qualified leaving rural area.

National Factors

Common source of stress originates from government policy, bureaucracy, and unpredictable climatic pattern such as floods or drought. In addition, other stressors include seasonal conditions, weather dependency for crop, and loss of crops due to disaster, disease, or pests. Some of these may vary regionally in large countries with wide geographical areas such as the United States, Africa, Australia, China, and India.

Local Factors

Although local factors vary across countries, most farmers encounter them, perhaps with varying severities. The working conditions subject farmers to natural, chemical, and mechanical hazards. They include hazards due to direct exposure to natural dangers such as harsh climates, vector-borne disease, higher injuries due to manual work and machinery, and long working hours. Although there are no clearly established links, there are concerns regarding the psychological effect such as depression and suicidal ideation from exposure to organophosphorus chemicals and other pesticides commonly used in farming.

There is a wide variation in the use of mechanical devices in farming across different countries and mostly limited to few large farms. Although sophisticated mechanical devices are available, they are still not used widely in several countries like India due to small-scale individual farming and non-affordability.

Family Factors

It has been noted that the younger generation has less interest in farming as their career. They prefer regular income jobs even if it is paid less because of guaranteed income. This trend is very common in emerging markets like India. It has led to outmigration of youth to urban service sector causing overcrowding in urban areas and leaving the elderly who are often sick and single to cope themselves. The elderly prefer to remain in rural places as they cannot cope with new culture, busy urban life, and unfamiliar work. They feel culturally and psychologically isolated if they migrate to urban area and feel geographically and socially isolated if they remain in rural area.

This has contributed to conflicts, strain on interpersonal relationships, and poor communication within families. These are potentially stressful to both generations because parents want their children to continue farming, while children prefer to build their life in other occupations.

Farmer or Individual Level Factors

The most common source of distress at individual level includes finances and work. Financial worries include irregular and uncertain income, financial debt, the effect of government regulations, compliance with the bureaucracy and the amount of paperwork required. Farmers reported worrying about work including high workloads; time pressures; long working hours, especially during peak work seasons such as harvesting and calving/lambing; and difficulties with understanding new technology and solitary work. The most commonly reported symptoms include sleep problems, feeling irritable and despondent, fatigue, and high rates of stress.

Other areas which contribute to stress include physical health problems such as farm-related injuries, chronic physical symptoms, back problems and respiratory problems, acute and chronic poisoning, and zoonosis.

Management of Rural Stress

Barriers to Identification and Provision of Care in Rural Area

It is a challenge to identify stress-related illnesses and provide care in the rural population. A wide range of physical and psychological factors act as barriers. They can be broadly grouped into individual factors, illness factors, community factors, and service provision factors.

Individual Factors

Knowledge: Limited understanding of different manifestations of stress, not acknowledging that persistent stress manifests with symptoms, and not seeking professional help through health and social services are the consequences of poor knowledge. In addition even if recognized, people may not have enough understanding on the link between stress, mental illness, and availability of intervention.

Attitudes: Although stress is universal and experienced by all, rural people hide their problems as they take pride in their independence and self-efficacy. The culture of self-reliance and stoicism prevent seeking professional help for mental health manifestation of stress. They have tendency to underestimate their injury or illness and procrastinate, avoiding help. Unfortunately, the more psychologically distressed the less likely to seek help even if it is available. Individual factors such as suitability to mental illness may play a role in their tendency to see things in a fatalistic way, a sense of not having control and giving up. The consequences of this hidden stress including high suicide rates in farmers are well explained using the iceberg stress model (Lobley et al. 2004). Therefore, the services should be assertive in identifying those individuals.

Community Factors

Community factors that hinder rural patients to seek help from professionals include the following:

Confidentiality: Rural population understandably has greater concern about confidentiality due to small community.

Stigma: It is well known that the perceived stigma about mental illness and intervention has a negative impact on help-seeking behavior. Rural population has greater sensitivity to stigma of mental health problems, and consequently they do not seek help readily. Rural population see their stress-related symptoms as personal weakness or failure which prevents seeking help especially from mental health professionals. This is further reinforced by the traditional masculine attitude of farming.

Illness Factors

Subthreshold nature of symptoms: In the contemporary classificatory system, stress-related illnesses are diagnosed if cluster of symptoms meets the threshold criteria for a diagnosis. Although the psychological, cognitive, physical, and behavioral manifestations of stress are distressing and disabling, the symptoms may not meet criteria for a disorder in the current classificatory system due to subthreshold nature. Furthermore, the symptoms may not fit into a single diagnostic category even if the symptoms are severe because of a wide range of symptoms. It is also possible that the stress researchers did not systematically explore diagnoses using strict diagnostic criteria. Although they may not meet criteria of classificatory system, they are still a source of suffering and need intervention. The intervention need not be necessarily by highly specialized teams but with existing professionals in the health-care system with special skills.

Service Provision Factors

Mental health care provision is mostly based on established diagnoses. Necessity of having a diagnosis is not only important to seek secondary care but also necessary for insurance claims. Unfortunately as highlighted earlier, stress-related problems do not always meet diagnostic criteria due to subthreshold and wide range of symptoms. Furthermore, the services are divided and subdivided into smaller groups such as child and adolescent psychiatry, adult services, older adult service, and drug and alcohol services, and they are located in urban areas. This makes it even harder to get the right and timely help. Therefore services should be made available under single roof so that they do not have to travel to different places.

Lack of trained staff: Professionals are heavily concentrated in urban areas; this is even more prevalent in developing countries. People have to travel several miles to get services with limited resource and scanty transport spending a whole day away from work and loss of wages. This can be particularly difficult in extreme weather conditions and in certain seasons due to work commitment and loss of earning.

High cost: Access to health care is limited due to high cost because of dispersed population, poor transport facilities, and limited resources.

Interventions

The stress research can be broadly grouped into sociological and biomedical. Sociological research focuses on the origin of stress (stressors), while biomedical approach focuses on the clinical aspect of the stress. Sociological approach is helpful for policymaking and for developing preventive measures. Biomedical research focuses on consequences of stress such as effects of stress on health and well-being and is useful in helping our thinking around various treatment modalities.

Management of stress-related problems in the rural setting can be broadly categorized into general and individual. The general measures are directed at making the rural environment less stressful by reducing stressors and improving service provision. The individual measures are aimed to improve once ability to cope with immediate stressful experiences and develop mastery.

The mode of intervention is guided primarily by extrapolating from general stress literature and to a limited extent by specific research on rural population.

General Measures

Nature of help needed: The nature and range of rural stressor and the manifestation are such that medical approach to manage rural stress is not sufficient and in some circumstances may not be appropriate. The stressors such as financial, administrative, and price regulation are better addressed by policy change and practical help rather than psychiatric approach.

Counter Urbanization Measure
Due to lack of urban facilities such easy access to education, health-care facilities, entertainment, etc., people understandably prefer to live in urban area unless there are incentives to live in rural areas. Most western developed governments subsidize services in rural areas and provide transport facilities and other infrastructures to rural locations.

Integrate Mental Health Care into Primary Care
Patients with stress-related problems do not feel comfortable to see a specialist due to small community, but they may accept help from general health services. Therefore, breaking barriers between mental and physical health care provisions

facilitates seeking help, reduces stigma, and is also cost-effective. This can be achieved by integrating mental health care into already well-established primary care system. This is also a solution to limited availability of specialists as majority of stress-related illness can be managed in primary care.

Dealing with Subthreshold Symptomatology
Developing psychological therapy service (similar to Improving Access to Psychological Therapies service in the United Kingdom) which aims to address people's problems without diagnostic label could provide some answers to "sub-threshold nature of symptoms of stress."

Transport Problems
The rural population is scattered making specialist care expensive for the care provider and inaccessible for the rural population. Therefore, it is necessary to develop models using technology which works well in rural areas and also cost-effective. This has been recognized by various governments, and several rural health projects have been developed to promote mental well-being and prevent and treat illness. Some examples include mobile service units at a convenient location, for example, in market and mobile day care facilities. Use of advanced technology can also help, for example, telepsychiatry can be used for video conferencing, remote clinics, and remote one-on-one session.

Isolation and Social Support
Various forms of isolation such as geographical, psychological, social, and cultural have been identified as stressful factors in rural population. Similarly lack of social support and network contributes to poor work life balance. Perhaps, encouraging development of community centers by local council/local authorities will provide more opportunities for both structured and informal social interactions in rural communities, attempting to minimize the impact of social isolation (Findlay 2003).

Innovative Measures
Due to a wide variation in problems, availability of resources, and responsiveness of local government agencies, it is not possible to have a simple one-size-fit-for-all solution. Solutions suitable for one group may not be suitable for another group. Therefore, innovative measures are necessary to find suitable local solutions to local problems. In India and perhaps in most developing countries, distribution of commodities and fluctuation in prices are a major source of stress. Successful interventions for financial problems can be brought about more effectively by policy change and governmental intervention rather than individual: for example, in India, the dairy farmer is not as stressed as other farmers (agricultural farmers, livestock farmers) because there is well-established dairy marketing system which fixes the price and takes care of distribution system and hence there is no loss to farmer. This has been very well accepted by the farmers and the consumers. The same model can be adopted for other farming such as grains, vegetables, fruits, poultry, etc., for example, consumers will be happy to pay a fixed price if fresh fruits and vegetables are delivered by organized distribution system.

Funding

The funding of the services is greatly influenced by perception about the rural life among policymakers, politicians, and general public. Mental health policies and planning should have rural dimension with specific measurable targets for rural population.

Individual Measures

The individual stress management techniques are universal and similar to all, though their use may vary depending on the situation. They are usually directed at developing effective coping skills and appropriate use of social support system. There is large general literature available in area of stress management. Following is the extrapolation of it into rural area. It has been noted that most individual stress management advice is prescriptive with full common sense and good intention and with some empirical evidence. Therefore, the literature is not reviewed in depth and a list of intervention is provided.

Role of Medication

Although clinical manifestations of stress do not meet criteria for a diagnosis, medications and psychological intervention have a role in providing symptomatic relief and enhancing coping skills. Therefore approach to rural stress interventions should be pragmatic with judicial combination of social through policy change and biomedical by use of drugs and psychological intervention. The use of medications in rural setting is discussed in ▶ Chap. 20, "Psychopharmacology in Rural Settings" by C. Andrade.

Coping and Adaption

Weigel (1980) noted the following key characteristics of people who successfully coped with stress more than three decades ago which still hold relevance in individual stress management. They are summarized below:

- Self-awareness – they know their strengths, skills, and weaknesses. They practice time management and goal setting.
- Variety of reaction – they use many techniques depending on the situation such as problem-solving, assertiveness, relaxation, and exercise.
- Varied interests – they draw on many sources of personal satisfaction – hobbies or recreation.
- Active and productive – they make things happen and practice stress management when stressed, not just when things go well. In a sense they are continuous learners.
- Use support systems – no one can develop mastery and cope with all situations alone. They develop relationships with others to help re-establish competence in stress periods.

Similarly the leaflet produced by the North Dakota State University provides a good guide to stress management for farmers/ranchers (North Dakota State University 1998). The recommendation in the leaflet for successful management of stress by people has been expanded here (also refer above to conceptual issues related to stress).

1. Individual varies in their capacity to tolerate stress – individual stress level varies and can be improved by practicing and developing new skills.
2. Feeling in control – Use problem-solving coping to manage stressors which are within one's control and actively find solutions. People who can identify and accept the stressors which are beyond their control such as unpredictable weather, prices, and market fluctuation can cope better by preparing for it and by using prospective coping.
3. The stress level can also be reduced by changing one's attitude, perception, and meaning assigned to the stressful situation. An event is stressful only if the person perceives it so.

The following are some of the suggestions to improve coping with stressors. They are derived from cognitive and positive psychology, and many are actually common sense approach.

1. Control events – People find difficult to cope if they are overwhelmed by the multiple stressors present at that time. Therefore reduce accumulation of stressful situations at one time; thereby reduce the stress levels. Some techniques are as follows:
   Plan ahead – for example, repair the broken machinery during off time and get it ready for the time it is needed rather than waiting for the season to begin.
   Set priorities.
   Plan your time.
   Simplify life – reduce dependence on others excessively; reduce taking debts.
   Positive procrastination – postpone events which are not needed to be done immediately and plan at a time you can.
2. Control attitudes – how farmers view the situation is key to eliminating unwanted stress.
   See the bigger picture.
   List the stressful situation – segregate them into events where you have control and those in which you have no control. Change the situation where you have control and accept where there is no control.
   Use problem-focused coping instead of emotion-focused coping by actively solving problems for stressors where you have control and prospective coping where you can anticipate the stress.
   Take the stressful situation as opportunity.
   Focus on the positive aspect of stress – what has been achieved than what is not and what was successful over what failed.
   Set realistic goals.
3. Control responses.
   Relax.
   Take care of physical health.

Notice signs of stress – they include physical, psychological, behavioral, and
   cognitive symptoms and seek help from professionals.
Share burden.
Delegate.
Speak to a confidant.
Balance work, homelife, and pleasure.
Give up events which are beyond your control – learn to let go.

The combined approach is more effective than individual approach. As stated
above there are several reports, online guides, and leaflets. The following report by
Joseph Leonard (2015) is one such example. It provides a comprehensive approach
which can be adopted to local requirements.

Joseph Leonard, in his report on stress management in farming in Ireland for
Nuffield Ireland Farming scholarships, concludes the following. He has arrived at
these conclusions after a series of interviews and case studies from the United
Kingdom, the United States, and New Zealand to France and discussions and
interviews with Nuffield Scholars and officers.

The findings appear to be reflective of stress of farmers universally across cultures
with minor variation; therefore, the conclusions are presented here without any
changes:

- Managing stress involves the ability to cope with or lessen the physical and
  emotional effects of everyday pressure and challenges. These challenges can be
  often seem insurmountable at first, but with the right support and advice and by
  tackling them in structured ways, they can be successfully overcome.
- By giving young people an insight into what sort of challenges they will face in
  starting out in farming, they will be better equipped to overcome these issues.
  An ounce of prevention is better than a pound of cure.
- Succession planning and intergenerational relationships need to be tackled in
  an open manner early in the career of the succeeding generation to avoid
  conflict and familial stress.
- Realistic goal setting and proper planning help to identify a clear path to
  achieving ones ambitions. This clarity of vision helps to insulate against stress
  caused by unforeseen events.
- Farmers are more likely to talk about their problems to support personnel
  who have strong rural connection/empathy. Therefore support services would
  be better able to engage with farmers in need if they were manned by farmers or
  rurally based staff/volunteers.
- Developing a "farmer experience" leads debate in the media and in education will
  resonate far more strongly with the farmers than one lead by non-farming
  examples.
- Managing stress is not an independent risk; rather it is part of one's overall
  personal and business development. Following practices that limit stress will
  overlap with farm business viability while at the same time help to strengthen
  personal relationship which enhances success and "quality of life."

The report emphasizes safety, self-care, prevention, and developing training modules to recognize signs of distress and best methods to respond to farmer's needs. The popular training modules include Mental Health First Aid and safeTALK which are used in the United Kingdom, Australia, and Canada.

Conclusion and Future Direction

The assumptions about rural area influence policy decisions, service provision, and research priorities. Although significant population (20–50%) reside in villages/countryside, the clinical services concentrate in urban areas. In general mental health research has focused on the urban poor more than the rural poor. Similarly stress-related research focused less on rural compared to urban population.

It is difficult to generalize the research findings across cultures due to the lack of uniformity in defining rural and different meanings and characteristics of rural, many studies do not even define rural, and a mere urban rural differentiation is not enough to critically evaluate rural difficulties.

There are contradictory findings regarding direct mental health morbidly. Direct measure with exception of schizophrenia does not find any significant differences, while indirect measures of mental health morbidity are disproportionally high in rural area. The stress is hidden in rural area due to factors inherent to rurality and diagnostic approach.

Research should include a standard diagnostic category to suffering to enhance the chances of applying the findings to a wider community. A mere list of symptom patient report is not good enough to understand rural stress and develop an interventional model.

Social scientists and mental health professionals have a role to change the outdated perception about rural designing studies which tap into the hidden stress and develop specific intervention strategies for local needs.

Cross-References

▶ Organization of Mental Health Services in Rural Areas
▶ Psychopharmacology in Rural Settings

References

Allen RC (2000) Economic structure and agricultural productivity in Europe, 1300–1800. Eur Rev Econ Hist 4(1):1–25
Allen J, Balfour R, Bell R, Marmot M (2014) Social determinants of mental health. Int Rev Psychiatry 26(4):392–407
American Psychiatric Association (2013) Diagnostic and statistical manual of mental disorders (DSM-5®). American Psychiatric Publishing, Arlington
Bartlett D (1998) Stress: perspectives and processes. McGraw-Hill Education (UK). Open University Press, Buckingham and Philadelphia

Breslow RE, Klinger BI, Erickson BJ (1998) County drift: a type of geographic mobility of chronic psychiatric patients. Gen Hosp Psychiatry 20(1):44–47

Cannon WB (1963) The wisdom of the body. The Norton Library, New York

Carver CS, Connor-Smith J (2010) Personality and coping. Annu Rev Psychol 61:679–704

Cox T (1993) Stress research and stress management: putting theory to work, vol 61. HSE Books, Sudbury

Findlay RA (2003) Interventions to reduce social isolation amongst older people: where is the evidence? Ageing Soc 23(5):647–658

Folkman S (2013) Stress: appraisal and coping. In: Encyclopedia of behavioral medicine. Springer, New York, pp 1913–1915

Goffin A (2014) Farmers' mental health: a review of the literature. ACC Policy Team, Wellington

Gray JS (2011) Rural Mental Health Research White Paper for National Institute of Medicine. School of Medicine & Health Sciences. Centre for Rural Health 501, N. Columbia

Gregoire A (2002) The mental health of farmers. Occup Med 52(8):471–476

Gregoire A, Thornicroft G (1998) Rural mental health. Psychiatr Bull 22:273–277

Hiott AE, Grzywacz JG, Davis SW, Quandt SA, Arcury TA (2008) Migrant farmworker stress: mental health implications. J Rural Health 24(1):32–39

Holmes TH, Rahe RH (1967) The social re-adjustment rating scale. J Psychosom Res 11:213–218

Hoyt DR, O'Donnell D, Mack KY (1995) Psychological distress and size of place: the epidemiology of rural economic stress. Rural Sociol 60:707–720

Hoyt DR, Conger RD, Valde JG, Weihs K (1997) Psychological distress and help seeking in rural America. Am J Community Psychol 25:341–347

ILO (International labour organisation) (2017) Agriculture; Plantations; other rural sector. http://www.ilo.org/global/industries-and-sectors/agriculture-plantations-other-rural-sectors/lang%2D%2Den/index.htm. Accessed on 19 Sept 2017

Jones P, Hawton K, Malmberg A, Jones JW (1994) Setting the scene: the background to stress in the rural community, causes, effects and vulnerable groups. In: Rural stress: positive action in partnership. National Agriculture Centre, Stoneleigh

Lazarus RS (1966) Psychological stress and the coping process. McGraw-Hill, New York

Lazarus RS, Folkman S (1984) Stress, appraisal, and coping. Springer, New York

Leonard J (2015) Stress Management in Farming in Ireland. A report for Nullified Farming Scholarships Sponsored by Peter Daly Trust, Dairymaster and LIC

Letvak S (2002) The importance of social support for rural mental health. Issues Ment Health Nurs 23(3):249–261

Lobley M (2005) Exploring the dark side: stress in rural Britain. J R Agric Soc Engl 166:3–10

Lobley M, Johnson G, Reed M, Winter M, Little J (2004) Rural stress review. Centre for Rural Policy Research, University of Exeter

Monk AS, Thorogood C (1997) Stress, depression and isolation and their effects in the rural sector: are the younger rural generation equipped to overcome these problems? Part II. J Rural Manage Hum Resour 1(2):9–20

Nicholson LA (2008) Rural mental health. Adv Psychiatr Treat 14(4):302–311

North Dakota State University (1998) Stress management for farmers/ranchers. Accessed on 19 Sept 2017. https://www.ag.ndsu.edu/publications/kids-family/farm-stress-fact-sheets-stress-management-for-farmers-ranchers

O'Brien DJ, Hassinger EW, Dershem L (1994) Community attachment and depression among residents in two rural Midwestern communities. Rural Sociol 59:255–265

Paykel E, Abbott R, Jenkins R, Brugha T, Meltzer H (2003) Urban-rural mental health differences in Great Britain: findings from the National Morbidity Survey. Int Rev Psychiatry 15(1–2):97–107

Reddy DN, Mishra S (2008) Crisis in agriculture and rural distress in post-reform India. In: R Radhkrishna (ed) India development report, 2008: 40–53. Oxford University Press, New Delhi

Sarafino EP, Smith TW (eds) (2014) Health psychology: biopsychosocial interactions. Wiley, Hoboken

Selye H (1956) The stress of life. McGraw-Hill, New York

Selye H (1974) Stress without distress. Cox and Wayman, Philadelphia: Lippincott

Selye H (1976) The stress of life. McGraw-Hill, New York

Simkin S, Hawton K, Fagg J, Malmberg A (1998) Stress in farmers: a survey of farmers in England and Wales. Occupational and environmental medicine 55(11):729–734

Weigel RR (1980) Helping farmers handle stress. Journal of Extension 18(2):37–40

Wilkinson RG, Marmo M (Eds.) (2003) Social determinants of health: the solid facts 2nd edition international Centre for Health and Society World Health Organization Reginal office for Europe Scherfigsvej 8 Copenhagen DK2100 Denmark

WHO (World Health Organization) (2004) ICD-10: International statistical classification of diseases and related health problems. 10th revision, 2nd edition Geneva: World Health Organization

WHO (World Health Organization) and Calouste Gulbenkian Foundation (2014) Social determinants of mental health. Geneva, World Health Organization

Wolff HG (1953) Stress and disease. Springfield. IL, US: Charles C Thomas

Intellectual Disability in Rural Backgrounds: Challenges and Solutions

7

Mahesh M. Odiyoor and Sujeet Jaydeokar

Contents

Abstract

Intellectual disability (ID) or mental retardation is defined as a condition of arrested or incomplete development of the mind, which is especially characterized by impairment of skills manifested during the developmental period, skills which contribute to the overall level of intelligence, i.e., cognitive, language,

M. M. Odiyoor (✉) · S. Jaydeokar
Learning Disability, Neurodevelopmental Disorders, and Acquired Brain Injury Services, Centre for Autism, Neurodevelopmental Disorders and Intellectual Disability (CANDDID), Cheshire and Wirral Partnership NHS Foundation Trust, Chester, UK
e-mail: mahesh.odiyoor@nhs.net; sujeet.jaydeokar@nhs.net

© This is a U.S. government work and not under copyright protection in the U.S.; foreign copyright protection may apply 2020
S. Chaturvedi (ed.), *Mental Health and Illness in the Rural World*, Mental Health and Illness Worldwide, https://doi.org/10.1007/978-981-10-2345-3_28

motor, and social abilities (International Classification of Diseases ICD 10, World Health Organisation 1992). This is characterized by significant reduced ability to understand new information and learn new skills, i.e., impaired intelligence (usually taken as an IQ below 70), a reduced ability to cope independently (impaired social functioning), and starting before adulthood with a lasting effect on development (DOH, 2001).

There is recognition that the prevalence of mental health issues in people with ID is much more than the general population. Epidemiological data, available globally, on Intellectual disability population is very limited and at times not reliable. There is a wide variation in the availability of mental health services across the world and access to good mental health care in the rural areas of some countries could be difficult. Many of these countries have limited resources for health provision, and mental health services of people with ID are not always seen as priority. Hence, it would be essential to explore issues specific to people with ID when it comes to mental health services in rural areas and explore potential solutions.

Keywords
Intellectual disability (ID) · Learning disability (LD) · Mental retardation · Mental health · Rural settings · Challenging behaviours

Introduction

Intellectual disability (ID) or mental retardation is defined as a condition of arrested or incomplete development of the mind, which is especially characterized by impairment of skills manifested during the developmental period, skills which contribute to the overall level of intelligence, i.e., cognitive, language, motor, and social abilities (International Classification of Diseases ICD 10, World Health Organisation 2004). This is characterized by significant reduced ability to understand new information and learn new skills, i.e., impaired intelligence (usually taken as an IQ below 70), a reduced ability to cope independently (impaired social functioning), and starting before adulthood with a lasting effect on development (DOH 2001). The American Association on Intellectual and Developmental Disabilities (AAIDD) defines it as "limitations both in intellectual functioning and adaptive behaviour" (Schalock et al. 2007). Adaptive behavior assesses conceptual skills (e.g., language, money, and time concepts), social skills (e.g., interpersonal skills and social problem-solving), and practical skills (e.g., activities of daily living and occupation).

The International Classification of Functioning, Disability and Health (ICF, World Health Organisation 2002) provides a standard language and framework for the description of health and health-related states and incorporates the concept of disability and functional adaptation to disability. This classification system helps us understand what a person with a health condition can do in a standard environment (their level of capacity), as well as what they actually do in their usual environment (their level of performance).

Conceptualizing models of disability is complex as the word "disability" is used interchangeably with different meanings. When considering "disability" it is useful to consider both "medical" and "social" aspects that contribute to the disability of an individual. Medical aspect of disability mainly focuses on the "impairment of functioning," causes of it, and resultant handicap. However, the resultant disability is not always just down to the underlying impaired functioning and resultant handicap but also due to number of social factors. The medical model, focusing on the impairment and resultant handicap, views an individual's disability as a feature of the person, directly caused by disease, trauma, or other health condition, which requires medical care provided in the form of individual treatment by professionals. The social model of disability, on the other hand, sees disability as a social construct and not at all an attribute of an individual. The medical model calls for medical or other treatment or intervention, to "correct" the problem with the individual, whereas the social model demands a political response, since the problem is created by an unaccommodating physical environment brought about by attitudes and other features of the social environment. On their own, neither model is adequate, although both are partially valid. ICF contends that a more useful model would be the "biopsychosocial model" combining the best of both models.

Conceptual evolution of terminology used to describe intellectual disability over the years meant that recognition of ID in general population can be variable. The term ID is characterized by a very heterogenous population of differing abilities. The recognition of intellectual disabilities in low- and middle-income (LAMI) countries can be further influenced by other characteristics; these include educational level and cultural aspects that could influence the current measures of intellectual functioning as well as by socio-environmental factors impacting person's adaptive capacity. These factors make the identification of mild and moderate intellectual disabilities particularly challenging (Mercandante et al. 2009; Parmenter 2008). Intellectual disability being commonly seen as a social disorder in some of these regions, its recognition in health service provision and more so in mental health service provisions, at times, has remained peripheral. The provision of the mental health service in the LAMI countries and especially in the rural areas has its challenges; this is even more so when it comes to the mental health service provision for those with ID.

Epidemiological data, available globally, on intellectual disability population is very limited and at times not reliable. There is a recognition that the prevalence of mental health issues in people with ID is much more than the general population. Availability of reliable epidemiological data around prevalence of mental illness in people with ID globally is complicated by various factors including poor recognition of ID, diagnostic difficulties including diagnostic overshadowing in people with ID, high prevalence of comorbid physical health conditions in people with ID, over-reliance on the social model of disability, etc. Recognition of the levels of complexity in the diagnosis and management of mental disorders in this population has led to development of specialism in this field in countries like the United Kingdom. However, such specialism, in service provision is not seen across the world. There is a wide variation in the availability of mental health services across the world, and access to good mental healthcare in some of the rural areas of LAMI countries could be difficult. Many of these countries have limited resources for health provision, and mental health services of people with ID are not always seen as priority. Hence, it would be essential to explore issues specific to people with ID when it comes to mental health services in rural areas and explore potential solutions.

Epidemiological Considerations

Studies on the prevalence of ID across the world give variable figures. Here, we are looking at the prevalence of those accessing health services; the variation in prevalence rates would depend on the definition of ID as well as other factors like process for identification. A global prevalence of intellectual disability is around 1–3% (Harris 2006; King et al. 2009); a meta-analysis estimated the prevalence to be 10.37/1000 population (Maulik et al. 2011). There is also a variation in the prevalence rates among different populations groups. The prevalence is higher in males in both adult and children/adolescent population. The prevalence of ID is also almost two times more in low- and middle-income (LAMI) countries compared to high-income countries. The prevalence was noted to be highest in urban slums/mixed rural-urban settings followed by rural settings (Maulik et al. 2011). It is interesting that the prevalence was found to be lower in rural settings compared to other settings; this could be a result of problems with recognition of ID in some of the settings. There is also countrywide variation among the prevalence rates.

The causes of intellectual disability could be categorized into antenatal, perinatal, and postnatal conditions. Antenatal conditions include genetic conditions, e.g., Down syndrome, fragile X, infections, substance misuse, malnutrition, or pre-eclampsia; perinatal conditions include premature birth, complications during childbirth, and neonatal infections; and postnatal conditions include infections such as meningitis, severe head injury, extreme malnutrition, exposure to toxic substances, etc. Etiological considerations could also explain the variation in the prevalence of ID in various settings and countries. Specific etiological factors are more easily recognized in people with moderate to severe ID. However, in a significant proportion of people with ID, the causes are unknown. A number of factors such as poor antenatal screening methods (Dave et al. 2005) and birth-related infections and injuries due to poor maternal and child healthcare facilities (Tao 1988) could be the cause for the disparity of the prevalence rates in LAMI countries compared to higher-income countries. A study in Pakistan identified a strong association of intellectual disabilities with lack of maternal education (Durkin et al. 1998).

It is acknowledged that the prevalence of mental health problems in people with ID is much higher than the general population (Cooper et al. 2007; Royal College of Psychiatrists 2013). There are a number of considerations that could explain high prevalence of mental ill health in people with ID. Biological factors like genetic conditions with behavioral phenotypes; high prevalence of epilepsy and other medical conditions; mental health conditions associated with certain syndromes, e.g., dementia and Down syndrome; etc. contribute toward the high prevalence in this population. Psychological factors like trauma, social isolation, stigma, life experiences, etc. also contribute as predisposing, precipitating, as well as perpetuating factors. Similarly social circumstances including rejection, isolation, abuse, etc. contribute to the high level of mental ill health in this population.

Challenges

People with intellectual disabilities experience a number of challenges. These are seen in various degrees across cultures and irrespective of economies; it would be important to understand these issues as their impact would be manyfold in rural settings compared to urban settings.

Social Exclusion

People with ID continue to remain marginalized from the society in spite of community care and a move away from institutions in many of the countries. Poor access to healthcare has a number of underlying reasons, and most of them are sociological reasons. In most of the countries, people with ID are one of the most marginalized sections of the society. When it comes to service provisions, most of the services are not designed or delivered with people with ID in mind. Social exclusion takes many forms, e.g., lack of appropriate opportunities for social inclusion to deliberate exclusion. There are very limited opportunities for people with ID in terms of community engagement and fulfilling life. Where provisions are available, they are limited to few and not always of good quality. In most situations, families are left to care for people with ID into their adulthood and beyond.

Families with disabled children have higher costs as a result of the child's disability coupled with diminished employment prospects. Their housing needs may not be adequately met. There is little evidence of a flexible and coordinated approach to support by health, education, and social services, and there is significant unmet need for respite for families through the process of short breaks. Young disabled people, at the point of transition to adult life, often leave school without a clear route toward a fulfilling and productive adult life. Carers can feel undervalued by public services, lacking the right information and enough support to meet their lifelong caring responsibilities. Social isolation remains a problem for too many people with learning disabilities. Employment is a major aspiration for people with learning disabilities, but very few people are in work. Most people remain heavily dependent on families or social security benefits. Where adequate social care systems are not available, this causes extra and lifelong burden on families with significant economic impact for both families and societies. These issues are multifold in LAMI countries. Disenfranchisement of the people with ID and their families meant that it had adverse impact on the health of people with ID. It also meant that there was poor recognition of carer burden and its impact on their mental health as well as of economic impact on the family and society. Socioeconomic status impacts on cognitive, social, attention, and emotional development in typically developing children (Conger and Donnellan 2007; Kishiyama et al. 2009). This may be affected by parenting.

Stigma continues to be a major issue for people with ID. There is a lot of stigma associated with mental illness in most of the countries. Similarly, there is a huge amount of stigma associated with ID which in turn may lead to social exclusion. In many countries, ID is considered as a social condition. This can in turn lead to poor recognition of health issues and poor investment in the service provision for this marginalized group. People with ID can be the victim of bullying, harassment, and hate crimes that are not uncommon. These issues along with poverty of environment, limited choices for daytime activities, problems accessing transport, limited one-to-one attention, and restricted social networks are associated with increased incidence of mental health problems (Ali et al. 2012, 2015; Cooney et al. 2006; Jahoda and Markova 2004).

Choice and Control

Many people with learning disabilities have little choice or control in their lives. Exposure to social disadvantages such as lone parent family, poor family functioning, lack of parental educational qualifications, income poverty, and households with no paid employment have been shown to be associated with mental health problems in children and young people with learning disabilities (Emerson 2015; Emerson and Hatton 2007). Advocacy services are patchy and inconsistent. There is an over-reliance on families to advocate on behalf of people who lack the capacity. Housing can be the key to achieving social inclusion, but the number supported to live independently in the community, for example, remains small. Many have no real choice and receive little advice about possible housing options. Day services frequently fail to provide sufficiently flexible and individual support. Some large day care centers offer little more than warehousing and do not help people with learning disabilities undertake a wider range of individually tailored activities. Provision of any day services varies widely globally and is nonexistent in many areas, especially rural areas of LAMI countries. Finances can be a significant challenge, and measures devised in developed countries like the United Kingdom such as "direct payments" to support people with health problems have been slow to take off for people with learning disabilities, while in some of the LAMI countries, families bear the burden with huge financial implications for them as well as for the society.

Ethnicity

The needs of people with learning disabilities from minority ethnic communities are too often overlooked. In the United Kingdom, key findings from the study by the Centre for Research in Primary Care at the University of Leeds identified prevalence of learning disability in some South Asian communities can be up to three times greater than in the general population. It also identified that diagnosis is often made at a later age than for the population as a whole and parents receive less information about their child's condition and the support available; social exclusion is made more severe by language barriers and racism, and negative stereotypes and attitudes contribute to disadvantage. Carers who do not speak English receive less information about their support role and experience high levels of stress; and agencies often underestimate people's attachments to cultural traditions and religious beliefs (Waqar 2001).

Health Including Mental Health of People with ID

Health Inequalities
Health problems such as epilepsy, visual impairments, hearing impairments, gastro-esophageal reflux disorder, constipation, diabetes, osteoporosis, contractures, mobility and balance impairments, injuries, thyroid dysfunction, eczema, asthma, obesity, and pain are very common in children, young people, and adults with ID and markedly more so than in the general population (Cooper et al. 2015; Hermans and Evenhuis 2014; McCarron et al. 2013). A number of factors such as the underlying cause of the person's ID, socioeconomic factors, lifestyle, socio-environmental factors, etc. in their own right or in combinations may contribute to this.

The substantial healthcare needs of people with learning disabilities often go unmet. They can experience significant health inequalities compared with the general population, experience avoidable illnesses, and die prematurely. Several studies demonstrated that between 72% and 94% people with learning disabilities had one or more unmet health need (Alborz et al. 2005). Although their life expectancy is increasing, it remains much lower than for the rest of the population (McGuigan et al. 1995; Patja 2000). The standardized mortality ratio has been found to be 8.4 for people with severe intellectual disabilities in the United States and 4.9 and 3.18, respectively, for people with intellectual disabilities of all levels in Australia and England. Additionally, in England it was noted that the life expectancy at birth was 19.7 years lower than for people without learning disabilities (Decouflé and Autry 2002; Durvasula et al. 2002; Glover et al. 2017). It is well recognized that people with intellectual disabilities have higher levels of health needs than the general population (Wilson and Haire 1990; Beange et al. 1995; Kapell et al. 1998; Cooper and Bailey 2001), and these are often unrecognized and unmet (Whitfield and Russell 1996; Lennox and Kerr 1997). This contributes to ongoing health inequality, chronic ill health, and premature death. Many biological, psychological, social, and developmental factors, as well as life experience, contribute to this inequality. Multiple comorbidities combined with communication difficulties can contribute to under-recognition of health problems, including mental health problems, or recognition of health problems only at a late stage of the disease process making recovery less likely. This is further compounded by diagnostic overshadowing, with problems being attributed to the person's learning disabilities and so being ignored, rather than being recognized and addressed. Lack of identification or misdiagnosis of underlying health conditions could lead to mental health difficulties in this population.

The life expectancy of people with ID has increased across the world through the advances in public health policies and in the medical sciences. Still, people with ID have higher risks of dying prematurely due to associated medical and genetic conditions. However, in countries such as India, there were no reliable estimations of the mortality and life expectancy rates available for the ID population. A study (Lakhan and Kishore 2016) found that mortality rate did not differ statistically between rural and urban adults and that age was highly associated with the mortality rate. For every 1-year increase in the age of the population, the mortality rate was found to increase by 3.3 and 3.0 persons per 100,000 in rural and urban ID adults, respectively (Patja 2000). At the same time, people with ID have little choice and control when it comes to accessing health. People with ID are sometimes assumed to lack capacity to make decisions about their healthcare. Multiple morbidities cause difficulties for health, social care, and education staff in identifying that the person with learning disabilities may potentially have additional health needs, seeking help for the person, and also introduce diagnostic complexities for healthcare professionals. Societal attitudes toward disability also meant that people with ID, sometimes, were not thought to have quality of life worth needing active interventions. "Do not resuscitate" signs are sometimes seen because medical staff thought their lives not to be worth saving. Services like "advocacy" and "ID liaison nursing" that were seen in some of the developed countries are not there in most of the LAMI countries. In spite of having advocacy services, liaison nursing services, ID health facilitators, etc. in developed countries, ID population continues to experience stigma, lack of reasonable adjustment, and poor access to health services compared to general population.

Access to Mental Health Services

People with ID also experience access barriers in using health services and is a major issue for them. This is seen across cultures and economies and the problem is more significant in LAMI countries. A number of factors affect "access" to healthcare for people with ID; complexity of the factors contributing to overall health outcomes was well captured by Gulliford et al. (2001) adapted by Alborz et al. (2005) for people with ID (Fig. 1). There are a number of considerations for mental health service provisions:

- Wider determinants of health including personal characteristics, such as genetic makeup, age, aspects of an individual's physical or social environment, and personal lifestyle choices, would have an impact on mental health as well as on access to mental health services.
- Identification of need is significantly impacted in people with ID, who rely, to a greater or lesser extent, on the skills of a third party in recognizing/interpreting their behavior as indicating distress or illness.
- Organization of healthcare provision that considers reasonable adjustments and includes the availability of appropriate services to meet wide ranging personal circumstances including support in obtaining appointments, providing escort or transport, and facilitating communication with health professionals.

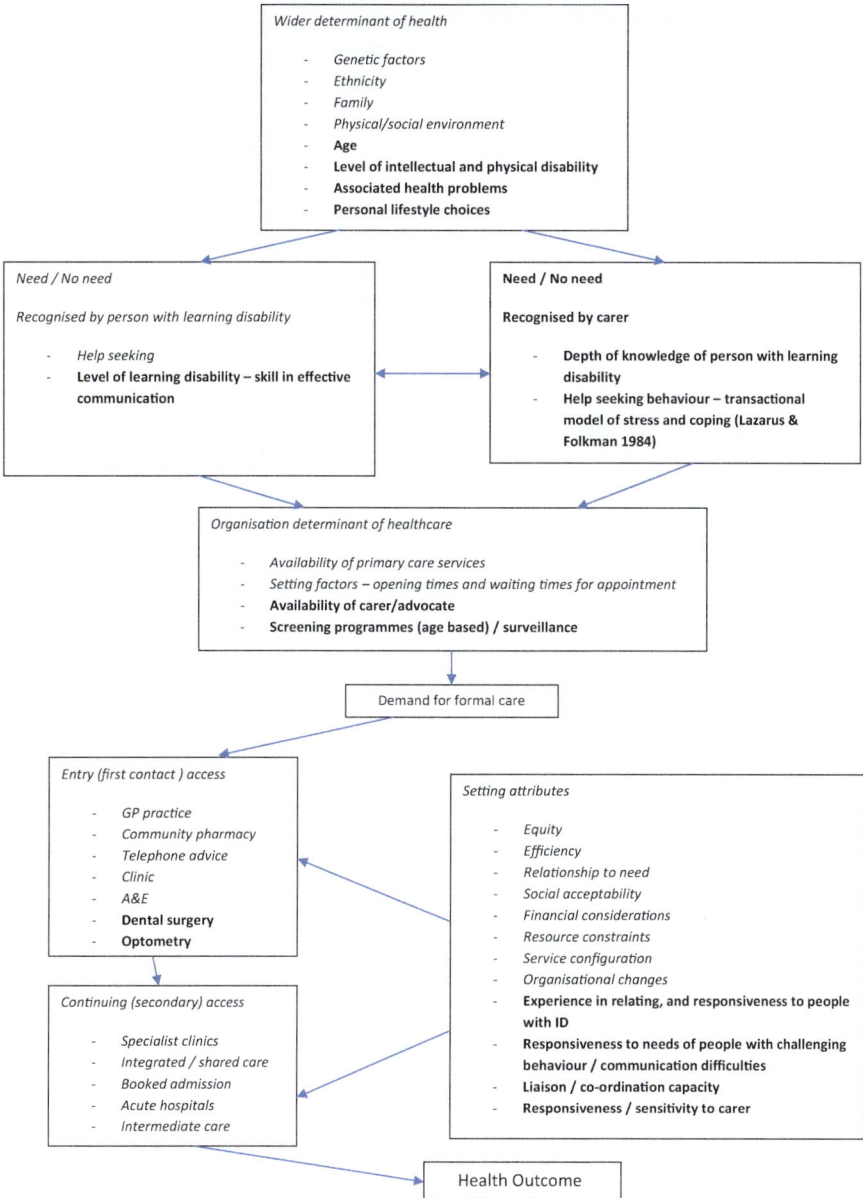

Fig. 1 Access to healthcare for people with ID (*Italics* indicates items taken from "Access to health care," Guildford et al. 2001; **Bold** indicates items added to adapt the model for people with ID by Alborz et al. 2005)

- Access to regular health screening/surveillance.
- Access to entry (first contact) healthcare services where individuals may refer themselves and require no professional assessment to determine access; these are the most frequently accessed services and provide a "gateway" through which people may "gain access" to secondary or "continuing" health services.
- Access and availability of continuing healthcare which is usually provided on referral from a health professional. The long-term health problems experienced by many people with intellectual disabilities meant that they were more likely to use these services than non-disabled patients. Health professionals themselves are therefore more likely to be involved in detecting additional symptoms, and problems, and making referrals to other services.

Issues of health provision for people with ID start at the health education level. As part of the health education curriculum, there is very little by way of specific understanding, teaching, or training regarding the health issues of people with ID. Hence health professionals qualify with having a very limited knowledge and skills in managing health issues in people with ID. It also meant that they were ill equipped to facilitate reasonable adjustments to meet the challenges needed in supporting people with ID accessing health services. Where management is protocol based, most protocols or care pathways are designed for single conditions and so are less relevant for a person with multiple morbidities. The fact that people with ID can present with multiple comorbidities adds further challenges with regard to recognition and diagnosis of conditions as well as impact on treatment. Multiple conditions can result in multiple treatment options increasing the chances of both medication-medication interactions and medication-disease interactions with adverse consequences. As people with learning disabilities may have difficulties expressing any new symptoms or side effects they are experiencing, regular monitoring is particularly important. Additional training and support is necessary where deliberate deviations from single-disease protocols may well be necessary in the person's best interest and highly individualized care is typically needed.

Mental Illness
It is generally acknowledged that there is a higher prevalence of mental illnesses in people with ID compared to the general population (Deb et al. 2001; Cooper et al. 2007). The rates are even higher when it comes to people with ID and autism. One of the common presentations leading to a referral to mental health services is challenging behavior. Emerson (1995) describes challenging behavior as a social construct defined by social impact, without any implication of the underlying process (Emerson et al. 1999). They suggest that challenging behaviors may represent atypical presentations of core symptoms of psychiatric disorder and may occur as secondary features or psychiatric disorders may establish a motivational basis for the expression of these behaviors.

Prevalence rates of mental health problems in population-based studies for adults with learning disabilities (excluding problem behaviors) are reported to be from 14.5%, when also excluding ADHD, autism, dementia, and personality disorder, and people aged 65 and over (Deb et al. 2001), to 43.8%, in adults with moderate to profound learning disabilities only (Bailey 2007). The largest prevalence study among adults in the United Kingdom found that 28.3% of adults had current mental health problems (excluding problem behaviors, or 40.9% including them) (Cooper et al. 2007). Prevalence rates of mental health problems including problem behaviors in children and young people have been reported in the range of 30% (Birch et al. 1970; Rutter et al. 1970) to 50% (Dekker and Koot 2003). Emerson and Hatton (2007) reported a rate of 36% in children and young people (aged 5–16 years) with ID in the United Kingdom, compared with 8% in children without ID participating in the same surveys: the children with ID accounted for 14% of all children with a mental health problem. Some specific types of mental health problems are notably more common in people with ID than in other people. Indeed, prevalence rates of mental health problems for children and young people with learning disabilities have been reported to be higher than for other children and young people for 27 out of 28 ICD-10 diagnostic categories, and statistically significantly so for 20 of these 28 comparisons (Emerson and Hatton 2007).

The challenge, when it comes to mental health problems in people with ID, is not just the high incidence rates but also the endurance of these problems. Studies on common mental health problems in adults with ID reported that 15% of adults with mild learning disabilities compared with 3% of the cohort without ID met definition for chronic depression (Collishaw et al. 2004; Maughan et al. 1999). Another UK longitudinal study of a 1946 birth cohort also included people with mild learning disabilities (Richards et al. 2001). The people with mild learning disabilities at ages 15, 36, and 43 years were found to have significantly more anxiety and depressive symptoms compared with other adults.

Studies that look at the difference in the incidence of mental illness in people with intellectual disabilities in urban vs rural populations have come up with interesting findings. A study in the United Kingdom did not identify any difference in the incidence between these two categories (Kiani et al. 2013), whereas a study in Denmark showed that birth in an urban environment is associated with an increased risk for mental illness in general and for a broad range of specific psychiatric disorders (Vassos et al. 2016).

Despite the high prevalence of mental health problems, they are often not recognized in people who have ID. This can be due to presumptions around the person's behavior and symptoms being attributed to their ID or changes in their presentation not being noticed by carers. This can result in prolonged distress for the person with ID. There is also an enormous gap between our knowledge of effective approaches to managing challenging behaviors and the routine availability of such approaches.

Generic mental health services often do not meet the needs of people with an ID either in community settings or in hospitals (Bouras and Holt 2000). A number of factors could contribute to this including staff training and understanding of the mental health needs of people with ID, lack of resources (people with ID could be perceived as taking away resources that are already scarce from the needs of general population with mental illness), lack of reasonable adjustments, and vulnerabilities of people with ID when they access general psychiatric settings.

Challenges Specific to Rural Settings

Background health disparities exist among individuals living in rural and urban contexts in terms of access to healthcare and overall mortality. Residing in a rural location is associated with socioeconomic disadvantage and reduced access to essential services such as housing, education, transport, and healthcare. Issues like substance abuse are also more prevalent in rural areas of some of the countries like the United Kingdom. This will have a direct impact on the prevalence of ID in the population but also on prevalence of mental illnesses in rural population of people with ID. These disparities are typically greater for youth with disabilities living in rural areas, who face additional barriers in receiving health and support services specific to their disability. Families face many different struggles such as identifying needs and/or appropriate supports, accessing those supports, and then locating providers to help implement those supports (Boehm and Carter 2016). Parents are typically the ones responsible for coordinating the care needed by children with a disability; however, with numerous barriers present, families are not provided adequate support to care for a child with disabilities. When comparing urban and rural areas, barriers to access do differ in terms of availability, but analysis revealed more similarities existed among parents from both contexts. (Walker et al. 2016). Primary carers for people with ID living in remote rural areas experience high demand care commitments that may require them to be available 24 h 7 days a week and reduce their access to formal or respite support. This could leave them little time to engage in other occupations by limiting opportunity to develop occupations. This could then lead to an impact on lifestyle and occupational roles, wellness and health, engaging quality supports, and societal and community engagement and have an impact on their vision for the future (McDougall et al. 2014).

It is not clear whether researchers consistently consider geographic impact for people with intellectual disability. A failure to consider geographic disadvantage potentially limits the applicability of research findings to a significant proportion of the community (Wark 2018). Three main themes of "funding," "training," and "access to services" act as impediments to community workers supporting people with an intellectual disability in rural settings. By identifying these impediments to supporting people with an intellectual disability in the community, both services and government funding bodies have the ability to plan to overcome both current and future problem areas (Wark et al. 2014).

Demographic Changes

Welfare for the disabled is becoming an important issue in countries such as China, and care for people with intellectual disability is challenging because of the inadequacies in formal support and the social service system. Rural families strive to provide care through a set of arrangements and bear tremendous stress in the process. Family care for people with intellectual disability in rural areas has been increasingly challenged by the forces of labor migration, demographic changes, and the ever-growing processes of commoditization. The role of the state has to be strengthened in welfare provision to balance the weakened family care ethos in transforming societies (Pan and Ye 2015).

Experience of aging in older adults with learning disability resident in rural locations can be problematic (Wark et al. 2013). Capacity of certain rural areas to support meaningful choice-making can be limited due to constraints of access to key services, including community-based aged care, generic and specialist health services, and both supported disability and aged-care residential options (Wark et al. 2015).

Service Models

The key elements within a good service model include focus on prevention, early detection, specific interventions, training, and making reasonable adjustments. The National Institute for Health and Care Excellence (NICE) in the United Kingdom summarized the mental health challenges in people with ID and provided a guideline on prevention, assessment, and management (NICE 2016).

Prevention and Early Detection

Many of the antenatal factors causing intellectual disability could be avoided through education, improving nutrition, and better standards of obstetric practice. It is important that public health policies of various countries account for issues specific to rural settings when it comes to preventable causes of intellectual disabilities that could be influenced through public health initiatives. Issues like alcohol and substance abuse in rural settings not just have an impact on the mental health of the abuser but could have consequences for like fetal alcohol syndrome. Genetic screening and genetic counselling could facilitate early detection with appropriate intervention. The care system involving education, training, counselling, and rehabilitation should reach the community more efficiently. Dave et al. (2005) trialled a model of population screening including a doorstep approach through community health volunteers to facilitate identification. The detection skills of the community health volunteers were enhanced through training and supervision by trained medical officers who also formed a vital link to gaining genetic counselling and follow-up. People with high-risk genetic factors detected through this method were referred to the Centre for Research in Mental Retardation (CREMERE) for cytogenetic and metabolic investigations. This community approach gave a better opportunity to gain acceptance from the community and to modify previously unacceptable practices (Dave et al. 2005). In addition to genetic screening, education and training of primary healthcare workers and educational staff in identification of intellectual disabilities would be of great benefit in early detection and intervention.

At the strategic level, any policy initiatives would need to be underpinned by quality epidemiological data. Unfortunately, quality of available epidemiological data is quite varied and not always reliable. To ensure appropriate public health policies as well as to ensure adequate service provisions, it would be important to understand the extent of the problem in rural settings; this would need identifying gaps in provisions and analysis of needs. This would allow development of innovative solutions to meet the needs of people with ID.

Improving Access to Healthcare

It is imperative that there is a concerted effort to improve access to healthcare for people with ID. In the United Kingdom most evidence on interventions designed to improve access described health check programs (Roy et al. 1997; Paxton and Taylor 1998). These primary care-based checks involved GPs, practice nurses, and/or community nurses in learning disability. Other primary care nurses and health workers can be trained to explore the health needs of people with ID. This has the potential to uncover high levels of unmet need and can be an effective method in overcoming barriers raised by difficulties in identifying or communicating health need by people with learning disabilities or their carers. Organizational barriers include communication problems between people with learning disabilities and health professionals; adequate resourcing, training, and managing cultural needs have to be better addressed.

Training and education are key aspects for improving access to healthcare for people with ID. Training and education would need to be considered at various levels. But before considering training and education, as a health services policy, one would need to consider a question of generalization versus specialization: whether mental health services for people with learning disabilities should be delivered by a specialist mental health team or by generic mental health services. This question might be irrelevant in countries where there are poor healthcare resources but would be essential consideration to understand the complexity of underlying issues. Services for people with ID have evolved differently in different parts of the world; services in the United Kingdom have mainly evolved as specialist mental health services for people with ID. Some of the European countries have specialist ID physicians who look after both mental and physical health of people with ID. Generalized services are the ones where needs of people with ID are met through generic main stream services. In such services, it is assumed that people with ID would access same services as anyone else and would have access to all the services as anyone else. Mainstreaming or generalization of services for people with ID has its advantages, and it makes sense to have such as approach where resources for specialist services are not available. However, success of such services is dependent of the skills and competencies of the professionals involved in identifying and managing complex issues experienced by people with ID. It would be appropriate to consider service model where there is some element of specialism available that supported general mainstream services in dealing with mental health issues in people with ID.

Access to mental health services for people with ID could be improved through training and education at various levels. At health education policy level, curriculums for various health professional trainings including medical and nursing students should include ID-specific training to improve their skills and competencies in identifying and managing mental health problems in this population. It would help to have some specialist training programs available at postgraduate levels for professionals to further develop their skills. Education and training initiatives should also involve general practitioners, primary healthcare workers, community health workers, outreach services, and families and carers. Training provision at these various levels would help improve access to mental health services. Where specialist services are not available or where availability of such services is limited, "hub and spoke" model would help improve access. Training initiatives should also involve training of educational staff in supporting children and young adults with ID in educational settings; this would have much more positive impact in the longer term on the mental health of these individuals.

Telehealth Technology

Innovative solutions like telehealth technology could be a way forward for making assessment and treatment options for families in rural and underserved regions of the country to manage the disparity in accessing specialty care due to distance, poverty, and transportation. A study in Baltimore was able to establish that telemedicine is a replicable and valid method for a comprehensive neurodevelopmental examination for children and young people with special needs and that it was accepted by practitioners and families. This enabled improved access to care for families with children and young people with special healthcare needs and early diagnosis helped to improve the care provided to the this group (Menon et al. 2016).

In a study using telehealth technology to teach three teachers in a rural school district, the basics of a mindfulness-based procedure Singh et al. (2017) were able to establish that telehealth may be an effective approach to providing training and therapy to caregivers in remote locations that cannot readily access specialist services. The authors used telehealth technology to teach teachers meditation on the soles of the feet (SoF) procedure. The students were able to use the procedure to downregulate their emotions associated with the earliest precursors to verbal and physical aggression.

Telepsychiatry has also been noted to have improved outcomes for older patients with a wide range of psychiatric disorders and a high rate of medical comorbidity. The group included people with intellectual disabilities with other comorbid conditions (Steinberg et al. 2014).

Rehabilitation of Individuals: Specialist Schools, Supported Employment, and Support to Families

Achieving good outcomes for people with ID with mental health needs is not just dependent on specialist mental healthcare provisions but also on the ability of statutory bodies to deal with the social determinants of health including mental health. Dealing with such social determinants should be considered at every level including at education, family, independence, and employment. Educational institutions including special school provisions have a key role in identifying and supporting those with mental health needs; that role also includes preparing individuals, to the best of their abilities, for adulthood. Family support is also a key factor; with training and support, family members could support people with ID and mental health needs better where access to healthcare provision was limited.

It is important for people with ID to be able to express choice throughout their lives. Similarly having opportunities for meaningful employment is very important if people with intellectual disability are to become truly self-determined. However, people with ID and their families have low expectations for competitive employment, if considered at all. There are examples of supported employments or employment through reasonable adjustments; however, these are far and few and are not a norm.

Advocacy and Making Reasonable Adjustment

Training and making reasonable adjustment are key elements to improving access to healthcare for people with ID. Diagnostic delays and diagnostic overshadowing are common in people with ID. Most healthcare services are not designed to or geared up for delivering appropriate healthcare for people with ID. Lack of training, with resultant lack of skills and competencies, is a common factor in these situations. Improving access also meant that the services were able to make reasonable adjustments.

Training has another key component. Some of the developed countries have health workers and professional groups that were specialized in the management of healthcare needs including mental healthcare needs of people with ID. Such specialism does not always exist in LAMI countries. Access to healthcare and mental healthcare is poor in some of the rural regions of LAMI countries; having expertise to deliver services for ID population is even scarcer in these regions. In such situations, innovative approach is needed, e.g., hub and spoke model, implemented in Goa where the clinical expertise are sited in the hub but the Anganwadi workers are trained to provide outreach services to families in remote rural parts. A similar model but with training to families and carers was successfully tested in rural Pakistan; a system successfully tried out training volunteers from families with ID in evidence-based interventions, who worked under the supervision of specialists to provide care and work with "family networks" (Hamdani et al. 2015).

Conclusion/Ways Forward/Future Considerations

Delivering mental health services for people with ID in rural areas across the world poses significant challenges, and any service delivery model would need to take into consideration a number of issues. Sociological issues like social isolation and social exclusion along with overreliance on social model of disability, societal attitudes, and double stigma of having ID and mental health issues meant that access to healthcare for these marginalized individuals could be extremely limited or nonexistent. Where such services were to be available, they are affected by geographical factors and issues of lack of adequacy of resources.

Access to mental health services to people with ID in rural settings is significantly affected by resource constraints. Investment in specialist mental health services in LAMI countries is limited or nonexistent. There is also lack of available expertise or specialism in managing mental health issues of those with ID and autism. Services are not geared up to meet the complex needs and are not always designed to provide reasonable adjustment where needed. There are additional considerations like lack of reliable prevalence data, problems of poor early identification, poor investment in public health preventative strategies, etc. which meant that there was poor recognition and investment in commissioning services for rural population. Lack of multi-disciplinary professional input along with overreliance on biological model of therapeutic intervention meant that the outcomes for those with ID and mental health needs were not optimum. Such services also lacked the ability to take into account social determinants of health in delivering therapeutic interventions. Lack of consideration in dealing with social determinants of health or lack of ability to deal with such factors due to poor infrastructure and resources in rural settings, especially in LAMI countries, meant that people with ID with mental health issues experienced poor quality of life with resultant further adverse impact on health.

Having considered challenges, there are examples of innovative good practice, and there is a need for upscaling such practices across wider footprints. Tackling some of these issues could be considered at various levels. There is marked need to get reliable epidemiological data to understand the extent of problem experienced by those with ID with mental health problems and the burden of care experienced by their families. This needs to be further supported by analyzing economic impact of such a burden; this would give a clear idea of the gaps in the provision and help develop an overall sustainable model. Such an analysis would need to be underpinned by human rights of those with ID and mental health issues and their families. A human rights-based approach should underpin all the service delivery models, outcome that those services were trying to achieve, training of professionals and other stakeholders, and social care provision. There has been significant positive development in some of the developed countries when it comes to stigma and mental health; however, it still continues to be a problem. This is even more so in LAMI countries. There would be a need for systematic campaign against stigma in relation to ID and mental health. This would need political backing, and organizations like the Royal College of Psychiatrists have a role in lobbying to influence the political will.

Improving mental health of people with ID in rural settings would need a public health and social care approach. It is not just about having specialist services to meet the needs of this population but also about having preventative and public health strategies. Preventative public health programs to tackle some of the preventable etiological causes of ID, better screening programs for early detection and support, public health initiatives to improve overall health of people with ID, and improved education and social care provisions in rural setting would have a positive impact on the health including mental health of people with ID.

Finally, provision of specialist mental health services in rural settings would need some innovative solutions. There is a lot that could be learned from pockets of good practices around the world. Lack of enough resources and specialist services to spread over wider and thinly spread rural population could benefit by "hub and spoke" model where a specialist health service provides training and support to outreach primary healthcare workers in rural settings. Teaching and training of mainstream health services and that of families and carers could have considerable positive impact. Such models are successfully tried in some countries.

References

Alborz A, McNally R, Glendinning C (2005) Access to healthcare for people with learning disabilities: mapping the issues and reviewing the evidence. J Health Serv Res Policy 10 (3):173–182

Ali A, Hassiotis A, Strydom A, King M (2012) Self stigma in people with intellectual disabilities and courtesy stigma in family carers: a systematic review. Res Dev Disabil 33:2122–2140

Ali A, King M, Strydom A, Hassiotis A (2015) Self-reported stigma and symptoms of anxiety and depression in people with intellectual disabilities: findings from a cross sectional study in England. J Affect Disord 187:224–231

Bailey NM (2007) Prevalence of psychiatric disorders in adults with moderate to profound learning disabilities. Adv Ment Health Intellect Disabil 1:36–44

Beange H, McElduff A, Baker W (1995) Medical disorders of adults with mental retardation: a population study. Am J Ment Retard 99:595–604

Birch H, Richardson S, Baird D, Horobin G, Illsley R (1970) Mental subnormality in the community: a clinical and epidemiological study. Williams and Wilkins, Baltimore

Boehm TL, Carter EW (2016) A systematic review of informal relationships among parents of individuals with intellectual disability or autism. Res Pract Persons Severe Disabil 41:173–190

Bouras N, Holt G (2000) The planning and provision of psychiatric services for people with mental retardation. In: Gelder MG, Lopez-Ibor JJ, Andreasen NC (eds) New Oxford textbook of psychiatry. Oxford University Press, Oxford, pp 2007–2012

Collishaw S, Maughan B, Pickles A (2004) Affective problems in adults with mild learning disability: the roles of social disadvantage and ill health. Br J Psychiatry 185:350–351

Conger R, Donnellan M (2007) An interactionist perspective on the socioeconomic context of human development. Annu Rev Psychol 58:175–199

Cooney G, Jahoda A, Gumley A, Knott F (2006) Young people with learning disabilities attending mainstream and segregated schooling: perceived stigma, social comparisons and future aspirations. J Intellect Disabil Res 50:432–445

Cooper S-A, Bailey NM (2001) Psychiatric disorders amongst adults with learning disabilities: prevalence and relationship to ability level. Ir J Psychol Med 18:45–53

Cooper S-A, Smiley E, Morrison J, Allan L, Williamson A (2007) Prevalence of and associations with mental ill-health in adults with intellectual disabilities. Br J Psychiatry 190:27–35

Cooper S-A, McLean G, Guthrie B, McConnachie A, Mercer S, Sullivan F et al (2015) Multiple physical and mental health co-morbidity in adults with intellectual disabilities: population-based cross-sectional analysis. BMC Fam Pract 16:110

Dave U, Shetty N, Mehta L (2005) A community genetics approach to population screening in India for mental retardation – a model for developing countries. Ann Hum Biol 32:195–203

Deb S, Thomas M, Bright C (2001) Mental disorder in adults who have a learning disability, 1: prevalence of functional psychiatric illness among 16–64 years old community-based population. J Intellect Disabil Res 5:495–505

Decouflé P, Autry A (2002) Increased mortality in children and adolescents with developmental disabilities. Paediatr Perinat Epidemiol 16:375–382

Dekker MC, Koot HM (2003) DSM-IV disorders in children with borderline to moderate intellectual disability. II: child and family predictors. J Am Acad Child Adolesc Psychiatry 42:923–931

Department of Health (2001) Valuing people: a new strategy for learning disability for the 21st century. The Stationery Office, London

Durkin MS, Hasan ZM, Hasan KZ (1998) Prevalence and correlates of mental retardation among children in Karachi, Pakistan. Am J Epidemiol 147:281–288

Durvasula S, Beange H, Baker W (2002) Mortality of people with intellectual disability in northern Sydney. J Intellect Develop Disabil 27:255–264

Emerson E (1995) Challenging behaviour: analysis and intervention in people with learning disabilities. Cambridge University Press, Cambridge, UK

Emerson E (2015) The determinants of health inequities experienced by children with learning disabilities. Public Health England, London

Emerson E, Hatton C (2007) Mental health of children and adolescents with intellectual disabilities in Britain. Br J Psychiatry 191:493–499

Emerson E, Moss S, Kiernan C (1999) The relationship between challenging behaviours and psychiatric disorders in people with severe developmental disabilities. In: Bouras N (ed) Psychiatric and behavioural disorders in developmental disabilities and mental retardation. Cambridge University Press, Cambridge, UK, pp 38–48

Glover G, Williams R, Heslop P, Oyinlola J, Grey J (2017) Mortality in people with intellectual disabilities in England. J Intellect Disabil Res 61(1):62–74

Gulliford M, Morgan M, Hughes D, Beech R, Figeroa-Munoz J, Gibson B et al (2001) Access to health care: report of a scoping exercise for the National Co-ordinating Centre for NHS Service Delivery and Organisation R & D (NCCSDO). NCCSDO, London

Hamdani SU, Minhas FA, Iqbal Z, Rahman A (2015) Model for service delivery for developmental disorders in low-income countries. Peadiatrics 136:1166–1172

Harris JC (2006) Intellectual disability: understanding its development, causes, classification, evaluation and treatment. Oxford University Press, New York, pp 42–98

Hermans H, Evenhuis HM (2014) Multimorbidity in older adults with intellectual disabilities. Res Dev Disabil 35:776–783

ICD-10: International statistical classification of diseases and related health problems:tenth revision, 2nd ed. World Health Organisation (2004)

Jahoda A, Markova I (2004) Coping with social stigma: people with intellectual disabilities moving from institutions and family home. J Intellect Disabil Res 48:719–729

Kapell D, Nightingale B, Rodriguez A, Lee JH, Zigman WB, Schupf N (1998) Prevalence of chronic medical conditions in adults with mental retardation: comparison with the general population. Ment Retard 36:269–279

Kiani R, Tyrer F, Hodgson A, Berkin N, Bhaumik S (2013) Urban–rural differences in the nature and prevalence of mental ill-health in adults with intellectual disabilities. J Intellect Disabil Res 57(2):119–127

King BH, Toth KE, Hodapp RM, Dykens EM (2009) In: Sadock BJ, Sadock VA, Ruiz P (eds) Comprehensive textbook of psychiatry, 9th edn. Lippincott Williams & Wilkins, Philadelphia, pp 3444–3474

Kishiyama M, Boyce W, Jimenez A, Perry L, Knight R (2009) Socioeconomic disparities affect prefrontal function in children. J Cogn Neurosci 21:1106–1115

Lakhan R, Kishore TM (2016) Mortality in people with intellectual disability in India: correlates of age and settings. Life Span and Disability 19(1):45–56

Lennox NG, Kerr MP (1997) Primary health care and people with an intellectual disability: the evidence base. J Intellect Disabil Res 41:365–372

Maughan B, Collishaw S, Pickles A (1999) Mild mental retardation: psychosocial functioning in adulthood. Psychol Med 29:351–366

Maulik PK, Maya N, Mascarenhas C, Colin D, Mathers C, Dua T, Saxena S (2011) Prevalence of intellectual disability: a meta-analysis of population-based studies. Res Dev Disabil 32(2):419–436

McCarron M, Swinburne J, Burke E, McGlinchey E, Carroll R, McCallion P (2013) Patterns of multimorbidity in an older population of persons with an intellectual disability: results from the intellectual disability supplement to the Irish longitudinal study on aging (IDS-TILDA). Res Dev Disabil 34:521–527

McDougall C, Buchanan A, Peterson S (2014) Understanding primary carers' occupational adaptation and engagement. Aust Occup Ther J 61(2):83–91

McGuigan SM, Hollins S, Attard M (1995) Age-specific standardized mortality rates in people with learning disability. J Intellect Disabil Res 39:527–531

Menon D, Singh V, Lipkin P (2016) Improving access to specialty care for underserved children with neurodevelopmental disorders using telemedicine. Ann Neurol 80:S386–S386

Mercandante MT, Evans-Lacko S, Paula CS (2009) Perspectives of intellectual disabilities in Latin American countries: epidemiology, policy and services for children and adults. Curr Opin Psychiatry 22:469–474

NICE guideline [NG54] (2016) Mental health problems in people with learning disabilities: prevention, assessment and management

Pan L, Ye J (2015) Family care of people with intellectual disability in rural China: a magnified responsibility. J Appl Res Intellect Disabil 28(4):352–366

Parmenter TR (2008) The present, past and future of study of intellectual disabilities: challenges in developing countries. Salud Publica Mex 50(2):S124–S131

Patja K (2000) Life expectancy of people with intellectual disability: a 35-year follow-up study. J Intellect Disabil Res 44:590–599

Paxton D, Taylor S (1998) Access to primary health care for adults with a learning disability. Health Bull 56:686–693

Richards M, Maughan B, Hardy R, Hall I, Strydom A, Wadsworth M (2001) Long-term affective disorder in people with learning disability. Br J Psychiatry 170:523–527

Roy A, Martin DM, Wells MB (1997) Health gain through screening – mental health: developing primary health care services for people with an intellectual disability. J Intellect Develop Disabil 22:227–239

Royal College of Psychiatrists (2013) People with learning disability and mental health, behavioural or forensic problems: the role of in-patient services. Royal College of Psychiatrists, London

Rutter M, Tizard J, Whitmore K (1970) Education, health and behaviour. Longman, London

Schalock RL, Luckasson RA, Shogren KA, Borthwick-Duffy S, Bradley V, Buntinx WHE, Coulter DL, Craig EM, Gomez SC, Lachapelle Y, Reeve A, Snell ME, Spreat S, Tasse MJ, Thompson JR, Verdugo MA, Wehmeyer ML, Yeager MH (2007) The renaming of mental retardation: understanding the change to the term intellectual disability. Intellectual and Developmental Disabilities 45(2):116–124. PubMed PMID: 17428134

Singh NN, Chan J, Karazsia BT, McPherson CL, Jackman MM (2017) Tele-health training of teachers to teach a mindfulness-based procedure for self-management of aggressive behavior to students with intellectual and developmental disabilities. Int J Dev Disabil 63(4):195–203

Steinberg SI, Gallop R, Syed I, Shraddha J, Mohammed AA, Singh H, Bogner HR (2014) Telepsychiatry for geriatric residents in rural nursing homes. Alzheimer's & Dementia: Journal of Alzheimer's association 10(4): 768

Tao KT (1988) Mentally retarded persons in the People's Republic of China: review of epidemiological studies and services. Am J Ment Retard 93:193–199

Vassos E, Agerbo E, Mors O, Bocker PC (2016) Urban-rural differences in incidence rates of psychiatric disorders in Denmark. Br J Psychiatry 208(5):435–440

Walker A, Alfonso ML, Weeks K, Telfair J, Colquitt G (2016) "When everything changes:" parent perspectives on the challenges of accessing care for a child with a disability. Disabil Health J 9 (1):157–161

Waqar A (2001) Learning difficulties and ethnicity. Centre for Research in Primary Care, University of Leeds, England

Wark S (2018) Does intellectual disability research consider the potential impact of geographic location? J Intellect Develop Disabil 43(3):362–369

Wark S, Hussain R, Edwards H (2013) Rural and remote area service provision for people aging with intellectual disability. J Policy Pract Intellect Disabil 10(1):62–70

Wark S, Hussain R, Edwards H (2014) Impediments to community-based care for people ageing with intellectual disability in rural New South Wales. Health Soc Care Community 22(6):623–633

Wark S, Canon-Vanry M, Ryan P, Hussain R, Knox M, Edwards M, Parmenter M, Parmenter T, Janicki M, Leggatt-Cook C (2015) Ageing-related experiences of adults with learning disability resident in rural areas: one Australian perspective. Br J Learn Disabil 43(4):293–301

Whitfield ML, Russell O (1996) Assessing general practitioners' care of adult patients with learning disabilities: case control study. Qual Health Care 5:31–35

Wilson D, Haire A (1990) Health care screening for people with mental handicap living in the community. BMJ 301:1379–1381

World Health Organisation (2002) Towards a common language for functioning, disability and health. ICF WHO/EIP/GPE/CAS/01.3

Mental Health of Rural Women

8

Prabha S. Chandra, Diana Ross, and Preeti Pansari Agarwal

Contents

P. S. Chandra (✉)
Department of Psychiatry, National Institute of Mental Health and Neurosciences,
Bangalore, Karnataka, India
e-mail: chandra@nimhans.ac.in; prabhasch@gmail.com

D. Ross · P. P. Agarwal
Department of Psychiatry, NIMHANS, Bangalore, India
e-mail: rossd810@gmail.com; preeti_pansari@yahoo.com

© Springer Nature Singapore Pte Ltd. 2020
S. Chaturvedi (ed.), *Mental Health and Illness in the Rural World*, Mental Health and
Illness Worldwide, https://doi.org/10.1007/978-981-10-2345-3_12

Abstract

Women from rural areas form a unique group due to the realities of rural living. The various challenges rural women face include gender disadvantage, poverty, poor physical health, roles of caregiving, and being women farmers. Besides gender disadvantage, exposure to intimate partner violence (IPV) is also one of the main risk factors for common mental disorders (CMDs) in rural women. Suicide, depression, anxiety, perinatal disorders, somatization, and substance abuse are few of the disorders that rural women often face. Challenges preventing rural women from approaching mental healthcare include an absence of women-friendly mental healthcare services, lack of trained mental health professionals in rural areas or those who understand the unique needs of rural culture, stigma associated with help seeking, poor knowledge about treatment facilities, lack of accessibility to transport, and the high costs of mental health services. Factors that contribute to enhanced vulnerability include older age, widowhood, poverty, and living in areas of armed conflict. HIV infection and comorbid physical illnesses also add to risk. Rural women's well-being is strongly associated with their physical health, inner well-being, economic security, rural identity, household and family well-being, and community relations. Psychological therapies tailored to the needs of rural women have reported improvement in mental health. Rural women are more accepting of help from their own community-level workers and peer volunteers. Mental health delivery, therefore, has to be tailored to the norms and realities of rural women.

Keywords

Women · Rural · Mental health · Mental health interventions · Well-being

Introduction

Health and illness are determined by various factors. These include individual behavioral factors, economic and psychosocial factors, and their complex reciprocal relationships. The mental health definition used in the WHO 1981 report on the social dimensions of mental health states that: "Mental health is the capacity of the individual, the group and the environment to interact with one another in ways that promote subjective well-being, the optimal development and use of mental abilities (cognitive, affective and relational), the achievement of individual and collective goals consistent with justice and the attainment and preservation of conditions of fundamental equality" (WHO 1986). A broadened health field concept consists of the social model of health, in which human biology and healthcare are integrated (Raeburn and Rootman 1989).

This definition has several advantages in relation to women's mental health because it:

- Emphasizes the various situations that determine mental health
- Goes beyond the biomedical understanding and incorporates the personal
- Affirms the vital role of social surroundings and the living environment
- Underlines the significance of fairness and equity in ascertaining mental health

Women need to be able to control their own lives and thereby their health by autonomy, decision-making power, and access to independent income, and absence of this can affect their physical and mental health including their susceptibility to communicable and noncommunicable diseases (Okojie 1994). Across the life cycle, women have various concerns which affect their psychological well-being, and this does not limit to reproductive roles alone. Inadequate sources of data and an overdependence on a purely biological model often hinder the true assessment of women's mental health needs.

According to the International Labour Organization (ILO), women are half the world's population, who receive one-tenth of the world's income, account for two-thirds of the world's working hours, and own only 100 of the world's prosperity (ILO 2009).

Women constitute and help in building the foundation in the rural and national economies. Rural women, most of whom depend on natural resources and agriculture for their livelihoods, make up over a quarter of the total world population. In developing countries, rural women represent approximately 43% of the world's agricultural labor force, which rises to 70% in some countries. A large responsibility for food security lies with rural women as they mainly work in areas of food production and related fields. Interestingly, the world observes *International Day of Rural Women* on *15th October*, such is the relevance of women who live in rural areas. This new international day, established by the United Nations General Assembly, recognizes "the critical role and contribution of rural women, including indigenous women, in enhancing agricultural and rural development, improving food security and eradicating rural poverty."

Unlike what we may imagine, rural women are a heterogenous group. They dwell in areas that are dissimilar in their topography, population, economy, and progress. However, there are some aspects which are common around the world: lower populations, larger distance from cities, segregation, close communal connections, social isolation, a culture of self-competence, and fewer financial resources and also less workforce. All of this influences the prevention and treatment of the healthcare needs of rural women. The broad sociocultural and lifestyle factors that characterize rural life are also known to influence both physical and mental health aspects. These include geographic barriers, distance, lack of transportation, poor job benefits of an agrarian society, and inadequate funding for health which affect access to both medical and mental health services.

Significant health disparities exist between rural and urban women. In comparison with urban women, rural women have less access to healthcare and have poorer health consequences. Rural women encounter numerous restrictions in accessing low-cost, satisfactory health services (e.g., clinics, hospitals, reproductive health/family planning, and counseling). Apart from issues of affordability, they also face constraints on their movement and a lack of access to conveyance or the means like a mobile phone to contact transport. Rural areas often have less healthcare providers, of which very few cater to women (Kim Tjaden 2015).

Economic and social change in most countries is driven by rural women. They also play an essential and active role in protecting the environment. Regardless of their role restrictions, they continue to be farmers, producers, investors, caregivers, and consumers and have a vital role in ensuring nation's food and nutrition security, obliteration of rural poverty, and cultivating their family's well-being. Despite this in many places around the world, they continue to face significant challenges as a result of gender-based compartmentalization and prejudices that deny them fair access to opportunities, resources, assets, and health services.

Challenges Faced by Rural Women

Life in rural areas is thought to be idyllic, free from stress and strain, relaxing, and resulting in more calmness and peace. However, as mentioned in the previous section, the challenges of rural life are many, and rural women often face a different set of difficulties than urban women.

In rural communities, women's mental health is affected by the reinforcement of traditional gender roles, which leads to fewer women in areas of employment or higher education.

Given below is a list of challenges that rural women face that affects their mental health:

Gender Disadvantage

Gender is a concept which has its origin in biology and yet is susceptible to the weaknesses in the social environment which is characterized by violence, prejudices, and disenfranchisement. Gender influences people's power over their health determinants, including their economic and social status and access to essential resources and treatment in society.

Gender disadvantage is clearly seen in the form of intimate partner violence (IPV), lack of autonomy in decision-making, lack of support, being married, and bearing children during adolescent age. This is more prominent in the rural areas where women often are tied down to their traditional gender-specific roles such as being a wife, mother, and caretaker of their families.

Gender disadvantage and exposure to IPV have been found to be main risk factors for common mental disorders in women (Patel et al. 2006). Surveys have also shown an association between reproductive and sexual complaints (such as the complaint of vaginal discharge) and common mental disorder. Rural women's marital history is found to be a main factor for gender disadvantage and for common mental disorder, especially, if the woman is widowed or divorced. This could be related to social isolation and stigma which comes with being widowed or divorced. The increased risk in married women is partly because of them having to cope with multiple roles leading to potentially more restricted lives in their marital homes. The very low age at marriage for women may also be another factor. Pregnancy and childbirth are major causes of death for women and children in the world's poorer countries especially in the rural areas. The early age of marriage that disrupts education and the lack of proper employment altogether restrict women's participation in societal development (Patel 2006).

Rural women's environment is more stressful and disadvantaged than that of men. This is characterized by features such as lesser access to transport, large number of women in the elderly age groups with health issues, economically disadvantaged status, difficulties in getting employed, greater share of family and caring responsibilities, lack of property rights, patriarchy, unseen labor in agricultural areas, and less role in decision-making. Ancillary factors associated with depression among rural women consist of isolation due to geographical location, lack of social support, as well as access to mental health services; weather problems; and waning farm economy which gives an uncertain income and the absence of social, educational, and child care resources (Mulder et al. 2000).

Poverty

Rural poverty rates are consistently higher and more persistent than urban poverty. Probst et al. (2005) found that while both urban and rural women were more likely to be depressed than men, further the rural women were more likely to be depressed than their urban counterparts (Probst et al. 2006).

Poverty is pervasive throughout the world, particularly in rural areas. In 2002, the percentage of population in developing countries living on US $1 or less per day was 19.2%. The agricultural sector still holds the major share of the economy in many developing countries and is a vital element in reducing poverty in rural areas, especially for women. Two aspects of poverty, time poverty and hunger, have clear gender dimensions, particularly in rural areas. The concept of time poverty talks about families with unavailability of time which keep them away from attaining well-being levels due to long working hours, because of sheer necessity and not by choice. Time poverty focuses on time spent doing house activities as much as time spent on productive activities (Sandys 2008).

In developing countries poverty also causes undernourishment and malnourishment. This is caused by inadequate income and lack of knowledge of food resources within the household. Because of discrimination, most women and girls do not eat healthily, which may further affect their mental health. Poverty in rural areas is also related to changes in the natural environment and by floods, droughts, deforestation, and other natural disasters (Sandys 2008).

Poor Physical Health

Because rural women are less frequently socially active, isolation and lack of social interaction often affect rural women's motivation to maintain physical health. Rural women must be strong to survive their living conditions, but even stronger to maintain positive physical and mental health. Rural women endure physical labor as part of farming and other related activities. Often many develop chronic problems such as diabetes, cancer, hypertension, heart disease, stroke, and lung disease and experience more disability and morbidity due to this, further affecting their mental health. Despite this many are unwilling to go for preventive check-ups with physicians for their physical complaints. This may be due to lack of knowledge about the significance of early detection and prevention measures. Many of the health factors connected to their depression were linked to mental health factors directly or indirectly (Valverde et al. 2011).

Occupation

There are important differences among women in rural areas based on class, age, marital status, ethnic background, race, and religion. The main source of income for rural women is agriculture and related activities. Rural women play critical and diverse roles in agricultural production in the rural economy of developing countries as unpaid family workers, as own-account farmers, and as full-time or part-time wage laborers on large farms and plantations. Although women make a major contribution to agriculture production, this is underreported in all developing regions because women's work is often unrecognized or is considered as part of "housework."

Rural women's labor in rural production becomes invisible and their work is consistently undervalued. At the same time, rural women play an important and time-consuming role in the reproductive economy. While productive labor results in goods or services that have monetary value in the form of a paid wage, reproductive labor is associated with the private sphere and involves anything that people have to do for themselves that is not for the purposes of receiving a wage (i.e., cleaning, cooking, having children). Unpaid work is in the form of collecting water and fuelage, cooking, cleaning, child-rearing, and caregiving for the old, the sick, and the disabled. In all developing country regions, this work is critical for family. Rural women often work longer hours, under arduous circumstances, and without adequate access to appropriate technologies and infrastructure which further increase their workloads and constrain their contributions (Sandys 2008).

In developing countries many of the employees in the textile and electronic industries are young, single migrant women from rural areas. Large numbers of women are moving from rural to urban areas to take up employment in many South and Southeast Asian countries. Small numbers are engaged in the construction industry as unskilled laborers. Many of whom are young, unmarried, or divorced, with only basic education further increasing social isolation and vulnerability which contribute to poor mental health (Sandys 2008).

Women Farmers

Role overload and invisibility are two primary stressors of rural farm women: role overload (i.e., working on the farm, in the household, or outside the home) and *invisibility* or lack of recognition for the work that they do at home or in the farm put a huge strain on their roles and responsibilities and give them no time or space for self-care or even for some kind of support. This has been linked to causing feelings of anxiety and depression as they are unable to meet their friends or attend religious or social functions due to role overload. Rural farm women also face stressors consisting of reduced farm work, managing conflict within family members as they all work with each other, unpredictable work due to weather changes, and irregularity of income (Mulder et al. 2000).

Caregiving

Caregiving responsibilities for aging parents, as well as sick relatives, often fall on the shoulders of rural women. Extended families often live close to each other in rural areas. The illness of any family members puts a time and labor overload due to caregiving needs, which falls on older women and girls. Ruralites tend to take care of their own family and resist having outsiders take care of them (Mulder et al. 2000). Thus caregiving chores in a family become an added responsibility for the rural woman. The burden of caregiving may prevent them from caring for their own physical and emotional health. The absence of day care or respite centers also adds to the caregiving burden.

Migration from Rural Areas and Its Impact on Conflicting Gender Roles

Migration is another cause for many changes in the rural women's role. These movements happen from rural to rural areas, from rural to urban areas, and from country to country, as women move for work. Armed conflict or natural disasters also force migration of women through human trafficking and forced labor. Migration generally results in the redistribution of tasks and responsibilities among those left behind, and there is evidence of a strong impact on gender relations. Migration can be an empowering experience for women as it brings them more freedom and power to make decisions which affects both those left behind and those women who migrate themselves (Sandys 2008).

Female-headed household is a new phenomenon which has occurred as an answer to the changes in the migratory patterns. One-third of the households in sub-Saharan Africa are permanently headed by women, either widows or women who are single, divorced, or separated from their partners (Sandys 2008).

With men moving away for work or other reasons, women have takeover male tasks and responsibilities thereby challenging gender-based norms. Despite this so-called empowerment, women even in single households face challenges, such as time shortage, inadequate access to resources, property restrictions, and less power in decision-making.

Gender-Based Violence

Violence against women is a pervasive violation of human rights and a major hurdle for gender equality. Women in rural areas experience violence within their families and communities. Such violence can be exacerbated during armed conflict and natural disasters. Violence occurs in many forms and different levels, including domestic violence, early and forced marriages, absence of nutritious food and adequate healthcare, harmful traditional practices such as female genital mutilation, forced prostitution, rape, and sexual violence. Exploitative working conditions and human trafficking are also other forms of indirect violence (Sandys 2008).

Lifetime exposure to IPV is a major risk factor for common mental disorders in women (Patel et al. 2006). Numerous risk factors for violence have been identified for rural women which include women's isolation and lack of social support, community attitudes that tolerate and legitimize male violence, and high levels of social and economic disempowerment and poverty. The under-resourcing of rural areas often makes women's access to shelters and legal services difficult, which in turn allows for women to be exploited as they have no avenues to turn to for help. Violence results in victimization increasing risk for AOD disorders and axis I mental disorders in rural women (Boyd 2003).

Several forms of violence have been reported from rural areas and around the world. From 5.1% in rural areas in the USA (Hillemeier et al. 2008) to 30% and 23% in Vietnam (Vung et al. 2008) and India (Kermode et al. 2007), respectively. Some forms of violence which are more culturally determined such as honor killing have also been reported in rural areas (Patel and Gadit 2008).

Female Genital Mutilation (FGM)

"Female genital mutilation" (FGM), or "female genital cutting" (FGC) or "female circumcision," is a cultural practice which occurs in many parts of the world, predominantly in Africa. It is found in both rural and urban areas despite religious background. Every year two million girls undergo this practice which causes severe psychological trauma along with serious physical and sexual complications.

It is estimated that 98% of Somali women and girls have undergone some form of genital mutilation. About 90% have been subjected to the most drastic form (type III or pharaonic circumcision). FGM/FGC has been found to be much more in traditional rural societies than in urban areas. A WHO study involving 1744 women aged 15–49 years in Somalia found that 90% of women prefer that FGM/FGC be continued. Several projects were designed region-wise to fight FGM, but the tradition continues (Diop et al. 2004 World Bank).

Special Populations in Rural Areas

Certain groups of rural women maybe particularly vulnerable to mental health problems and may need interventions tailored to them. Described here are some specific vulnerable groups.

Widows in Rural Areas

Across the world regions, in rural societies, death of the husband pushes women into tremendous difficulties, poverty being one of them. As widows, they lose their right to inherit property and often maybe unsupported by husband's family. Eviction from their homes, violence, and loss of household possessions are something they may have to endure during this period.

Due to the HIV/AIDS pandemic, the percentage of households headed by widows in rural Zambia increased from 9.4% to 12.3% between 2001 and 2003. Both widows from patrilineal and matrilineal villages were equally likely to lose their rights to land.

While both men and women in rural areas have poor mental health following widowhood, in some cultures, women bear a double burden of stigma as well as loneliness and are hence more vulnerable (Sandys 2008).

Women in Conflict Areas

Gender equality gets shattered during periods of war and armed conflict and post-conflict restoration. During this time women are abused and face much trauma due to the various kinds of abuse, death of their family members, loss of property, and other material things. War and conflicts demand role changes for both men and women, and this puts a strain on their responsibilities especially more on the women as she may be forced to become the main breadwinner due to the absence or death of her husband and face safety issues both for herself and her children (Sandys 2008).

Elderly Women

In the absence of geriatric mental health specialists in rural communities, older persons who require mental healthcare rely on physicians when they face emotional problems (Colenda et al. 2002). Sometimes rural physicians are the only source of mental healthcare available or the single professional with whom they feel comfortable sharing their emotional or mental distress (Merwin et al. 2003). The number of specialist doctors in rural areas is very few; the group practice arrangements are stretched very thin, allowing only limited physician contact; and hence many rural elderly face medical and mental healthcare crises (Judd and Malcolm 2002). They have to depend on their family members for living arrangements as no other ways are feasible. This often puts a strain on other family members as many rural women are working and trying to balance with caregiving tasks along with other family roles and responsibilities. Counseling, family meetings, and support group services have been found to be helpful to caregivers of older persons, yet these are not easily accessed (Zarit and Zarit 2012). Urban models of best mental health practice are not made available to older rural communities due to various restrictions and budgetary cuts.

Adequate social support has been found to be a protective factor for the mental health of rural women, especially links with family and religious groups (Letvak 1997).

Social support was also found to be an important predictor of mental health outcomes such as depression, anxiety, positive affect, and negative affect in older rural adults who were 65 and older. McCulloch (1995) and Okwumabua et al. (1997) interviewed 96 African American men and women over 60 to compare rural and urban differences in depression. A significant relationship between levels of depression and social support was found wherein (p <0.015) social support helped in reducing the effects of loneliness and depression which was provided mainly by family, neighbors, friends, and church groups (McCulloch 1995; Okwumabua et al. 1997).

Among rural women with newly diagnosed breast cancer, spiritual or church groups and family members were the main sources of social support which helped them to cope with their diagnoses (Koopman et al. 2001).

The rural elderly woman may actually be better integrated with society compared to her urban counterpart, given the opportunities for larger social networks in rural communities in some countries.

Women with Disability

Women and girls with a disability face considerable discrimination and violence across all the regions of the world. Approximately 26% of women living in rural areas have disabilities. Living in rural areas with a disability adds to the risk of having depression. Depression prevented 30% of women with disabilities from being active, compared to 8% of women without disabilities (US Department of Health and Human Services 2000). Research suggests that depression is more prevalent in rural areas, and barriers to accessing mental health services are more pervasive (Probst et al. 2005, 2006). In Australia, 9.5% of the population were women and girls with a disability. Almost 700,000 women and girls with a disability live in rural and remote Australia. Australian people with disability are half as likely to be employed as people without disability and more likely to be living in poverty (Broderick 2012).

There are two types of problems that women with disability may face in rural settings: the absence of required facilities and a lack of access to essential services. Specialized services such as surgery, physiotherapy, dialysis, chemotherapy, and counseling may not be present in rural areas; hence, the disabled person must travel to urban centers which becomes difficult often requiring assistance of another person. Ramps, elevators, sloping curbs, automatically opening doors, grab bars in washrooms, and washrooms that can accommodate wheel chairs are mostly available in urban areas. The rural disabled person must hence depend on family and friends for help in moving around (Jennisen 1992).

Women and girls with disability are also susceptible to many forms of violence including sexual violence, increasing their vulnerability to mental health problems.

Adolescent Girls

Adolescents living in rural areas are often seen as living in a safe world. Research indicates that substance abuse and use are serious problems among adolescents in rural areas (Hall et al. 2008). Rural adolescents show lower academic achievement and a higher rate of dropping out of school when compared to their non-rural peers (Roscigno and Crowle 2001). According to Dunn et al. (2008), rural adolescents in the USA are exposed to significant risk factors such as isolation, early use of alcohol or drugs, and early exposure to sexual activities (Dunn et al. 2008; Reeb and Conger 2011).

Rural young women with emotional problems may face multiple issues. These issues consist of being identified easily, confidentiality and privacy issues, ostracization for deviation from community norms, discrimination against people from other countries and other sexual orientation, sexism, racism, and lack of spaces and activities designed for girls and young people.

Women Living or Affected by HIV

Globally, HIV infection among women influences their mental health and quality of life. Most HIV-positive women live in sub-Saharan Africa, but the epidemic is affecting multitudes of women in South and Southeast Asia and in Eastern Europe and Central Asia.

HIV/AIDS puts pressure on the assets of rural households of women, affects jobs, and earnings, thereby affecting their nutrition and health. Stigma, poor physical health, lack of disclosure, and being a widow or caregiver add to risk for mental health problems (Nyamathi et al. 2011).

Prevalence and Types of Mental Health Disorders

While mental health concerns of rural women remain largely unaddressed due to deficient services and difficult accessibility, studies have reported common mental disorders to be highly prevalent among this vulnerable population. The prevalence of different psychiatric disorders in this underprivileged populace across geographical boundaries is as follows.

The Epidemiologic Catchment Area (ECA) study conducted in the USA assessed prevalence rates of psychiatric disorders and compared rural and urban areas. The rural lifetime prevalence was found to be 32%, slightly lower than the urban (34%) (Mulder et al. 2000). Overall, it was found that the rate of mental disorders was highest in poor people, those who were least educated, *women*, young people, and *rural* communities.

The WHO-based assessment instrument for mental health systems (AIMS) used in 20 countries showed underrepresented rural population among outpatient users in 12 countries, more so for women (Saxena et al. 2007).

Common Mental Disorders (CMDs)

Overall CMDs are two to three times higher in women than men. In rural India the prevalence of common mental disorders was 10.7% among married women aged between 15 and 39 years. Older age, lower education, lower standard of living index (SLI), exposure to intimate partner violence (IPV), husband's unsatisfactory reaction to dowry, and husband's alcohol intake and tobacco consumption were independent risk factors for CMDs in the rural areas (Shidhaye and Patel 2010).

Depression

Unipolar depression is twice more common in women and more persistent than in men and is the most common psychological disorder in rural women (Hillemeier et al. 2008). Some of the risk factors responsible for this gender difference that have been studied include gender disadvantage and the nature and extent of multiple roles often assumed by women. In Central Virginia, 41% of women reported depressive symptoms in a primary healthcare setting (rural medical center), while the prevalence was 20% in the urban areas (Mulder et al. 2000).

In a study in the outpatient department in a rural African hospital, 30% of the women had depressive disorder (Chemali et al. 2013). In Afghanistan, nearly 73% reported depression, anxiety or PTSD, and gender violence (Cardozo et al. 2005). Psychological complaints were found to account beyond 40% of all patient visits to physicians in rural family practice (Mulder et al. 2000). Also, in comparison to urban women, rural women were found to have higher rates of chronic physical illnesses associated with depression leading to more disability.

In a multicentric study conducted in Bangladesh, women above 18 years had higher prevalence of mental disorders compared to men, and the rural population had higher rates compared to urban. Depression and somatoform disorders were the commonest conditions (Firoz et al. 2006).

In Vietnam, rural women had a higher prevalence of depression (6.8%), almost double of that in men (3.9%). Only 5% of those with mental distress sought healthcare at facilities where mental healthcare services were available. Women had more symptoms such as "headache, poor appetite, sleeping badly, easily frightened, having trouble thinking, feeling unhappy, crying more than usual, having thought of ending life, feeling tired all the time, being easily tired" (Bao Giang et al. 2010).

Depression in the Perinatal Period

Mental health problems in the perinatal period are an important public health issue. In a study at Missouri, USA, the overall prevalence of any psychiatric disorder was found to be 43.5% among rural residents, while major depressive disorder was found to be 13.9% among rural women. It was found that pregnant women from rural communities were less likely to receive treatment than their urban counterparts, despite having similar prevalence rates of psychiatric disorders (Cook et al. 2010).

In rural Tamil Nadu, India, the incidence of postpartum depression was found to be 11%. Risk factors identified were low income, birth of a daughter when a son was desired, interpersonal relationship difficulties with mother-in-law and parents, adverse life events during pregnancy, and lack of physical help (Chandran et al. 2002).

In a document by Courtney Pendray in 2012, high maternal mortality rate and pregnancy rate of Afghani women along with reported violence made prevalence of mental disorders very high in the perinatal period. Lack of access to health services related to maternity care was identified as the main factor which results in higher mental distress in this population (Pendray 2012).

Somatization Disorders

Somatization or somatoform group of disorders have been identified as disorders with bodily symptoms which are an expression of psychological distress. This type of stress has been invariably high among the rural women who are less educated and less expressive of their problems and have low socioeconomic status. Consistently studies have found higher prevalence of somatic symptoms in the population with these demographics.

Women from rural areas with psychological distress often present to primary care settings with bodily symptoms such as headaches, backaches, insomnia, fatigue, and abdominal pain more than depressive symptoms unlike their urban counterparts with depression. This could be as attention is paid to physical complaints and is more acceptable while mental complaints have stigma attached to them (Mulder et al. 2000).

Reporting of depression is also higher among women with somatic complaints in rural settings (Simmons et al. 2007). Similarly, the overlap of depression, anxiety, and somatization is not only visible clearly in primary care clinics but also in villagers with high PTSD due to past trauma (Löwe et al. 2008).

Possession syndrome or conversion disorders are also found among rural women. Dissociation has been found to be the indicator of stress in the mind. Purposes behind possession are to express the unmet desires and the need for power and prestige. It gives women a chance to liberate themselves from the clutches of patriarchy during the brief episodes of trance like state. It also gives them an opportunity to express their true selves and yet remain unharmed from the outburst, thereby protecting them from the social norms that otherwise suppress them (Sered 1994).

Anxiety Disorders

Anxiety as a symptom of another mental illness or an independent syndrome is common in women worldwide. In a study done in the rural households of the landlocked state of Tennessee, USA, women were found to have higher levels of anxiety and depression than urban women (Mulder et al. 2000).

A study conducted in the mountain villages of Chitral District in Pakistan showed a conservative estimate of 46% of women suffering from anxiety and depressive disorders (Mumford et al. 1996). In another study from the same region, 25% of women had anxiety, and 17% had a combination of anxiety and depression (Dodani and Zuberi 2000).

In the rural region of Sichuan Province of China, 12.4% (236) of the women expressed significant psychological distress with a score of more than 15 on the CES-D scale (Center of Epidemiological Studies-Depression Scale). MINI international neuropsychiatric interview administered to those with >15 on CES-D scale revealed that 11% had anxiety disorder and 49.8% qualified for major depressive episode. Unemployment and poverty were found to be associated with these disorders in this rural region (Qiu et al. 2016).

A cross-sectional study done in rural Brazil among 182 rural women farm workers from the state of Rio Grande do Sul reported 61% (111 of 182) of women having work-related mental disorders. The most common being anxiety spectrum of disorders with a rate of 37.9% (69 of 182), followed by acute stress reaction 36.8% (67), nonorganic disorder of the sleep-wake schedule 30.2% (55), depressive disorder 18.1% (33), and panic disorder 4.9% (9). Those who reported physical discomfort with work had higher prevalence of mental disorders (Cezar-Vaz et al. 2015).

Suicide

Women outnumber men in the total burden of suicidal morbidity and mortality due to greater gender-related vulnerability factors (Vijayakumar 2015). Suicide rates for rural women are usually underreported, often to protect the immediate family or avoid religious implications of suicide. An Indian survey reported that 72,000 women had suicide-related deaths in 2010 of which women aged 15–29 years had 56% suicide death rates higher compared to men (40%). The commonest mode used was pesticides due to its easy availability. Rural women had 1.08% higher risk of suicide. Death rate due to suicide was found to be 20.4 /1 lakh for women, with a higher rate in southern states (Patel et al. 2012).

Women in rural areas mainly work as farmers themselves or help their spouses who are farmers. Worldwide the rate of farmer suicide is high and has been a matter of grave concern for governments and policy makers. Data regarding suicide rates in women farmers is scanty. Hence, not much has been reported regarding the psychological issues of widows whose husbands have committed farmer suicide.

The World Health Organization (WHO) reported higher suicide rate among the elderly in Asian countries such as China, Hong Kong, Japan, South Korea, Malaysia, and Singapore in both rural and urban populations (Suh and Gega 2017). Suicide rates are two to three times higher among rural women and among the elderly five times more than urban population in China.

In China, almost 70% of the population lives in rural areas. Suicide rate in women has been on the rise in China compared to the rest of world. China finds a bimodal peak of suicide rate in women (one at 15–24 years and the other in the elderly) unlike the rest of the world and is high among rural women compared to their urban counterparts (Zhang et al. 2002).

In Nepal, suicide is the second leading cause for death among women in the reproductive group (Pradhan et al. 2011). High rates of trafficking in rural women to meet the commercial sex demand have been reported as one of the causes for suicide. Polygamy, alcoholism, and violence from men are additional risk factors for suicide among young women (Simkhada et al. 2015).

In Cambodia, the Royal University of Phnom Penh conducted the first-ever large-scale mental health survey in 2011 in 9 provinces (rural regions included) and estimated the female suicide attempt rate to be 5.5% versus 1.7% in men (Dr. Tanja Schunert 2012).

Economic polarization, marital problems, traditional gender role, traditional values and practices, low self-esteem, lack of education, lack of safety net for elderly, poverty among senior citizens, stigma and/or insufficient knowledge of mental illness, social isolation and disconnection, difficulty in accessing services, and ready access to lethal means of suicide have been explained as some of the reasons for higher suicidality among rural women in East Asian countries.

Substance Use

In India, dependence on chewable type of tobacco is highly prevalent and results in life-threatening diseases like oral cancer. Women folk of the rural community abuse it more than men, and it is a regular pastime as it is an accepted social behavior. The ITC (International Tobacco Control) study is a multisite survey which was initially carried out among 4 countries (UK, USA, Canada, Australia) in 2002 which was later extended across more than 25 countries consisting of both developing and developed countries. The Bangladesh survey – ITC study –reported that smokeless tobacco was used more in rural than urban areas and more in women than men (Nargis et al. 2015).

In a primary care outpatient sample of 600 rural South Africans, 10.7% of women were identified as hazardous drinkers, and 0.3% of women met criteria for probable alcohol dependence or harmful drinking as defined by AUDIT (Peltzer 2006).

Men and women differed on perceived barriers/facilitators and need for alcohol treatment. Women differed from men on measures of treatment affordability, accessibility, acceptability and report of social support, illness severity, comorbidities, and demographic characteristics. Rural women differed from urban women on measures of treatment affordability, accessibility, and report of illness severity and comorbidities (Small et al. 2010). In a US based study for 267 women in rural area, victimization was identified as significant factor for occurrence of mental illness (Boyd 2003).

Data related to substance use among rural women is, however, not available from many parts of the world (Table 1).

Table 1 Prevalence studies of psychiatric disorders among rural women from different parts of the world

Sl no.	Country	Population	Findings	Risk factors
1	Malawi (Local Governance Performance Index in Malawi: Selected findings on Health, November 2016)	15 districts; 8100 respondents from north/south and central regions	84% of the population live in rural areas. Women suffer from mental health issues more than men. 24% of females reported a depressed feeling most or all of the time	Education, income, and geographical location had an impact on the percentage
2	Malawi (Stewart et al. 2010)	501 mothers from a child clinic	Weighted prevalence: current depressive episode (minor or major) was 30.4%, and current major depressive episode was 13.9%	CMD associated with poverty, relationship difficulties, HIV infection, and infant health problems
3	Ethiopia (Deyessa et al. 2008)	3016 randomly selected women between 15 and 49 years	4.4% was the 12-month prevalence of depression among all women Higher divorced and widowed Holding traditional values around childbirth was protective factor in postnatal depression	Factors: khat chewing habit, seasonal job, living in village
4	Ethiopia (Chemali et al. 2013)	Women patients; 226 in the period 2006–2008	Depression MDD, 30%; psychosis 12%; BPAD, 19%	Rural women have difficulty to access basic mental health services
5	Afghanistan (Trani and Bakhshi 2013)	5130 households from December 2004 to July 2005. Women compared with nondisabled men	Women were 3.35 and 8.57 times more likely to experience mild or moderate mental distress disorders. Widowed or divorced women were 1–12 times more at risk of experiencing mild or moderate levels of mental distress disorders	Disabled women were a particularly vulnerable group at higher risk of experiencing mental distress disorders

(continued)

Table 1 (continued)

Sl no.	Country	Population	Findings	Risk factors
6	Nepal (Khattri et al. 2013)	146 women of 261	30.8% women (45/146) found to have psychiatric problem	Risk factor: caste (dalit)
7	Nepal (Suvedi et al. 2009)	1496 deaths in women of reproductive age. National survey in 8 districts	Women: suicide third highest cause of death; 27.6% is suicide rate. 16% death of women in reproductive age is due to suicide	Mental health problems as one of the causes for high rate of suicide
8	Nepal (Kohrt and Worthman 2009)	316 households	36.5% anxiety in women (n = 130) whereas 20.4% of men (n = 186)	Women have greater risk of anxiety which is moderated by social support
9	Nepal (Dørheim Ho-Yen et al. 2006)	426 postpartum women evaluated, EPDS done	Rural (n = 102); prevalence is 3.9% (low and not significant)	
10	Africa: Study done in African Somali women (Knipscheer et al. 2015)	66 women who had undergone FGM	20% had PTSD symptoms; 33% had depression; and 30% had anxiety symptoms	Substance use also was found as a mechanism of coping
11	India: The great universe of Kota Study done in rural region (Carstairs and Kapur 1976)	3 South Indian communities (agriculturists: Brahmins, Bants, and fishermen, Mogers)	40% of the women had one psychiatric symptom. 35% had neurotic symptoms. In 55% of widows and >40 years of age in women showed psychiatric illness. 24% had somatic symptoms, 15% insomnia, and 13% forgetfulness with poor concentration. Anxiety, depressive, and psychotic symptoms were about 1–3%	Women especially of the Bant and Moger community were found to have higher psychiatric rates than men

(continued)

Table 1 (continued)

Sl no.	Country	Population	Findings	Risk factors
12	India: Door-to-door study in rural India (Nandi et al. 2000)	2 culturally similar villages near Calcutta two decades (1972 and 1992) apart	Mental morbidity more in rural women (138–147 per 1000) than men (73–86 per 1000) Rural women affected with mental illness decreased from 146.8 to 138.5 per 1000 over 20 years. (economic independence) Hysteria, phobia rate decreased but depression increased.	Economic independence – working women/females increased
13	India: National Mental Health Survey of India 2015–2016 (Gururaj et al. 2016)	12 states, i.e., 2 each from different parts of India. Rural India represented 492 clusters out of total 720 surveyed	Females overall were 52.3% of all respondents. Two times higher prevalence of neurosis and stress-related disorder compared to men. Lower prevalence of depression and suicide in rural women compared to urban and metro	

Challenges in Accessing Mental Healthcare for Rural Women

Mental health resources are scarce, but they are also inequitably distributed: between countries, between regions, and within local communities. Need and access tend to vary inversely, and often those with highest need have least access to care. Due to stigma and discrimination, people worldwide are hesitant to seek help or even to accept that their difficulties maybe related to a mental disorder. Specific problems that rural women might face include the following.

Lack of Women-Friendly Mental Healthcare Services in Rural Areas

Due to the meager mental health services, many mental health problems in rural areas go underreported (Letvak 1997). Those suffering from a mental illness bear the burden of limited access to healthcare, a scarcity of resources, and traditional cultural belief systems. As a result they earn lower incomes, utilize lower levels of insurance coverage, have higher rates of chronic disease, higher infant mortality rates, and use less preventive health screening (Coburn and Bolda 1999).

The typical rural life has various sociocultural and lifestyle factors that affect physical and mental health in similar ways. Access to both medical and mental health services is affected by various factors such as inadequate funding, long distances, lack of transportation, and ecological barriers (Mulder et al. 2000). It is hence recommended that a combined public health approach that addresses both physical (including gynecological) and mental health at the same clinic maybe more appropriate for rural women than a specialist psychiatric service.

Privacy and confidentiality and availability of women health professionals would be an important requisite, which may be lacking.

Lack of Professionals with Training in Handling Rural Mental Health Problems

Rural populations have inadequate access to care, since mental health professionals in most countries tend to be concentrated around cities. The main reason for this is that services do not use strategies to deliver care equitably to all groups. Rural areas have serious shortages of mental health professionals. This means that rural residents must receive psychological services from primary care physicians who may be unprepared to identify and treat mental illness (Mulder et al. 2000). In the absence of trained mental health professionals, ruralities have no choice but to seek help from informal support systems such as ministers, self-help groups, and family or friends. While support from family and friends are important, they may be inadequate in giving the complete care and support of a professional.

Those with mental illness may encounter difficulty in accepting social support from families and friends for fear of maintaining anonymity of their mental health problems. Absence of mental health resources in rural communities makes this a major health concern.

Another barrier is created by the professionals themselves who being educated and trained in the cities may not necessarily understand the rural way of life which prevents them from relating to the rural problems (Mulder et al. 2000).

Apart from the "help-seeking" barriers which are "cultural," in rural areas, most mental health services are planned with the urban problems in mind. These programs designed for meeting the mental health needs of the urban population cannot be replicated for the rural population as they may not be relevant to the rural needs and problems (Mulder et al. 2000).

Here the role of women community health workers becomes very important. Many countries have women community health workers who are part of the same social milieu and hence understand the context in which women live and work. Rural women find these workers (like the ASHA workers in India) easier to relate to and are more willing to share their problems with them.

Culture, Stigma, Help-Seeking Patterns, and Treatment Preference

Rural culture emphasizes staunch self-reliance, and hence seeking help for mental health issues may be considered as a vulnerability which is looked down upon. In the rural areas outsiders are viewed with distrust, and this attitude prevents them from seeking help from people outside their intimate inner circle. A high value is placed on self-reliance, and rural residents may fear that community members will find out that they required assistance for emotional difficulties which would directly conflict with that value (Mulder et al. 2000).

Rural women may present to primary care settings with somatic complaints such as headaches, backaches, insomnia, fatigue, and abdominal pain and may not reveal symptoms of depression to their primary care providers. This may prevent them from getting treatment for their mental health problem. Women in rural areas may also prefer more integrative approaches to care. Hence, approaches that integrate the body (physical), mind (mental), and soul (spiritual and social) are more easily accepted, especially from people in their own community (Mulder et al. 2000).

Since maintaining confidentiality is a main concern in rural areas due to the stigma attached, there is often a reluctance to seek professional help (Howland 1995; Mulder et al. 2000).

Poor Knowledge About Facilities and Benefits

Utilization of healthcare services within the rural community is influenced by many factors such as the stigma attached with mental illness, lack of understanding about mental illnesses and their treatments, lack of information about treatment facilities, and the incapacity to pay (Mulder et al. 2000).

Lack of Access, Transport, and Services

Inpatient psychiatric services do not exist in many rural areas, and therefore the access to these services is limited or even nonexistent (Mulder et al. 2000). Most often mental health services in the rural areas fall on the government, and these mainly cater to those who are severely mentally ill. Hence, many rural residents may have to cover long distances to reach an appropriate care provider or at times may receive help from a mental health professional with less experience or training.

Another barrier to access treatment is the lack of proper transportation or difficulties in traveling to treatment facilities, especially for women due to geographical isolation. Rural women often have to depend upon public transportation to travel which may be sparse or with limited trips to their destination. Lack of communication facilities like not having a telephone also impedes access to mental health services (Mulder et al. 2000).

Another major problem in rural areas is the absence of quality inpatient care for severely mentally ill people. Hence patients should travel to cities for inpatient psychiatric care. Post discharge they do not have access to day care or half way homes in the community unlike in the urban settings and hence must settle for expensive hospitalization charges when relapses occur. Rural residents may also not be aware of the various monetary or pension benefits that maybe available to them. Due to the high levels of poverty in rural areas, funding in these areas is limited as tax benefits are minimal, causing understaffing of health care centers (Mulder et al. 2000).

High Costs of Mental Health Services

One of the major barriers which impedes using mental health services is the cost factor. Sixty-eight percent of women rural residents did not receive healthcare because of the high cost. In a study on 281 rural minority households, majority of the women reported their inability to pay as the main reason for not seeking physician's help (Mulder et al. 2000).

Absence of health insurance is a significant hurdle to receiving treatment, and in rural areas many are not covered under insurance schemes as they may be employed in the farming sector which is not an organized sector.

What Is Important for Quality of Life and Psychological Well-Being Among Rural Women?

While there has been interest in understanding factors that contribute to distress and poor mental health among rural women, there is a small body of research that has also focused on their well-being and quality of life (QOL).

Most people think of rural life as idyllic with good social networks and fresh air. Does that translate into a good quality of life for women living in rural areas? Studies among women in Africa including rural Malawi and South Africa have found several factors that contribute to quality of life of rural women. These include both essential needs, such as living conditions and nutrition and those addressing complex aspects of life including social networks, being accepted by the community, safety within and outside the household, and inner peace of mind and well-being (Greco et al. 2015).

Interestingly, studies done in Bangladesh, where microcredit has been used to enhance women's autonomy and economic stability, found that improving rural women's material conditions alone did not enhance well-being. In fact, studies have found that in patriarchal societies, addressing differential power within the household, decreasing domestic violence, and addressing needs of widowed and separated women are more important than or indeed as important as providing economic empowerment. Factors such as interpersonal discord and the well-being of children seemed to have a larger influence on women's well-being rather than being part of an income generation program (Ahmed et al. 2001).

Studies from countries where patriarchy is common found that gender-based identities within a household determine well-being. Higher levels of household work are associated with lower levels of well-being among women who disagree with patriarchal notions of gender roles, while the opposite is true for women who agree with patriarchal notions of gender roles. Importantly, this pattern holds only when a woman strongly identifies with patriarchal or egalitarian notions of gender role (Seymour and Floro 2016).

Singh et al. (2014) found that in India, rural women enjoyed performing household chores and undertaking agricultural activities and animal husbandry work. They also endorsed enjoying dual roles and revealed that they would like to work outside home too. Age and education were significant factors that affected their experience of well-being. Those with higher level of education experienced greater subjective well-being and better psychological well-being and projected more positive relations with others as well as compared to those with lower level of education. Women in the youngest age group showed higher levels of subjective happiness and well-being and reported better personal growth and positive relations with others as compared to their older counterparts (Singh et al. 2014).

In a meta-synthesis of qualitative studies on well-being among women in rural Australia, the four important themes related to psychological well-being included connectedness, a feeling of belonging, the ability to cope with adversity, and the women's rural identity.

Women reported a strong sense of belonging and a spiritual connection to the land as enhancing their sense of well-being. On the other hand, they also reported that the gendered rural identity of having to be stoic and strong as well as self-reliant was at odds with the desire to have social connections and help in times of crisis.

For women being part of a larger network, sharing of feelings and emotions and a sisterhood is important which may get compromised in rural and remote areas. This maybe particularly true for women who migrate into rural areas from other countries and may not know the language of the host country (Harvey 2007).

Resilience of rural women in Australian farms were found to be in their positive view of self and viewing themselves to be partners in farming. Along with this, the meaningful role they played along with relative autonomy and active decision-making helped in retaining hope and connecting to their place (King et al. 2009).

An important finding in all the studies related to well-being was the heterogeneity of factors related to it based on age, immigrant status and ethnicity, nature of work, and sexual identity. Single women and those who were separated or widowed constituted a vulnerable group, and their well-being was often determined by economic security, land ownership, and connectedness to and support from the community.

Measurement of Quality of Life in Rural Women

A few tools to specifically assess quality of life and psychological well-being of rural women have been developed keeping in mind the unique issues they face, which other quality of life measures may not be able to pick up. However, they have not been used as much as they should. These tools focus on specific areas that may influence QOL in rural women such as land ownership, distribution of chores, literacy, and birth control (Khatun et al. 1998).

A study from rural Africa which aimed to develop a method for measuring and assessing quality of life and well-being using focus group discussions with women in rural Malawi found physical strength, inner well-being, household well-being, community relations, and economic security as important factors (Greco et al. 2015).

The Well-Being of "Left Behind Women"

In countries like China and several countries in sub-Saharan Africa and Asia, an increase in migratory flow and the absence of husbands or male members of households have led to women taking over traditional male tasks and responsibilities. The women left behind potentially face difficulties, such as increased workload, poor access to information and resources, and inability to make decisions because of social norms and restrictions on their ownership of property. The impact of male migration is more in rural women if social support systems and services are poor. Female heads of households may have to deal with greater challenges than male heads of households in meeting the needs of their households, because of lower economic and social status, lack of control over agricultural income, and a heavy workload that may reduce their overall productivity.

Despite these problems, male migration may improve the status of women left behind in rural areas, including increased empowerment and an opportunity to acquire new skills and capacities leading to better self-esteem and independence.

A study among women in rural Kerala in India also provides insights into the processes shaping women's lives in the context of male migration. It documents both the constraints and the opportunities provided by male migration. While some women established or maintained their own households and gained increased autonomy as well as responsibility, others lived with extended family and were subject to strict supervision and regulation and had to cope without help from their husbands mediating between them and the extended households (Gulati 1993).

In China, Zhou et al. (2007) found that less than half of the *left-behind* women reported positive subjective well-being. The *left-behind* women lacked the rights to speak and make decisions, and as a result, their rights and benefits were more easily violated. Guo et al. (2013) analyzed the subjective well-being of 134 left-behind women in rural areas of northern Jiangsu Province in China and found that more than

Table 2 Factors influencing well-being among rural women

Physical well-being	Inner well-being
Being able to do physical work	Being happy and satisfied with life
Having enough food to eat	Having peace of mind
Being able to avoid diseases	Having control over personal matters
Being able to space births	Being free from oppression
Good health	Living without shame
	Having knowledge
	Having good conduct
Household and family well-being	**Community relations**
Living free from domestic violence	Feeling safe and comfortable in the village
Living in a satisfactory house	Being able to join community groups
Being able to take care of children and husband	Avoiding social exclusion and discrimination
	Being respected
Being able to educate the children	Being able to access services
Economic security	**Rural identity**
Owning assets	Sense of spiritual connection with the land
Having control over money	Power relations within the household, specifically patriarchy
Being able to access business opportunities	
Being able to rely on safety nets	
Being able to cope with shocks	

20% of them reported low subjective well-being. Desires for more children, lower family income, and less social chances for social interaction were the main factors that contributed to poor well-being (Table 2).

Mental Health Interventions for Rural Women

While there has been much research on the prevalence and risk factors associated with mental health problems among rural women, the literature on interventions aimed at enhancing mental health or managing mental health problems is limited to a few community-based trials, mainly in LAMI countries.

We describe here some of these interventions which have a reasonable level of evidence. The Thinking Healthy Programme which was used among mothers in a rural sub-district of Pakistan included an intervention which used techniques of active listening, collaborating with the family, and guided discovery and homework and encouraged health workers to use it in their routine maternal and child healthcare (Rahman 2007). Both health workers and depressed mothers found the intervention relevant and useful. In a cluster, randomized trial involving nearly 400 mothers in each arm, the Thinking Healthy Programme, delivered by community health workers, resulted in halving the rate of depression in prenatally depressed women compared with those receiving enhanced routine care. The women who received the intervention also reported less disability and better overall and social functioning, all of which were sustained till 1 year (Rahman et al. 2008).

In a qualitative study that examined the delivery of the above program through peer volunteers (PVs), the women found the intervention to be acceptable, and the peer volunteers found it easy to deliver. The PVs also called "barefoot therapists," which appeared to understand cultural and local nuances, were easily accepted by the family and were considered trustworthy (Atif et al. 2016). In rural areas, models such as these might be acceptable and worth examining, especially for common mental disorders among women. This first-level care may lead to a better pathway to more specialist care, especially where stigma related to mental health problems may prevent women from seeking help from existing healthcare facilities.

Community mobilization through participatory women's groups in poor rural areas aimed mainly at reducing the neonatal mortality rate was also found useful in decreasing depression in a trial conducted in a tribal and rural population in three districts in eastern India (Tripathy et al. 2010).

The reduction in depression was attributed to improvements in social support and problem-solving skills through participation in women's groups. In meetings, information was shared about the difficulties encountered by mothers, and practical ways to collectively address them were discussed, resulting in better problem-solving skills.

However, in rural Bangladesh, a similar trial using participatory groups among pregnant women did not yield positive results or decrease in depression. The authors attribute this to other social and contextual factors in patriarchal societies such as rural Bangladesh, which might override the gains achieved through group participation among pregnant women.

Smaller trials using mindfulness-based interventions have reported improvement in mental health among women reporting perinatal grief related to stillbirths. CBT has been found to be useful in decreasing antepartum depression among rural women in the USA.

Interestingly, most of these trials have been among rural women in the perinatal period. One reason for this maybe the finding that the perinatal period provides an opportunity for behavior change, and any intervention during this period also has a long-term impact on the woman and her newborn.

Interventions that specifically address mental health issues among nonpregnant women are few. Kermode et al. (2007) used a peer-facilitated participatory action group (PAG) to enhance the mental health of widows in rural Northeast India. The intervention consisted of 10 PAG meetings involving 74 IDU (injecting drug use) widows. The women in the study reported significant improvement over the course of intervention on their quality of life. Mental health and somatic symptoms also improved significantly.

Atmiyata (meaning shared compassion) is a rural community-based mental health intervention program on promoting well-being and reducing distress with low-cost interventions. It ran from 2013 to 2015 with help from a local NGO and covered 41 villages in Nasik District in Maharashtra, Western India. It utilized 323 community-based volunteers (CBV) who were trained by professionals and provided smartphones with the "Atmiyata app" which contained mapping tool, films, training, and educational materials and linked welfare benefits of nearly 14,000 adults. CBVs avoided medical jargons and used simple counseling techniques, behavioral activation, health management, and financial advice. This study demonstrated the usefulness of a low-cost model for community-based mental healthcare intervention with public-private partnership (DMHP and NGO) (Shields-Zeeman et al. 2017).

A qualitative research was carried out among two groups of women from four villages. Focused group discussions were conducted separately for Group 1: the intervention group with 180 self-help group members (SHG) received group counseling and stress management. Group 2: 160 was the control group who did not receive any mental health intervention. Eighty-six percent reported better quality of sleep and reduction in aches and pains in Group 1. Hence, combining SHGs who focus on microcredit along with mental health interventions can be used as models in the future in other parts of the world (Rao et al. 2011).

How should We design Interventions for Mental Health Problems Among Rural Women?

Interventions for alleviating psychological distress among rural women should be acceptable, be preferably delivered by people from the same community or at least by those who are aware of the social and cultural norms of the rural community, be participatory, and involve simple cognitive behavioral techniques or mindfulness training. They should not enhance stigma; rather the approach should be one of inclusiveness. Interventions should consider the realities of rural women such as dependence on agriculture and its related problems, social isolation, and lack of empowerment. They should use strengths of the women such as spirituality, relationship to the land and nature, and existing social support systems.

Given the poor resources in most rural settings, several models of task shifting have been used and are recommended. Task shifting involves shifting delivery of the task from professionals to less qualified health workers or creating a new work force with specific training for a particular task. A recent trial among women with depression and HIV in rural Karnataka has found that such interventions delivered by trained basic nurses are feasible and acceptable (Reynolds et al. 2016). Recent systematic reviews have also indicated effective use of task shifting for the delivery of psychosocial interventions for common mental disorders in resource-constrained settings (Hoeft et al. 2016).

Mobile phone health interventions (mhealth) have shown promising results for rural women in particular (Reynolds et al. 2016; Shields-Zeeman et al. 2017). Mobile applications are portable and allow care anywhere, anytime, regardless of patient geography and transportation barriers. These are low cost and can directly connect healthcare providers with patients. Text-based mobile messages and direct interventions are possible. More trials that are gender specific are needed regarding acceptability and risks of mhealth interventions among women from rural areas.

Conclusion

Rural women with mental health problems are a unique group with factors that are quintessentially related to rural life. Social isolation, a culture of self-reliance, and a high work burden both at home and outside, coupled with poor access to both physical and mental health services, place rural women at a very vulnerable position. Prevalence of some psychiatric disorders and the risk factors related to them are seen to be higher. Cultural and social issues such as patriarchy, traditional roles, low levels of economic independence, poor access to communication, and gender-based violence are some of the important risk factors. Some aspects of rural life such as belongingness to the land and self-reliance contribute to their well-being; however, there seem to be more risks than protective factors such as stigma, a fear of disclosing mental health problems, and poor access. On the other hand, world over, health services in rural areas have not geared themselves up to meet the mental health needs of rural women. There have been a few attempts at developing and testing interventions especially for this group, but the research is meager. More interventions that meet the mental health needs of rural women need to be designed and tested. Finally it is prudent to remember that one rural area is not like another, and any intervention for mental health has to be tailored to the norms and realities for women in different rural communities as one size may not fit all!

Cross-References

▶ Methodological Issues in Epidemiological Studies on Mental Health in Rural Populations
▶ Yoga and Traditional Healing Methods in Mental Health

References

Ahmed SM, Chowdhury M, Bhuiya A (2001) Micro-credit and emotional well-being: experience of poor rural women from Matlab, Bangladesh. World Dev 29(11):1957–1966
Atif N, Lovell K, Husain N, Sikander S, Patel V, Rahman A (2016) Barefoot therapists: barriers and facilitators to delivering maternal mental health care through peer volunteers in Pakistan: a qualitative study. Int J Ment Heal Syst 10(1):24
Bao Giang K, Viet Dzung T, Kullgren G, Allebeck P (2010) Prevalence of mental distress and use of health services in a rural district in Vietnam. Glob Health Action 3(1):2025
Boyd MR (2003) Vulnerability to alcohol and other drug disorders in rural women. Arch Psychiatr Nurs 17(1):33–41
Broderick E (2012) Violence against women with a disability in rural Australia. Australian Human Rights Commission. https://www.humanrights.gov.au/news/speeches/violence-against-women-disability-rural-australia. Accessed 10 May 2017
Cardozo BL, Bilukha OO, Gotway CA, Wolfe MI, Gerber ML, Anderson M (2005) Report from the CDC: mental health of women in postwar Afghanistan. J Women's Health 14(4):285–293
Carstairs GM, Kapur RL (1976) The great universe of Kota: stress, change, and mental disorder in an Indian village. University of California Press
Cezar-Vaz MR, Bonow CA, da Silva MRS (2015) Mental and physical symptoms of female rural workers: relation between household and rural work. Int J Environ Res Public Health 12(9):11037–11049
Chandran M, Tharyan P, Muliyil J, Abraham S (2002) Post-partum depression in a cohort of women from a rural area of Tamil Nadu, India. Br J Psychiatry 181(6):499–504

Chemali ZN, Borba CP, Henderson TE, Tesfaye M (2013) Making strides in women's mental health care delivery in rural Ethiopia: demographics of a female outpatient psychiatric cohort at Jimma University Specialized Hospital (2006–2008). Int J Womens Health 5:413–9

Coburn AF, Bolda EJ (1999) Rural elderly and long-term care. In: T. C. Ricketts (Ed.), Rural health in the United States. New York: Oxford University Press, pp 179–189

Colenda CC, Mickus MA, Marcus SC, Tanielian TL, Pincus HA (2002) Comparison of adult and geriatric psychiatric practice patterns: findings from the American Psychiatric Association's Practice Research Network. Am J Geriatr Psychiatry 10(5):609–617

Cook CAL, Flick LH, Homan SM, Campbell C, McSweeney M, Gallagher ME (2010) Psychiatric disorders and treatment in low-income pregnant women. J Women's Health 19(7):1251–1262

Deyessa N, Berhane Y, Alem A, Hogberg U, Kullgren G (2008) Depression among women in rural Ethiopia as related to socioeconomic factors: a community-based study on women in reproductive age groups. Scand J Soc Med 36(6):589–597

Diop M, Ba F (2004) Female genital mutilation/cutting in Somalia. World Bank and United Nations PopulationFund. http://siteresources.worldbank.org/INTSOMALIA/Data%20and%20Refer ence/20316684/FGM_Final_Report.pdf. Accessed 20 Apr 2017

Dodani S, Zuberi RW (2000) Center-based prevalence of anxiety and depression in women of the northern areas of Pakistan. J Pak Med Assoc 50(5):138

Dørheim Ho-Yen S, Tschudi Bondevik G, Eberhard-Gran M, Bjorvatn B (2006) The prevalence of depressive symptoms in the postnatal period in Lalitpur district, Nepal. Acta Obstet Gynecol Scand 85(10):1186–1192

Dr. Schunert T, Khann S, Kao S, Pot C, Saupe LB, Dr. Lahar CJ, Sek S, Nhong H (2012) Cambodian mental health survey report. Phnom Penh, Cambodia: Royal University of Phnom Penh, pp 1–101

Dunn MS, Ilapogu V, Taylor L, Naney C, Blackwell R, Wilder R, Givens C (2008) Self-reported substance use and sexual behaviors among adolescents in a rural state. J Sch Health 78(11):587–593

Firoz AHM, Karim ME, Alam MF, Rahman M, Zaman MM (2006) Prevalence, medical care, awareness and attitude towards mental illness in Bangladesh. Bangladesh J Psychiatry 20(1):28

Greco G, Skordis-Worrall J, Mkandawire B, Mills A (2015) What is a good life? Selecting capabilities to assess women's quality of life in rural Malawi. Soc Sci Med 130:69–78

Gulati L (1993) In the absence of their men: the impact of male migration on women. Sage, New Delhi

Guo HJ, Zhang ZH, Xu J, Li XN, Li J (2013) A survey on subjective wellbeing among rural left-behind women in northern Jiangsu. Jiangsu Prev Med 24(3):60–61. (in Chinese)

Gururaj G, Varghese M, Benegal V, Rao GN, Pathak K, Singh LK, Mehta RY, Ram D, Shibukumar TM, Kokane A, Lenin Singh RK, Chavan BS, Sharma P, Ramasubramanian C, Dalal PK, Saha PK, Deuri SP, Giri AK, Kavishvar AB, Sinha VK, Thavody J, Chatterji R, Akoijam BS, Das S, Kashyap A, Ragavan VS, Singh SK, Misra R, NMHS Collaborators Group (2016) National mental health survey of India, 2015–16: mental health systems. National Institute of Mental Health and Neuro Sciences, Bengaluru. NIMHANS, Publication no. 130

Hall JA, Smith DC, Easton SD, An H, Williams JK, Godley SH, Jang M (2008) Substance abuse treatment with rural adolescents: issues and outcomes. J Psychoactive Drugs 40(1):109–120

Harvey MR (2007) Towards an ecological understanding of resilience in trauma survivors. J Aggress Maltreat Trauma 14(1–2):9–32. https://doi.org/10.1300/J146v14n01_02

Hillemeier MM, Weisman CS, Chase GA, Dyer A-M (2008) Mental health status among rural women of reproductive age: findings from the Central Pennsylvania Women's Health Study. Am J Public Health 98(7):1271–1279

Hoeft TJ, Fortney JC, Patel V, Unützer J (2016) Task-sharing approaches to improve mental health care in rural and other low-resource settings: a systematic review. J Rural Health 34:48

Howland RH (1995) The treatment of persons with dual diagnoses in a rural community. Psychiatry Q 66(1):33–49

ILO (2009) Global employment trends for women. International Labour Organization, Geneva

Jennisen T (1992) Health issues in rural Canada. Political and Social Affairs Division. http://publications.gc.ca/Collection-R/LoPBdP/BP/bp325-e.htm. Accessed 26 May 2017

Judd F, Malcolm H (2002) Psychiatry and rural general practitioners: keeping patients and doctors healthy.-editorial. Current Ther 43(11):7

Kermode M, Herrman H, Arole R, White J, Premkumar R, Patel V (2007) Empowerment of women and mental health promotion: a qualitative study in rural Maharashtra, India. BMC Public Health 7(1):225

Khattri J, Poudel B, Thapa P, Godar S, Tirkey S, Ramesh K, Chakrabortty P (2013) An epidemiological study of psychiatric cases in a rural community of Nepal. Nepal J Med Sci 2(1):52–56

Khatun M, Wadud N, Bhuiya A, Chowdhury M (1998) Psychological well-being of rural women: developing measurement tools. Retrieved from Dhaka

Kim Tjaden MD (2015) Health Disparities between Rural and Urban Women in Minnesota. Minn Med 98:40

King D, Lane A, MacDougall C, Greenhill J (2009) The resilience and mental health and wellbeing of farm families experiencing climate variation in South Australia. Report for the National Institute of Labour Studies Incorporated, Adelaide

Knipscheer J, Vloeberghs E, van der Kwaak A, van den Muijsenbergh M (2015) Mental health problems associated with female genital mutilation. BJPsych Bull 39(6):273–277

Kohrt BA, Worthman CM (2009) Gender and anxiety in Nepal: the role of social support, stressful life events, and structural violence. CNS Neurosci Ther 15(3):237–248

Koopman C, Angell K, Turner-Cobb JM, Kreshka MA, Donnelly P, McCoy R et al (2001) Distress, coping, and social support among rural women recently diagnosed with primary breast cancer. Breast J 7(1):25–33

Letvak S (1997) Relational experiences of elderly women living alone in rural communities: a phenomenologic inquiry. J N Y State Nurses Assoc 28(2):20–25

Local Governance Performance Index in Malawi: Selected findings on Health (November 2016). (Series 2016:7)

Löwe B, Spitzer RL, Williams JB, Mussell M, Schellberg D, Kroenke K (2008) Depression, anxiety and somatization in primary care: syndrome overlap and functional impairment. Gen Hosp Psychiatry 30(3):191–199

McCulloch BJ (1995) The relationship of family proximity and social support to the mental health of older rural adults: the Appalachian context. J Aging Stud 9(1):65–81

Merwin E, Hinton I, Dembling B, Stern S (2003) Shortages of rural mental health professionals. Arch Psychiatr Nurs 17(1):42–51

Mulder PL, Shellenberger S, Streiegel R, Jumper-Thurman P, Danda CE, Kenkel MB, . . . Hager A (2000) The behavioral health care needs of rural women. American Psychological Association

Mumford DB, Nazir M, Jilani F, Baig IY (1996) Stress and psychiatric disorder in the Hindu Kush: a community survey of mountain villages in Chitral, Pakistan. Br J Psychiatry 168(3):299–307

Nandi DN, Banerjee G, Mukherjee SP, Nandi PS, Nandi S (2000) Psychiatric morbidity of a rural Indian community. Br J Psychiatry 176(4):351–356

Nargis N, Thompson ME, Fong GT, Driezen P, Hussain AG, Ruthbah UH et al (2015) Prevalence and patterns of tobacco use in Bangladesh from 2009 to 2012: evidence from International Tobacco Control (ITC) Study. PLoS One 10(11):e0141135

Nyamathi AM, Sinha S, Ganguly KK, William RR, Heravian A, Ramakrishnan P et al (2011) Challenges experienced by rural women in India living with AIDS and implications for the delivery of HIV/AIDS care. Health Care Women Int 32(4):300–313

Okojie CE (1994) Gender inequalities of health in the third world. Soc Sci Med 39:1237–1247

Okwumabua JO, Baker F, Wong S, Pilgram BO (1997) Characteristics of depressive symptoms in elderly urban and rural African Americans. J Gerontol Ser A Biol Med Sci 52(4):M241–M246

Patel S, Gadit AM (2008) Karo-kari: a form of honour killing in pakistan. Transcult Psychiatry 45(4):683–694

Patel V, Kirkwood BR, Pednekar S, Weiss H, Mabey D (2006) Risk factors for common mental disorders in women. Br J Psychiatry 189(6):547–555

Patel V, Ramasundarahettige C, Vijayakumar L, Thakur J, Gajalakshmi V, Gururaj G et al (2012) Suicide mortality in India: a nationally representative survey. Lancet 379(9834):2343–2351

Peltzer K (2006) Prevalence of alcohol use by rural primary care outpatients in South Africa. Psychol Rep 99(1):176–178

Pendray C (2012) Letter to Minister of Finance, Afghanistan. Women's health in Afghanistan. https://www.jbpub.com/essentialpublichealth/skolnik/2e/docs/Maternal%20and%20Child%20Health/Womens%20Health%20in%20Afghanistan,%20Pendray.doc. Accessed 13 May 2017

Pradhan A, Poudel P, Thomas D, Barnett S (2011) A review of the evidence: suicide among women in Nepal. National Health Sector Support Program. Ministry of Health and Population, Kathmandu, p 117

Probst J, Laditka S, Moore C, Harun N, Powell M (2005) Depression in rural populations: prevalence, effects on life quality, and treatment-seeking behavior. South Carolina Rural Health Research Center, Columbia

Probst JC, Laditka SB, Moore CG, Harun N, Powell MP, Baxley EG (2006) Rural-urban differences in depression prevalence: implications for family medicine. Fam Med (Kansas City) 38(9):653

Qiu P, Caine ED, Hou F, Cerulli C, Wittink MN, Li J (2016) The prevalence of distress and depression among women in rural Sichuan Province. PLoS One 11(8):e0161097

Raeburn J, Rootman I (1989) Towards an expanded health field concept: conceptual and research issues in a new age of health promotion. Health Promot Int 3(4):383–392

Rahman A (2007) Challenges and opportunities in developing a psychological intervention for perinatal depression in rural Pakistan–a multi-method study. Arch Womens Ment Health 10(5):211–219

Rahman A, Malik A, Sikander S, Roberts C, Creed F (2008) Cognitive behaviour therapy-based intervention by community health workers for mothers with depression and their infants in rural Pakistan: a cluster-randomised controlled trial. Lancet 372(9642):902–909

Rao K, Vanguri P, Premchander S (2011) Community-based mental health intervention for under-privileged women in rural India: an experiential report. Int J Fam Med. Article ID 621426, 7. https://doi.org/10.1155/2011/621426

Reeb BT, Conger KJ (2011) Mental health service utilization in a community sample of rural adolescents: the role of father–offspring relations. J Pediatr Psychol 36(6):661–668

Reynolds NR, Satyanarayana V, Duggal M, Varghese M, Liberti L, Singh P et al (2016) MAHILA: a protocol for evaluating a nurse-delivered mHealth intervention for women with HIV and psychosocial risk factors in India. BMC Health Serv Res 16(1):352

Roscigno VJ, Crowle ML (2001) Rurality, institutional disadvantage, and achievement/attainment. Rural Sociol 66(2):268–292

Sandys E (2008) Rural women in changing World: opportunities and challenges

Saxena S, Thornicroft G, Knapp M, Whiteford H (2007) Resources for mental health: scarcity, inequity, and inefficiency. Lancet 370(9590):878–889

Sered SS (1994) Ideology, autonomy, and sisterhood: an analysis of the secular consequences of women's religions. Gend Soc 8(4):486–506

Seymour G, Floro MS (2016) Identity, household work, and subjective well-being among rural women in Bangladesh (Vol. 1580). Intl Food Policy Res Inst

Shidhaye R, Patel V (2010) Association of socio-economic, gender and health factors with common mental disorders in women: a population-based study of 5703 married rural women in India. Int J Epidemiol 39:1510. https://doi.org/10.1093/ije/dyq179

Shields-Zeeman L, Pathare S, Walters BH, Kapadia-Kundu N, Joag K (2017) Promoting wellbeing and improving access to mental health care through community champions in rural India: the Atmiyata intervention approach. Int J Ment Heal Syst 11(1):6

Simkhada P, Van Teijlingen E, Winter R, Fanning C, Dhungel A, Marahatta S (2015) Why are so many Nepali women killing themselves? A review of key issues. J Manmohan Mem Inst Health Sci 1(4):43–49

Simmons LA, Huddleston-Casas C, Berry AA (2007) Low-income rural women and depression: Factors associated with self-reporting. Am J Health Behav 31(6):657–666

Singh K, Kaur J, Singh D, Junnarkar M (2014) Socio-demographic variables affecting well-being: a study on Indian rural women. Psychol Stud 59(2):197–206

Small J, Curran GM, Booth B (2010) Barriers and facilitators for alcohol treatment for women: are there more or less for rural women? J Subst Abus Treat 39(1):1–13

Stewart RC, Bunn J, Vokhiwa M, Umar E, Kauye F, Fitzgerald M et al (2010) Common mental disorder and associated factors amongst women with young infants in rural Malawi. Soc Psychiatry Psychiatr Epidemiol 45(5):551–559

Suh GH, Gega L (2017) Suicide attempts among the elderly in East Asia. Int Psychogeriatr 29(5):707–708

Suvedi BK, Pradhan A, Barnett S, Puri M, Chitrakar SR, Poudel P et al (2009) Nepal maternal mortality and morbidity study 2008/2009: summary of preliminary findings. Family Health division, Department of Health Services, Ministry of Health, Government of Nepal, Kathmandu

Trani J-F, Bakhshi P (2013) Vulnerability and mental health in Afghanistan: looking beyond war exposure. Transcult Psychiatry 50(1):108–139

Tripathy P, Nair N, Barnett S, Mahapatra R, Borghi J, Rath S et al (2010) Effect of a participatory intervention with women's groups on birth outcomes and maternal depression in Jharkhand and Orissa, India: a cluster-randomised controlled trial. Lancet 375(9721):1182–1192

U.S. Department of Health & Human Services (2000) Healthy people 2010: understanding and Improving health and objectives for health, 2nd edn. U.S. Government Printing Office, Washington, DC

Valverde M, Garcia M, Bain SF (2011) Rural women's health: the correlation between physical and mental health

Vijayakumar L (2015) Suicide in women. Indian J Psychiatry 57(Suppl 2):S233

Vung ND, Ostergren P-O, Krantz G (2008) Intimate partner violence against women in rural Vietnam-different socio-demographic factors are associated with different forms of violence: need for new intervention guidelines? BMC Public Health 8(1):55

WHO (1986) Ottawa charter for health promotion. WHO, Geneva

Zarit SH, Zarit JM (2012) Mental disorders in older adults: fundamentals of assessment and treatment. Guilford Press, New York

Zhang J, Jia S, Wieczorek WF, Jiang C (2002) An overview of suicide research in China. Arch Suicide Res 6(2):167–184

Zhou QH, Zeng Z, Nie ZM (2007) A survey among rural left-behind women from Chongqing City, the China Women's University. Acad J 19(1):63–66. (in Chinese)

Suicide and Self-Harms in Rural Setting

9

With Special Reference to Farmers' Suicide

Prakash B. Behere, Himanshu Mansharamani, Aniruddh P. Behere, and Richa Yadav

Contents

P. B. Behere (✉) · H. Mansharamani
Department of Psychiatry, Jawaharlal Nehru Medical College, Wardha, Maharashtra, India
e-mail: pbbehere@gmail.com; himanshu.mansha@gmail.com

A. P. Behere
Helen DeVos Children's Hospital, Michigan State University, Grand Rapids, MI, USA
e-mail: aniruddhbehere@gmail.com

R. Yadav
Elmhurst Hospital Center, Ichan School of Medicine at Mount Sinai, New York, NY, USA
e-mail: drrichayadav@gmail.com

© Springer Nature Singapore Pte Ltd. 2020
S. Chaturvedi (ed.), *Mental Health and Illness in the Rural World*, Mental Health and
Illness Worldwide, https://doi.org/10.1007/978-981-10-2345-3_21

Abstract

Suicide is an important public health issue which is rising at an alarming rate both in the developed and developing countries since the past few decades. It leaves an abysmal impact on the survivors of the deceased individual. It is fairly known that people in rural areas face a multitude of problems which are relatively unknown to those living in urban areas. These include poor healthcare services, low standard of living, greater physical labor and exposure to hazardous situations, among others. Among the top risk factors for suicide is the agriculture industry as farmers totally depend on monsoon for farming. Suicide rates tend to be high in rural areas in part because there is greater access to pesticides, high rates of drug and alcohol abuse, and few healthcare providers and emergency medical facilities. Suicide among rural agrarian community is now a universal phenomenon. Several factors like financial constraints, altercation with any family member, and easy availability of pesticides have been attributed as the common reasons for suicide attempt among the rural population, and the farmer is gradually becoming an endangered species.

Suicide is an important, largely preventable public health problem. Varied measures have been taken by the government throughout all these years. Various schemes to prevent monetary stress in the agrarian community have been taken. Efforts to improve the mental health of people residing in rural areas have also been initiated.

This chapter gives an overview of suicide in rural areas and factors affecting it. Historical aspect of suicide have also been described, and focus has been put on farmer suicide, its causes, the current scenario, and the prevention strategies for the same.

Keywords

Farmers · Rural · Suicide · Self-harm

Introduction

Suicide is an important public health issue which is rising at an alarming rate both in the developed and developing countries since the past few decades. It leaves an abysmal impact on the loved ones of the deceased individual. Among the top risk factors for suicide is the agriculture industry (McCurdy et al. 2003).

Suicide has been defined by a variety of definitions for encompassing all of its various characteristics. Health Evidence Network (HEN) synthesis report given by the World Health Organization (WHO) on strategies for suicide prevention and their effectiveness provides us a definition for suicide encompassing its various characteristics as "an act with a fatal outcome that the deceased, knowing or expecting a potentially fatal outcome, initiated and carried out with the purpose of bringing about wanted changes" (Scott and Guo 2012).

According to the data given by the WHO, 1.4% of all the deaths worldwide were due to suicide. Suicide was the 17th leading cause of mortality worldwide in the year 2015. Another shocking statistic is that for every 1 adult committing suicide, there are approximately 20 others who attempted suicide (WHO 2014).

Since not all incidents end in suicide, there is another terminology that takes the stage – parasuicide. The World Health Organization's resource for nonfatal suicidal behavior case registration provides a definition for parasuicide, given by Platt et al. in 1992, as "an act with a nonfatal outcome in which an individual deliberately initiates a non-habitual behavior that, without intervention from others, will cause self-harm, or deliberately ingests a substance in excess of the prescribed or generally recognized therapeutic dosage, and which is aimed at realizing changes which the subject desired, via the actual or expected physical consequences" (WHO 2014). Self-harm, deliberate self-harm, attempted suicide, and nonfatal suicide are some of the terms that can be used interchangeably with parasuicide.

It is thought that the rural setup, having a close-knit cultural setup, is a safe haven for its people. But over recent years, studies have shown otherwise.

"Suicide rates tend to be high in rural areas in part because there is greater access to pesticide, high rates of alcohol abuse, and few healthcare providers and emergency medical facilities," says Julie Goldstein Grumet, PhD, director of prevention and practice at the Suicide Prevention Resource Center in Washington, D.C. "It's a lethal triad."

Suicide among rural agrarian community is now a universal phenomenon. Several factors like financial constraints, altercation with any family member, and easy availability of pesticides have been attributed as the common reasons for suicide attempt among the rural population, and the farmer is gradually becoming an endangered species.

Due to such severe state, the government appointed several inquiries to look into the causes of farmer suicide and farm-related distress in general in various regions of the world.

Schemes like 2006 relief package; agricultural debt waiver and debt relief scheme; Maharashtra relief package 2010; Kerala Farmers' Debt Relief Commission (Amendment) Bill, 2012, in India; agricultural subsidy in the USA; and income compensation in Japan among others have been initiated in the past to decrease the plight of farmers.

In this chapter, our focus will be on rural suicides. We will be discussing regarding farmer suicides.

Suicide Around the World

According to the World Health Organization (WHO), suicide in 2004 was the eighth leading cause of potential years of life lost worldwide among persons aged 15–44 years (WHO 2004).

The rate of suicide is highest in Eastern Europ, in countries such as Belarus, Estonia, Lithuania, and the Russian Federation. High rates of suicide have also been seen in Sri Lanka, according to the data obtained from the WHO Regional Office for Southeast Asia (Gururaj et al. 2001).

An interesting speculation which is now being hypothesized is that latitude and the daily amount of sunlight influence rates of suicide (Terao et al. 2002). Rates of suicide are higher in northern parts of Japan and in northern countries of Europe as compared to the southern countries. Suicide is now seen in almost all over the world (Behere and Behere 2008).

Low rates have been found predominantly in Latin America (notably Colombia and Paraguay) and few countries in Asia (e.g., the Philippines and Thailand). Haiti reported not even a single suicide in 2003. Countries in other parts of Europe, North America, and some parts of Asia and the Pacific tend to fall somewhere in between these extremes. Eighty-six percent of all suicides occurred in low and middle income countries (WHO 2002).

Farming is one of the oldest industries in the world and often portrayed as a happy way of life. Despite this, agriculture has higher rates of suicides than any other industry. Based on proportional mortality ratios, farmers were among the top ten occupational groups with highest proportional mortality rates in the UK. In a psychological autopsy study from India, mental illness was most often judged as an important risk factor for suicide by farmers. The farmers' suicide rate in the country in the year 2001 was 12.9 which was about one-fifth higher than the general suicide rate, which was 10.6 in that year. Economic factors such as indebtedness, crop failures, and acute financial loss or responsibility (e.g., marriage in family) were significantly associated with farmers' suicides in Vidarbha region of Maharashtra and rural Punjab. A qualitative study from Vidarbha region revealed that farmers perceived debt, addiction, environmental problems, low price for farm produce, stress, family responsibility, government apathy, etc., as the most significant reasons for farmers' suicides. Studies on this issue in India so far have focused on socioeconomic and farming-related risk factors (Behere and Bhise 2009, 2010a).

Historical Aspect

The story of suicide is probably as old as that of man himself. Through the ages, suicide has been described time and again in various contexts and has been condemned, glorified, bemoaned, and even romanticized. Be it the story of Greek heroes like Aegeus, Socrates, Lycurgus, Cato, Demosthenes, Zeno, or Seneca; or the famous Roman figures like Brutus, Mark Anthony, and Cassius; or the Egyptian princess, Cleopatra; or Samson, Saul, Abimelech, and Achitophel of the Old Testament; or the suicide bombers in the present world, the ubiquitousness of suicide transcends religion and culture (Evans and Farberow 1988).

An understanding of suicide in the Indian context needs a recognition of the diverse literary, religious, and cultural ethos of the subcontinent because tradition has rarely been present in the lives of people for as long as it has in India. Ancient Indian texts contain stories of valor in which suicide as a means to avoid shame and disgrace was mentioned a lot many times in these ancient texts.

Suicide has been mentioned in the great epics of Ramayana and Mahabharata. When Lord Rama departed this earth, there was a tide of suicide in his kingdom, Ayodhya. The sage Dadhichi sacrificed his life so that the Gods can use his bones in the fight against the demons.

The Bhagavad Gita condemns suicide for different reasons and states that such a death cannot have "shraddha," the all-important last rites. Brahmanical view had specified that those who attempt suicide should fast for a stipulated period. Upanishads, the Holy Scriptures, see suicide in a bad light and state that "he who takes his own life will enter the sunless areas covered by impenetrable darkness after death."

However, the Vedas permit suicide for religious reasons and tell that the best sacrifice was that of one's own life. Suicide by starvation, also known as "sallekhana," was linked to the attainment of "moksha" (liberation from the cycle of life and death) and is still practiced in today's world.

Sati, a woman who immolates herself on the pyre of her husband rather than living the life of a widow, and Jauhar (Johar), in which Rajput women killed themselves to avoid humiliation at the hands of the men of the invading Muslim armies, were practiced until as recently as the former half of the twentieth century. This was predominantly seen in the Indian state of Rajasthan.

Factors Leading to Farmer Suicide

It is fairly known that people in rural areas face a multitude of problems which are relatively unknown to those living in urban areas. These include poor healthcare services, low standard of living, greater physical labor and exposure to hazardous situations, among others.

The factors described under this have direct effect on suicide in all settings. Males are more prone to commit suicide than females, with exceptions in some cultures like India and China.

In European countries and the USA, the male/female ratio was found to be 3:1–4:1 (Cantor 2000), whereas in India and China, the ratio was found to be 1.8:2 and 1:1, respectively (Behere and Bhise 2009; Phillips et al. 2002).

The reason given for this in various studies is that males being the major workforce in rural areas and being the head of the family are subjected to more stressful conditions as compared to females which lead them to take such drastic steps, but in the Indian scenario, more stringent cultural laws for the females and bad practices such as dowry force the females to take such harsh steps.

1. Poverty
   An increased number of suicide rates have been noted in people living in poverty (Rehkopf and Buka 2006). It can be attributed to the lack of resources needed to lead a good life due to lack of funds. Loan is a stressor for farmers in rural areas (Behere et al. 2015)

2. Mental Illness
   Marked increase in suicidal ideations, completed suicide, and attempted suicide have been noted among people suffering with mood disorders (Nock et al. 2009; Posada-Villa et al. 2009; Almasi et al. 2009).

Scarcity of mental health services at the grassroot level further exacerbates this condition as the symptoms go unnoticed and the only chance to prevent the act of suicide is lost.

3. Substance Abuse Specially Alcohol

About 16% of persons completing suicide are substance abusers, which include either alcohol or other drug abuses.

People are more likely to commit suicide and harm themselves when they are under the influence of any substance or intoxicated (Manoranjitham et al. 2010; Schneider 2009).

Intoxication leads to faulty judgment and clouding of consciousness, causing a person to take such drastic steps.

4. Predisposing Factors

Factors such as genetic predispositions and environmental factors are deeply associated with greater suicide risk (Marušič and Videtič 2008; Brezo et al. 2008; McGuffin et al. 2001; Currier and Mann 2008).

Family history of mental disorders has stood out as a major predisposing factor for mental illnesses in majority of the cases.

5. Psychological Factors

Impulsivity, improper problem-solving skills, negative thoughts of the future, and improper coping mechanisms all lead to faulty decision-making.

Personality defects may also lead to suicide.

Stressors

1. Loneliness

Loneliness can be a stressor in some areas and occupations. Farmer may work for long hours in farms alone with no one to share their concerns and worries (Rasmussen et al. 2010).

Poor access to public transport can result in enforced isolation, or there may be long distances between settlements in countries such as Australia and Canada causing a person to feel outcasted and aloof (Stark et al. 2011).

2. Social-Cultural and Poor Crop

Many times, the crop is damaged because of natural calamities, which results in the loss of crop and the farmer gains nothing out of it. The hiked prices of machinery also add on to it (Behere and Bhise 2009). Hence, many farmers shift from farming to a non-farming job and may even change their place of residence from rural to urban areas in despair (Kinsella et al. 2000). The economy per say may be soaring, but agriculture is its Achilles' heel, the source of livelihood for millions of people but still suffering from a lot of problems and complications.

In a rural culture where joint family systems are central to way of life, it appears that the most common background stressors or precipitating issues were family interpersonal conflicts. A factor long assumed to be protective as far as suicide goes – the strong joint family system – appears to be the most common immediate trigger for suicide attempts in this type of population.

3. Social Exclusion

Many indigenous groups all over the world, such as Australian Aboriginal and Inuit in Canada, have reported cultural disruption due to forced education causing a disturbance in the natural cultural flow and values (Chandler and Lalonde 1998; Webster 2006; Hunter and Harvey 2002; Christian and Spittal 2008).

Many groups from rural areas all over the world have also reported feelings of exclusion from national processes, which may likely increase stress and despair (Wexler 2006).

Suicide risk has been seen in association with homosexuality in rural areas, as people in rural areas stigmatize it (Barton 2010; Hansen and Lambert 2011; Quinn and Quinn 2003; Crawford and Brown 2002).

Factors Affecting Support

1. Stigma

There is stigma towards mental illness; therefore, they do not access mental health services.

Compromised education level and lack of awareness in rural areas further worsen this situation, as lack of awareness about mental illnesses is present. This gap can be bridged by spreading awareness with the help of media (Behere et al. 2009)

2. Cultural and Religious Norms

In countries like India and Africa, many patients are first taken to faith healers, and if the disease is not cured, then they seek mental healthcare services, which increases the duration of untreated mental illness and hence a poor prognosis of the illness. (Judd et al. 2006; Rajkumar and Hoolahan 2004).

Visiting a psychiatrist still remains a taboo and is socially unacceptable at various places.

Methods like poojas and wearing a tabiz (amulet) or performing various rites are still considered more effective and popular methods for combating mental symptoms in rural areas.

3. Availability of Mental Health Services

In developing countries, rural areas are lacking in mental healthcare services, and still rural areas are not well connected with urban areas, where the services could be made available leading to a missing link which needs to be filled (Shi 1993).

Unavailability of easy access to mental health services further worsens the situation causing an increase in the magnitude of the problem.

4. Social Support

Social support in rural areas due to stigma towards mental illnesses and geographical isolation may not be available (Stark et al. 2011).

Mental symptoms are still considered a taboo and are classified as possessions or as a part of some supernatural phenomena which tend to further isolate the patient and cause cuts down of social support leading to abandonment of the patient by family members.

Gradually joint family system in India is breaking down to extended nuclear family and nuclear family system. In nuclear family, senior family support is lost.

Deliberate Self-Harm

Views on suicide and self-harm are different around the world. In many countries, suicide and self-harm are not seen as a crime, while in others they may be. Recently in India, suicide has been decriminalized, if the act has been done under the influence of mental illness (Tester and McNicoll 2004). This varied view about self-harm and suicide around the world affects the acceptance of it as well. In some traditions removing the burden from the world has been seen positively, which may ease the decision-making of self-harm and considering it as an ethical option. This may also increase the acceptability regarding self-harm and suicide in the community (Malmberg et al. 1999).

Likelihood of Death

Choice of method while self-harming and the emergency services available in the rural area are the factors affecting likelihood of death of a person committing self-harm (Stark et al. 2011; Behere and Bhise 2009).

Methods such as pesticide use, hanging, and drowning are particularly common in rural areas, which are lethal methods for self-harm and suicide and difficult to prevent. Firearm ownership is common in rural areas of some countries (Behere et al. 2015; Stark et al. 2006; Gunnell and Eddleston 2003).

Particularly in middle- and low-income countries, pesticide poisoning is a common method for suicide, as pesticides are easily available and poorly controlled in rural areas, and it results in high fatality rate (Gunnell et al. 2007; Behere and Bhise 2010).

Method chosen by an individual may also be affected by media depiction or knowledge which would have been gained by previous events that occurred in the area (Stark et al. 2011).

It has been seen in past studies that females use less aggressive methods for suicides as compared to males.

Other factors affecting likelihood of death in rural areas in middle- and low-income countries are unavailability of emergency services, transport facilities, low healthcare resources, lack of awareness, and monetary constraints.

Farmers' Suicide in Rural India

Suicide among rural agrarian community is now a universal phenomenon.

M. S. Swaminathan, the geneticist who was the scientific leader of India's Green Revolution 40 years ago, said: "The suicides are an extreme manifestation of some deep-seated problems which are now plaguing our agriculture."

In a country of 70 million farmers, 10 in every 100,000 farmers commit suicide. This is higher than the total national suicide rate (NCRB 2015).

In India, there is direct loan which is taken for farming and indirect loan which is taken for personal reasons.

Historical records depicting frustration, revolts, and high mortality rates among farmers in India, particularly cash crop farmers, date back to the nineteenth century. The high land taxes of the 1870s, payable in cash regardless of the effects of frequent famines on farm output or productivity, combined with colonial protection of usury, money lenders, and landowner rights, amounted to widespread penury and frustration among cotton and other farmers, ultimately causing the Deccan Riots of 1875–1877.

The British government enacted the Deccan Agriculturists' Relief Act in 1879, to restrict the interest rate charged by money lenders to Deccan cotton farmers, but applied it selectively to areas that served British cotton trading interests that is as per their own profit and not basically for the well-being of the farmers. Rural mortality rates, in predominantly agrarian British India, were alarmingly high between the 1850s and 1940s (IOSR).

The current trend is also not at all encouraging, and farmer suicide still remains a matter of concern as each year many farmers lose their lives due to suicide which is a preventable state.

There is no doubt that our agrarian community is facing a crisis. But questions arise on the exact nature and precise cause behind the deepening problem. Farmer suicides have largely been attributed to debt, drought, crop failure, poor facilities, or poor returns.

Such a high mortality rate seen in a small period points to the fact that farmers have been taking the drastic step regardless of a good rainfall year or bad and a good price year or a disappointing one. Why? Among the literature returned, numerous factors were identified and to varying degrees were investigated. As outlined below, indebtedness was considered as the overwhelming correlate of farmer suicides (Nagaraj et al. 2014).

Majority of farmers in India have low annual income. When there are additional or unexpected expenses, farmer gets trapped into a vicious cycle with ever-increasing amounts of debt, which causes distress.

It has been seen that most of the farmers attempting suicide were diagnosed with some or other psychiatric illnesses as compared to the general population. Farmers, in general, have higher psychological morbidity than general population as they face compounding effects of physical stressors, environmental changes, changing farm economics, lack of differentiation between professional and personal life with no definitive age of retirement that leads to psychological distress.

Farmers have struggles beyond indebtedness, with potentially stressful roles within their families as the head of the household, arguments with family members, alcohol abuse, and difficulties adjusting to changing statuses in their local communities. If they find themselves in difficulties, there is nowhere to go.

Determining a single defining cause for farmer suicides in India is impossible, especially considering the relative lack of extensive literature. What can be inferred

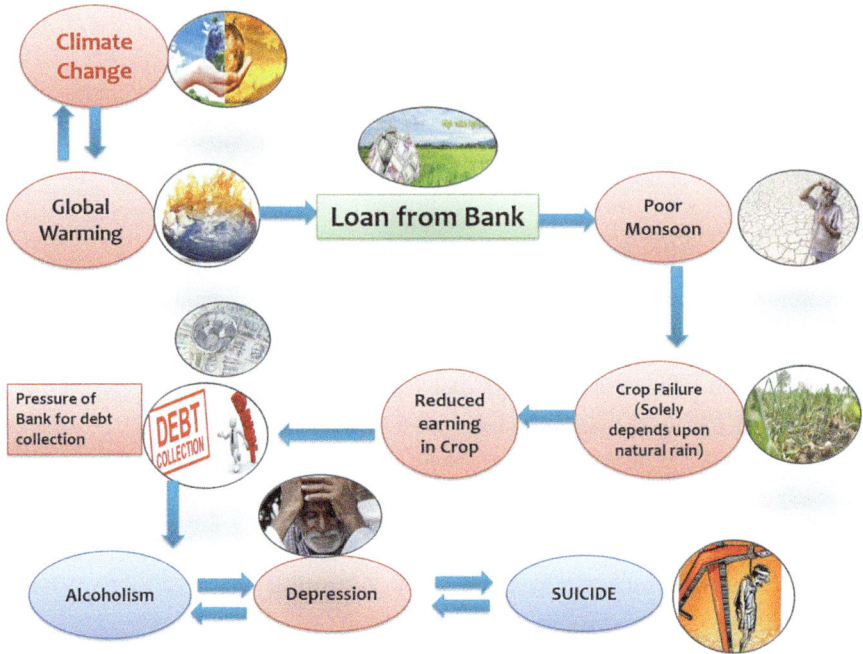

Image 1 An amalgamation of various factors leading to farmers' suicide in view of global warming

is that an amalgamation of factors has led to a picture of large-scale farmer indebtedness, which in combination with a volatile ecological climate and socioeconomic landscape has left hundreds of thousands, if not millions, of farmers vulnerable to a situation of such a crippling amount of debt and desperation that many have come to take their own lives (Image 1). In recent times, it was raised whether global warming causes farmers' suicide (Sarah 2017).

Among the states in India, the data showed Maharashtra (3,030), Telangana (1,358), Karnataka (1,197), Chhattisgarh (854), and Madhya Pradesh (516) lead the list (NCRB 2015). In India, national suicide rate is 10.6/100,000 population; similarly Maharashtra state is 14.2 and Telangana 27.7.

The numbers are still rising despite the efforts.

Current Scenario

A lot of rulings and schemes have been implemented by the government in the past for improving the problem of farmer suicide, but little has been achieved so far.

India has witnessed several droughts in the last few years. In 2014, as many as 5,650 farmers committed suicide. Between 1995 and 2014, a number of farmer suicides go as high as 296,438. A lack of rainfall has decreased food grain production significantly.

These figures are enough to prove that there is a severe agrarian crisis, but still there are no firm methods being used to combat rural suicides.

Despite their claims of being farmer friendly, the ground reality for farmers involves living a debt-ridden life and ultimately dying by committing suicide.

State governments then blame the center for not having transferred enough funds and that's why they are helpless. The center, on the other hand, puts forward the argument that states have not made proper utilization of the funds and have poor implication. Thus, the blame game goes on and on (Farmer Suicide and Response 2017).

Recently the Supreme Court expressed concern over farmers' suicide due to crop failure and asked the government to come up with a concrete "road map" within 3 weeks to eliminate the reasons for the agrarian crisis.

It has been pointed out many times that the government was working in a "wrong direction" in tackling the "real problem."

With all these researches, one more thing has been established, and that is paying compensation to the families of such victim's post facto is not the real solution.

Farmers take loan from banks and when they are unable to repay, they commit suicide. The remedy to the problem is not to pay money to farmers after suicide, but there should be schemes to prevent this. This came out as the verdict to PILs filed to draw the court's attention to the crisis.

Government Initiatives Globally: Are They Helpful?

How have agrarian policies heightened farmers' economic vulnerability?

Post liberalization, cutbacks in agricultural subsidies combined with the necessity to meet the international standards of quality increased the cost of input. There was also an increased demand for cash crops like cotton in the global market. Hence, a major part of the sector evidenced a government-encouraged shift from food grain to cash crop cultivation. However, due to an excess of such products in the market that caused saturation, prices fell making cash crops uneconomical.

This was the late 1990s – the time from when farmer suicides began to be recorded on a large scale. Production costs have steadily risen in the years since, but market prices have not seen a corresponding increase leading to a never-ending void.

Furthermore, public investment in agriculture gradually reduced – a move that has resulted in the lack of uniform modernization, irrigation facilities, communication and information systems, and storage facilities.

Credit supply to farmers diminished at the same time, forcing them to rely on moneylenders. Private companies and state-run Regional Rural Banks (RRBs) reject many applicants due to the lack of collateral or land titles.

The agricultural market structure has played a major role in denying farmers their actual returns. The Agriculture Produce and Market Committee (APMC) act was introduced to ensure fair prices to farmers. But corrupt middlemen have taken advantage of the provisions in the act to exert their influence on the supply chain. They exploited farmers by buying produce at far lower prices than deserved.

The coverage of crop insurance schemes is low. Not only is the area covered under these schemes is less, but even the amount offered to the farmers is dismal. The impact of the new schemes and plans on protecting them in case of crop failure is yet to be seen. The Government of India came out with new initiative called Farmers' credit card.

Farmers have been cultivating on smaller and smaller lands with depleting soil quality and groundwater levels. Farming will become completely unviable if there are no checks on environmental effects and their consequences. They do not even have many job options to rely on when cultivation fails.

It has been seen that the government hasn't done much to ensure jobs to the rural population. Many migrate to cities in search of labor work causing disruption in the urban-rural population ratio and opening doors for various other complications.

Besides economic, there are other reasons to be considered as well. It has been seen that though debt is the main cause of farmer suicides, others include social disrepute, health problems, addictions, marriage in the family, dispute with others, etc. In fact, many of the farmer loans are for marriages of the female members of the family. It is a social obligation to spend well on weddings, which sometimes even includes dowry, causing female infanticide in extreme cases.

Another pattern noticed in farmer suicides is more than 80% of them are male suicides. This probably has to do with the fact that the role of the breadwinner is traditionally that of a man's, and they are also usually the head of the family. The incapacity to fulfill that role adds to their mental stress.

A concrete reorientation of policy is crucial to stopping this depressing trend. Farmers in the current scenario are desperate for help. But taking their lives isn't the answer, a global farmer movement to demand for reform is. A revolution saved agriculture once, it can be done again.

The Real Problem

Despite so many schemes and efforts from the government, why is it that our country is not able to overcome farmer suicide?

The reason to why the magnitude of the problem is not decreasing is yet to be found.

It has been noted that a greater infrastructure of mental healthcare is needed in India. In areas hardest hit by low crop yields, field surveys have been conducted, and higher rates of alcoholism and depression have been noted.

It has been well established by now that loan forgiveness would only alleviate a symptom, rather than solving a more serious underlying problem.

Successive governments have focused more on control rather than prevention. It will take more than short-term measures and disproportionately implemented programs to wade through the crisis. The wave of farmer suicides could have been avoided if there were proper irrigation system, weather forecast information, competent crop insurance, and buffer stocks in cold storage facilities to distribute to the families in their time of need.

There are two basic aspects to this problem; one is to make the work environment stress-free for our farmers, and second is to provide them with adequate mental health services for referral and consultation.

Prevention

Suicide is an important and also a preventable public health problem. The problem is however a complex one and has been described time and again as a complex array of factors such as poverty, low literacy level, unemployment, family violence, breakdown of the joint family system, unfulfilled romantic ideals, inter-generational conflicts, loss of job or loved one, failure of crops, rising costs of cultivation, huge debt burden, unfulfilling marriages, harassment by in-laws and spouse, dowry fights, depression, chronic physical illness, alcoholism/drug abuse, and easy access to lethal means of suicide.

In 2000, the WHO launched the multisite intervention study on suicidal behaviors (SUPRE-MISS) which aimed to increase knowledge about suicidal behaviors and about the effectiveness of interventions for suicide attempters in culturally diverse places around the globe.

Early detection and adequate treatment of a primary psychiatric disorder is of paramount importance. In psychiatrically ill subjects, lithium, clozapine, olanzapine, antidepressants, and behavioral interventions such as dialectical behavior therapy (DBT) have been shown to have anti-suicidal efficacy.

Since the greatest predictor of completed suicide is the presence of a previous suicide attempt, investigations aimed at suicide attempters may prove to be effective in reducing suicide rates.

Psychological autopsies were adopted as a measure to find the cause for suicide in farmers in certain regions of India (Behere and Rathod 2006). It was also noted that in majority of the cases, the suicide completers had consulted a physician before the event and family of the victim were aware of their suicidal intent. This finding is similar to data from the west, where two-thirds visited their general practitioners in the month prior to death and 40% in the week before. This calls for adequate training of general practitioners in detection and referral of patients with common mental disorders, which may result in a significant decline in suicide rates as they are usually the first point of contact in such cases and not the psychiatrist. This may also have to be culturally sensitive; the higher rate of somatic symptoms, rather than cognitive symptoms, among depressed patients at times should also raise a suspicion in the mind of the treating physician.

The early identification and treatment of vulnerable populations with risk factors for suicide is one strategy. As there is a strong link between negative life events early in childhood and suicide risk, it is of utmost importance to identify populations that have been exposed to traumatic childhood experiences, such as sexual/physical abuse and domestic violence. The identification of such individuals requires a combined approach with active participation from teachers and school authorities, health professionals, and the judiciary. Primary prevention strategies include promoting positive health and instilling adaptive coping strategies in children and improving awareness among the parents, teachers, and healthcare professionals regarding child-rearing practices and early intervention for maladaptive coping behavior. At the community level, the establishment of social programs such as child and family support programs and programs aimed at achieving gender and socioeconomic equality can be of some help.

The need for a strategy which will raise awareness and help make suicide prevention a global priority has long been recognized. Such a strategy will need a comprehensive all-round approach that encompasses promotion, coordination, and support of activities to be implemented across the world at international, national, regional, and local levels. The program would need to be customized for populations at risk. For example, prevention programs aimed at children and young adults would have to address issues related to gender inequality, physical/sexual abuse, violence, and mental illnesses and stressors more common to that age group such as peer pressure and career. Strategies with empirical evidence in western literature such as the Universal, Selective, Indicated (USI) model, "gatekeeper training," and outpatient follow-up and emergency outreach may also be relevant to eastern part of the world. The USI model outlines "universal" preventive strategies for the population as a whole (e.g., restricted access to lethal means), "selective" strategies targeting at-risk individuals (e.g., psychiatrically ill, homeless, socially excluded groups, drug abusers), and "indicated prevention" strategies focusing on suicide attempters (e.g., outpatient contact and emergency outreach). Gatekeeper training focuses on skill development to enable community members such as teachers, coaches, and others in the community to identify signs of depression and suicide-related behaviors among youth and the population coming in their contact. It encourages individuals to maintain a high index of suspicion and to inquire directly about distress, persuade suicidal individuals to seek and accept help, and serve as a link for local referrals to fill the gap. Such approaches would also require a multidisciplinary team approach involving psychiatrists, general physicians, psychiatric nurses, psychiatric social workers, and nongovernmental organizations (NGOs) making it a holistic plan to tackle this problem.

The role of the media is becoming increasingly relevant. A delicate balance needs to be maintained between press freedom and responsibility of the press to minimize the harm to vulnerable individuals as they are easily influenced by whatever they are exposed to through these sources. The role of advocacy and legislature cannot be overemphasized and ignored. Laws restricting availability of lethal sources such as firearms have been advocated by the WHO. NGOs can play an important role by bridging the gap between the government and general masses. Sensitization of media regarding ethical broadcasting of farmers' suicide is also paramount.

In rural areas strategies like good financial schemes, supportive technical machinery, artificial techniques to combat failures due to weather change, alternate ways to earn money if crop fails, and adequate surveillance for mental health illnesses by appointing increased number of trained mental health professional in these far-fetched areas can help reduce the suicide number. Also spreading awareness about the mental illnesses and educating the masses about the warning signs such as changes in behavior or sleep cycle or personal hygiene will help bring this to the notice of a doctor soon and can help prevent suicide if adequate help is provided at the right time.

The task of suicide prevention is daunting. Although suicide attempters are at increased risk of completed suicide, about 10% of attempters persistently deny suicidal intent. This group may still be seen as vulnerable and prone to psychological distress (Behere et al. 2015). Though restricting availability of lethal means appears to be a possible and appropriate solution, some cases have come up where legislation was introduced to restrict sale of a pesticide and found no reduction in the overall suicide rate, but merely a change in the modes of suicide. Hence, the solutions to suicide prevention may prove to be more complex than the problem of suicide itself.

Conclusion

Suicide has been defined as "an act with a fatal outcome that the deceased, knowing or expecting a potentially fatal outcome, initiated and carried out with the purpose of bringing about wanted changes." The story of suicide originated with that of the man himself. Since then, suicide has been mentioned, described, and commented upon time and again in various contexts.

It is fairly known that people in rural areas face a multitude of problems which are relatively unknown to those living in urban areas. These include poor healthcare services, poor standard of living, greater physical labor and exposure to hazardous situations, social isolation, lack of awareness, and increased stigmatization, among others.

The rural agrarian community is at further more risk due to various stressors like physical stressors, environmental changes, changing farm economics, indebtedness, and lack of differentiation between professional and personal life with no definitive age of retirement leading to psychological distress.

A lot of rulings and schemes such as loan forgiveness have been implemented by the government in the past for improving the problem of farmer suicide, but little has been achieved so far. The basic aspect to this situation is to provide them with adequate mental health services for referral and consultation. The basic focus should be on preventing and not controlling suicide (Gibbens 2017).

The task of suicide prevention is daunting. Collaboration, coordination, cooperation, and commitment are needed to develop and implement an international and universal plan, which is cost-effective, appropriate, and relevant to the needs of the community. In India, suicide prevention is more of a social and cultural health objective than a usual exercise in the mental health sector. The time has come when the mental health professionals should adopt proactive and leadership roles in suicide prevention along with collaborations with government and NGOs and save the lives of thousands of human beings who deserve a life. Finally, there should be multiagency approach to farmers' suicide, i.e., mental health professionals, counsellors, economists, and agriculturist. This is not only the government's concern but everybody's concern (Behere and Bansal 2009).

References

Almasi K, Belso N, Kapur N, Webb R, Cooper J, Hadley S, Appleby L (2009) Risk factors for suicide in Hungary: a case-control study. BMC Psychiatry 9(1):45

Barton B (2010) "Abomination" – life as a bible belt gay. J Homosex 57(4):465–484

Behere PB, Bansal A (2009) Farmers' suicide in Vidarbha: everybody's concern. J MGIMS 4:iii–iiv

Behere PB, Behere AP (2008) Farmers' suicide in Vidarbha region of Maharashtra state: a myth or reality? Indian J Psychiatry 50(2):124

Behere PB, Bhise MC (2009) Farmers' suicide: across culture. Indian J Psychiatry 51(4):242

Behere PB & Bhise M. (2010a) Farmers Suicide in Rural India. "Comprehensive text book of community psychiatry in India" by Chavan BS, Gupta N, Arun P, Sidana A. Chandigarh

Behere PB, Bhise MC (2010b) Guest editorial farmers' suicides in central rural India: where are we heading? Indian J Soc Psychiatry 26:1–2

Behere PB, Rathod M (2006) Report on farmers' suicide in Vidarbha. Report submitted to Collectorate, Wardha

Behere PB, Bhise MC, Behere AP (2015) Suicide studies in India in developments in psychiatry in India: clinical, research and policy perspectives. Eds Savita Malhotra & Subho Chkarabarty. Springer

Brezo J, Klempan T, Turecki G (2008) The genetics of suicide: a critical review of molecular studies. Psychiatr Clin N Am 31(2):179–203

Cantor CH (2000) Suicide in the western world. In: Hawton K, Van Heeringen K (eds) The international handbook of suicide and attempted suicide. Wiley, Chichester

Chandler MJ, Lalonde C (1998) Cultural continuity as a hedge against suicide in Canada's first nations. Transcult Psychiatry 35(2):191–219

Christian WM, Spittal PM (2008) The cedar project: acknowledging the pain of our children. Lancet 372(9644):1132

Crawford P, Brown B (2002) 'Like a friend going round': reducing the stigma attached to mental healthcare in rural communities. Health Soc Care Community 10(4):229–238

Currier D, Mann JJ (2008) Stress, genes and the biology of suicidal behavior. Psychiatr Clin N Am 31(2):247–269

Evans G, Farberow NL (1988) The encyclopedia of suicide. New York, Facts on file

Farmers Suicide and Response of the Government in India -An Analysis Available from: www.iosrjournals.org. Last accessed on: 25 Aug 2017

Gibbens S (2017) Why these farmers are protesting with skills-interview by Sarah Gibbens Published in National Geographic August 2017

Gunnell D, Eddleston M (2003) Suicide by intentional ingestion of pesticides: a continuing tragedy in developing countries

Gunnell D, Eddleston M, Phillips MR, Konradsen F (2007) The global distribution of fatal pesticide self-poisoning: systematic review. BMC Public Health 7(1):357

Gururaj GA, Isaac MK, Latif MA, Abeyasinghe R, Tantipiwatanaskul P (2001) Suicide prevention-emerging from darkness, SEA/Ment/118; New Delhi, WHO/SEARO

Hansen JE, Lambert SM (2011) Grief and loss of religion: the experiences of four rural lesbians. J Lesbian Stud 15(2):187–196

Hunter E, Harvey D (2002) Indigenous suicide in Australia, New Zealand, Canada and the United States. Emerg Med Australas 14(1):14–23

Judd F, Jackson H, Fraser C, Murray G, Robins G, Komiti A (2006) Understanding suicide in Australian farmers. Soc Psychiatry Psychiatr Epidemiol 41(1):1–10

Kinsella J, Wilson S, De Jong F, Renting H (2000) Pluriactivity as a livelihood strategy in Irish farm households and its role in rural development. Sociol Rural 40(4):481–496

Malmberg A, Simkin S, Hawton K (1999) Suicide in farmers. Br J Psychiatry 105(2):103–105
Manoranjitham SD, Rajkumar AP, Thangadurai P, Prasad J, Jayakaran R, Jacob KS (2010) Risk factors for suicide in rural South India. Br J Psychiatry 196(1):26–30
Marušič A, Videtič A (2008) Suicide risk: where, why and how is it generated? Psychiatr Danub 20(3):262–268
McCurdy SA, Samuels SJ, Carroll DJ, Beaumont JJ, Morrin LA (2003) Agricultural injury in California migrant Hispanic farm workers. Am J Ind Med 44(3):225–235
McGuffin P, Marušič A, Farmer A (2001) What can psychiatric genetics offer suicidology? Crisis 22(2):61
Nagaraj K, Sainath P, Rukmani R, Gopinath R (2014) Farmers' suicides in India: magnitudes, trends, and spatial patterns, 1997–2012. Journal 4(2):53–83
National Crime Records Bureau [NCRB] (2015) Accidental deaths and suicides in India. Ministry of home affairs. Government of India, New Delhi
Nock MK, Hwang I, Sampson N, Kessler RC, Angermeyer M, Beautrais A, De Graaf R (2009) Cross-national analysis of the associations among mental disorders and suicidal behavior: findings from the WHO World Mental Health Surveys. PLoS Med 6(8):e1000123
Phillips MR, Li X, Zhang Y, Eddleston M (2002) Suicide rates in China. Lancet 359(9325):2274–2275
Posada-Villa J, Camacho JC, Valenzuela JI, Arguello A, Cendales JG, Fajardo R (2009) Prevalence of suicide risk factors and suicide-related outcomes in the National Mental Health Study, Colombia. Suicide Life Threat Behav 39(4):408–424
Quinn KT, Quinn K (2003) Establishing an association between rural youth suicide and same-sex attraction. Rural Remote Health 3:222
Rajkumar S, Hoolahan B (2004) Remoteness and issues in mental health care: experience from rural Australia. Epidemiol Psychiatr Sci 13(2):78–82
Rasmussen SA, Fraser L, Gotz M, MacHale S, Mackie R, Masterton G, O'Connor RC (2010) Elaborating the cry of pain model of suicidality: testing a psychological model in a sample of first-time and repeat self-harm patients. Br J Clin Psychol 49(1):15–30
Rehkopf DH, Buka SL (2006) The association between suicide and the socio-economic characteristics of geographical areas: a systematic review. Psychol Med 36(2):145–157
Schneider B (2009) Substance use disorders and risk for completed suicide. Arch Suicide Res 13(4):303–316
Scott A, Guo B (2012) For which strategies of suicide prevention is there evidence of effectiveness. World Health Organization, Denmark
Shi L (1993) Health care in China: a rural-urban comparison after the socioeconomic reforms. Bull World Health Organ 71(6):723
Stark C, Gibbs D, Hopkins P, Belbin A, Hay A, Selvaraj S (2006) Suicide in farmers in Scotland. Rural Remote Health 6(1):509
Stark CR, Riordan V, O'Connor R (2011) A conceptual model of suicide in rural areas. Rural Remote Health 11(2):article–art1622
Terao T, Soeda S, Yoshimura R, Nakamura J, Iwata N (2002) Effect of latitude on suicide rates in Japan. Lancet 360(9348):1892
Tester FJ, McNicoll P (2004) Isumagijaksaq: mindful of the state: social constructions of Inuit suicide. Soc Sci Med 58(12):2625–2636
Webster P (2006) Canadian Aboriginal people's health and the Kelowna deal. Lancet 9532:275
Wexler LM (2006) Inupiat youth suicide and culture loss: changing community conversations for prevention. Soc Sci Med 63(11):2938–2948
World Health Organization (2014) Preventing suicide: a resource for non-fatal suicidal behaviour case registration
World Health Organization. Global Burden of Disease. 2004. [Last cited in 2004]. Update. Available from: http://www.who.int/healthinfo/global_burden_disease/GBD_report_2004upda te_full.pdf
World Health Organization. Injuries, & Violence Prevention Department (2002) The injury chart book: a graphical overview of the global burden of injuries. World Health Organization, Geneva

Organization of Mental Health Services in Rural Areas

10

Vimal Kumar Sharma

Contents

V. K. Sharma (✉)
University of Chester, Chester, UK

Cheshire and Wirral Partnership NHS Foundation Trust, Chester, UK
e-mail: v.sharma@chester.ac.uk

© Springer Nature Singapore Pte Ltd. 2020
S. Chaturvedi (ed.), *Mental Health and Illness in the Rural World*, Mental Health and
Illness Worldwide, https://doi.org/10.1007/978-981-10-2345-3_14

Abstract

A large proportion of people live in rural areas of the world especially in the low- to middle-income countries. The services for mental disorders of rural population remain poor even in high-income countries. The treatment gap for mental illness is about 80% in low-income countries and perhaps even worse for rural people. The reasons of treatment gap are complex. Mental health is often given low priority despite constituting nearly one-fifth of the overall health morbidity. Low investment in mental health leads to limited mental health resources available to the country's health service. In low- to middle-income countries, these are concentrated in the cities, depriving rural communities of any specialist mental health services. Stigma, discrimination, poor literacy, and specific cultural belief toward mental illness restrict rural people to accept and access appropriate help.

Integration of mental health at primary care level is the only way forward to address this unmet need. Primary care health workers, however, lack in knowledge and skills to detect and manage mental disorders. The World Health Organization has come up with an ambitious mhGAP initiative to train primary care workers using its intervention guidelines. Recent mental health programs (PRIME) involving primary care workers and local organizations have shown some benefits as well as highlighted challenges in organizing services, especially in low-income countries to serve predominant rural communities.

Organizing services for people in the rural areas needs to involve local people by understanding their views and perception of mental disorders, an extensive public health education program addressing the issues of stigma and discrimination existing toward mental illness, investing in adequate mental health resources, but more importantly integrating mental health with general health. This can be achieved by training frontline health workers in diagnosing and managing mental disorders supported by specialists. Modern technology and tools using mobile- and computer-assisted methods could greatly assist frontline workers if well supported by telemedicine approaches by mental health specialists.

Keywords

Rural mental health · Mental health services · Global health · Mental health services · Mental health

Introduction

Nearly half (46%) of the world's population still resides in rural areas. The proportion of the rural population is lower (18–25%) in high-income (developed) countries such as North America and Europe and significantly higher in middle- (50%) and low-income countries (69%) as reported by World Bank data (World Bank 2015).

Mental health services available to people living in rural areas are generally inadequate. Lack of specialist resources available to this population especially in low- to middle-income countries is often blamed for the poor mental health service provision. However, a number of other equally important factors contribute to poor services for this population. Rural communities' stigma-based views toward mental illness and their understanding of causes and remedies of mental illness influenced by their cultural beliefs in many developing countries may have led to poor acceptance or even rejection of the Western ways of treating mental health problems. A planning of a comprehensive mental service provision to rural communities requires a deeper understanding of their needs, beliefs, and attitudes toward mental illness as well as a good deal of knowledge of existing ways and resources they routinely use in dealing with them.

Rural areas differ significantly from country to country and from one continent to the other. A single global model of services for rural areas will therefore not only be difficult to propose but equally be unwise to implement. A need-based, tailor-made service model for different rural communities will therefore be more advisable.

A general framework of mental health services addressing the needs and ways to address them can be proposed. But that should be modified and adapted from region to region depending on their circumstances.

This chapter outlines the existing mental services available to rural areas around the world and attempts to identify gaps in services and ways to organize services for rural communities.

What Is the Mental Health Service Provision in Rural Areas at Present?

North America

Mental health services are less than satisfactory in rural areas even in the most prosperous country in the world (Wang et al. 2005). Carpenter-Song and Snell-Rood (2016) reported a high rate of depression and suicide in rural communities without adequate care and treatment as compared to their urban counterparts. Substance misuse problems are now equally prevalent. Most people with mental health difficulties seek help from their primary care doctors who have insufficient training in mental health. Those who have more severe problems such as psychosis that require specialists' services or inpatient care have to travel long distances (Gamm et al. 2010).

Latin America

Limited resource to provide mental health services in Latin American countries has led to 70% treatment gap for mental disorders in these countries (Epstein 2017). This is even more so for rural and remote areas. In 1990, Pan American Health Organization (PAHO) outlined proposals for improving mental health services in the region under Caracas Declaration. The document emphasized on providing comprehensive and participatory community mental health services, taking account of human rights issues. Latin American countries have taken some steps in setting up necessary policies and strategies for delivering comprehensive mental health services to their communities, but just about a third of these countries have partially implemented such policies (PAHO 2016). Rural and remote areas are still deprived of adequate mental health services.

Europe

Rural communities in Europe are generally considered well off and less prone to mental health problems. A UK-based public health research (Riva et al. 2011) supports this view that the life expectancy and general well-being including mental health are better compared to urban dwellers. Another research report from Northern Ireland (Donaghy 2012) however paints a different picture. In the Northern Irish SWARD region, people living in rural areas had poor mental health and poor access to services, higher self-harm, and negative views toward mental illness. Changing economic realities have put a considerable financial pressure on farming communities. Media, as a result, routinely highlight a frequent and high suicide acts among European farmers.

Asia

China and India are the most populated countries in the world and have over half of their population still living in rural areas. Poverty, lack of mental health resources for rural communities, distances from mental health facilities, and stigma toward mental illness are the main barriers of mental health service provision for rural people. In rural China, primary care health professionals are the only people who can take care of their mental health, but they are ill equipped to deal with mental health problems due to their poor training in mental health (Ma et al. 2015). In India, despite of the government's District Mental Health Programme and Rural Health Mission, mental health of rural communities remains a neglected area (Kumar 2011; Sharma 2015). A high suicide rate by Indian farmers is well recognized. Financial debt, lack of adequate health care, and poor support system are the main reasons for farmers' suicide (Behere and Behere 2008).

Africa

Rural population in the African region is poorly served by mental health services, the main reason being that the limited mental health specialists either leave the region or settle in cities. Attempts have been made to integrate mental health services in primary care by training the frontline health workers in mental health (Jenkins et al. 2010). Poor literacy and specific cultural beliefs toward mental disorders and their treatments are other challenging issues in planning services for rural communities.

Negative attitude toward mental illness with a feeling of shame, cultural belief that mental illness is caused by bad deeds, long distance travel for treatment, and poor financial resources are found as main barriers in one of the Nigerian studies (Jack-Ide and Uys 2013).

Australasia

Indigenous population reside largely in rural Australia and poorly served by mental health services. There is a substantial high rate (twice compared to urban population) of suicide among young men in this population (National Rural Health Alliance 2017). Most specialists live in cities. A high suicide rate among farmers in New Zealand led to the study commissioned by Farmsafe (Walker 2012) and highlighted the stressors encountered by rural communities particularly farm and dairy workers and their families.

Major Gaps in Service Provision and Their Reasons

A broad evaluation of mental health services for the people living in rural areas in different parts of the world gives rise to some common themes responsible for the poor mental health services to their communities. These are outlined in the following section.

A Low Priority Given to Mental Health

Mental health generally receives lower priority in the overall health spending despite the fact that mental health constitutes a substantial proportion (20%) of overall health morbidity in any population. This is even worse for rural communities especially of low- (Monteiro et al. 2014) and middle-income countries, where a large proportion (over half) of people live in rural areas. Lack of governments' sustainable initiatives and programs on rural mental health leads to poor distribution of mental health-related resources and somewhat demoralization in health workers who are keen to provide services for mentally ill people.

Poor Resources

Mental health professionals are very few for rural areas even in high-income countries. They get scarce in low–middle-income countries and even worse in low-income countries. Figures produced by Mental Health Atlas (WHO 2015) show that whereas high-income countries spend around $57 per capita on mental health, the figures for low- to middle-income countries are below $2 per person per year and most of that is spent on inpatient facilities. Rural communities of low–middle-income countries are therefore left with hardly any mental health resource. The number of psychiatrists in high-income countries is over 7 per 100,000, whereas in low–middle-income countries, their number is less than 0.4 per 100,000, and some places have just one or two psychiatrists for the whole country. Distribution of these meager resources in low–middle-income countries is limited to cities, leaving rural population with hardly any specialist resource.

Poor Accessibility

As outlined above, rural population has no local mental health facilities. Travelling to cities to seek help for mental health problems is inconvenient and expensive. Accessibility to mental health services by rural population remains a problem even in high-income countries as there are very few specialists in rural counties (Gamm et al. 2010) and primary care physicians have little skill and knowledge needed to manage mental health problems. This is a huge problem for rural dwellers of low-income countries where distances are greater, transport facilities are poor, and affordability is very limited.

Social Attitudes and Acceptability of Services

Rural communities still have their shared cultural beliefs and misconceptions toward behavior resulting from mental illness as being caused by supernatural causes or bad spirits (Ngui et al. 2010). Poor literacy largely plays an important part in such deeply held beliefs. They often seek help from faith healers. People with mental illness are stigmatized believing they are being possessed by evil spirits (Mohit 2001). As a result they avoid getting help from mental health specialists even if that help is made available.

Lack of Mental Health Knowledge and Skills Among Frontline Health Workers

Health workers and primary care doctors serving rural communities are not equipped to identify and manage mental disorders (Sharma and Copeland 2009). Inadequate training on mental health is well recognized by the WHO and considered as an important reason for treatment gap for mental disorders (WHO 2010, 2016).

How to Organize Mental Health Services for Rural Areas

A proper planning of mental health services for people living in rural areas needs a multipronged approach. A thorough understanding of the population's needs is crucial for such planning. Thornicroft, Deb, and Henderson (2016) outlined that community mental health services should take account of socioeconomic context, individual- and population-based preventative approach, open and easy access, team-based approach, long-term and sustainable approach, and the cost-effectiveness of the services provided. The services should focus on strengths and aspirations of people with a focus on recovery and fulfillment (Jacobson and Greenley 2001), should strengthen from family to communities and organizations' support, and should apply evidence-based interventions with genuine involvement of people. The Mental Health Action Plan 2013–2020 (WHO 2013) set out its objectives for most countries around the world to implement in their service organization. These objectives are: (1) effective leadership and good governance in providing community-focused mental health services; (2) comprehensive, integrated, and responsive mental health and social care in the communities; (3) implementing mental health promotion and prevention strategies; and (4) developing and strengthening integrated information systems to gather evidence and carry out applied research in collaboration with academic institutes. organization.

Community Involvement: Users, Families, and Village Leaders

A broad understanding of rural communities' mental health needs is only possible by speaking to people, families, and key community leaders to find out what they know about mental health problems, their causes, and remedies. This knowledge would be valuable to develop appropriate public educational programs for mental health. Community involvement in organizing health services is a widely recommended essential approach (WHO 2001, WHO 2013, Thornicroft et al. 2016). User involvement is well advanced in the Western world and has made a real impact on effective service development (Simpson and House 2002).

Public Policies and Education About Mental Health

UK government's policy and strategy on "No health without mental health" (Department of Health 2011) has really taken a lead in making mental health a mainstream health issue. It sets out ambitious goal to consider mental health at parity with general health and emphasize on integrated approach of care for better health outcome. This approach is relevant to all countries, rich or poor, even more so where large proportion of population lives in rural areas.

Everyone in the communities should be made aware of mental ill health and its impact of peoples' lives. Teachers at school, employers at workplace, police, families, friends, and relatives in social interactions often notice change in people's behavior, performance, etc. Having some understanding through mental health public education that mental ill health may be attributable to the change of behavior can prompt them to get the right help at the earliest opportunity.

There should therefore be tailor-made public mental health education programs as a part of government's initiative by involving communities in formulating and delivering it. With advancing technology now available to rural people, media including social media could be used for this purpose. Public health education program should also address the issue of stigma and discrimination and misconceptions associated with mental disorders. A simple message should be effectively conveyed to people that mental illness is a health issue and people with timely and right care and treatment recover and function well.

Integrating Mental Health with Primary Care: Upskilling Existing Workers: Training and Education

Integrating mental health in general health at primary care level is the only solution to meet the mental health needs of population, especially that of the rural communities around the world. The World Health Organization (WHO 2010, 2016) in the Mental Health Gap Action Programme (mhGAP) highlighted that four out of five people with mental disorders in low–middle-income countries fail to receive treatment for their mental conditions. This proportion is even higher for rural communities in low-income countries which form nearly one third of the world's population. As a result, training of frontline health workers in mental health so that they can identify, diagnose, and manage mental health problems of rural people themselves as far as possible is the only answer. The mhGAP therefore has provided intervention guidelines (mhGAP-IG, WHO 2017) for nonmental health workers on identifying and managing priority conditions such as depression, psychosis, bipolar disorders, epilepsy, developmental and behavioral disorders in children and adolescents, dementia, substance misuse, self-harm, and other emotionally or medically unexplained complaints. Lund et al. (2012, 2016) developed a program to reduce treatment gap for mental disorders (PRIME) for low–middle-income countries and a detailed evaluation process. The program incorporates mhGAP-IG aimed at upskilling health workers. The PRIME package targeted community level (frontline workers), health-care facility level (health center), and organizational level (district health administration). Initial PRIME field trials in five countries, Ethiopia (Fekadu et al. 2016), India (Shidhaye et al. 2016), Nepal (Jordans et al. 2016), South Africa (Petersen et al. 2016), and Uganda (Kigozi et al. 2016), have shown some promising findings. Three of the five trials (India, Nepal, and Ethiopia) included mostly rural population. These studies identified various challenges of integrating mental health in primary care level. A sufficient length of mental health training, ongoing support from specialist, making medicines, and other resources available at primary care level are some of them.

The author of this chapter has long-standing interest in integrating mental health services in primary care (Sharma 2015) and in developing mental health assessment tools suitable for primary care, the Global Mental Health Assessment Tool (GMHAT) (Sharma and Copeland 2009). The primary care version-GMHAT/PC is a semi-structured, computer-assisted clinical assessment tool that is developed to assist health workers in making quick, convenient, and comprehensive standardized mental health assessments in both primary and general health care. The assessment program starts with basic instructions giving details of how to use the tool and rate the symptoms. The first two screens help in getting brief background details including present, past, personal, and social history including trauma, epilepsy, and learning disorder. The following screens consist of a series of questions leading to a comprehensive yet quick mental state assessment. They start with two screening questions about every major symptom complex followed by additional questions only if the screening questions are answered positively. The series of questions cover the following symptom areas: worries; anxiety and panic attacks; concentration; depressed mood, including suicidal risk; sleep, appetite, and eating disorders; hypochondriasis; obsessions and compulsions; phobia; mania; psychotic symptoms; disorientation; memory impairment; alcohol misuse; illegal drug misuse; personality problems; and stressors. The questions proceed in a clinical order along a tree branch structure. The GMHAT/PC has been widely tested and now being tried to detect and manage mental disorders in primary and general health settings in English, Hindi, Arabic, and Spanish (Sharma et al. 2004, 2008, 2010, 2013b; Krishna et al. 2009; Tejada et al. 2016). Further translations in various languages are in progress. The GMHAT team has also developed a 2- to 3-day mental health training program for frontline workers to provide knowledge and skills to identify, diagnose, and manage mental disorders at primary care level. The findings of field trials are promising and detailed in a book recently published by Indian Psychiatry Society (Behere et al. 2017). GMHAT/PC may prove to be a very useful clinical tool for frontline health workers in association with mhGAP-IG.

Making Treatments and Support Available at Community Level

All recent PRIME field studies' findings (Lund et al. 2016) highlighted that poor supply of medicines particularly in rural areas caused problems in treating people with mental illness. A need of ongoing support from mental health specialists was also considered necessary. These matters require a strong commitment at a higher governmental level.

Using Technology in Diagnosing and Managing Mental Health Problems

Technology such as mobile phones and tablets is increasingly getting affordable to people of even low-income countries. The mobile network is now available in the remote areas of these countries. It is therefore necessary to explore the ways to reach out to people living in rural and remote areas using smartphones and computers. Health workers following training may use tools such as GMHAT/PC that is now available in android app and can easily be used for mental health assessment using android phone.

Advancement of telemedicine is very promising. People from remote areas can easily communicate with specialists from any part of the world (Rathod et al. 2017). The governments should invest in technology to maximize the health-care benefits for rural and remote population.

Maximize Efficiency of Specialists: Supporting Roles Than Just Service Provider Role

The number of specialists in mental health available in low–middle-income countries and even in high-income countries such as the USA and Australia for rural population will always be insufficient in number to provide direct mental health care to sufferers. Frontline workers who provide general health care will be best suited to serve rural population for their mental health needs. Mental health specialists therefore have to take more and more roles of providing training and education on mental health and support them in managing people with mental health problems of rural areas. This should be a part of health strategy and policy of the country and not merely a voluntary gesture on specialist's part.

Human Rights and Legislation

In the twenty-first century, it is heartbreaking to see the neglect and mistreatment of people with severe mental illness in some rural communities. Patients' human rights are often ignored and treated without much respect or dignity (Ngui et al. 2010). It is important that in organizing and planning mental health services for the rural communities, patients' human rights issues are duly considered and health workers are trained in treating patients with mental illness and their relatives with respect and dignity. Mental health sufferers' rights should also be protected by appropriate legal framework. In a recent Lancet communication, Patel et al. (2016) highlighted that despite of high-level initiatives taken by Lancet in 2007 (Patel et al. 2008) with a follow-up in 2012 (Patel et al. 2011) and by WHO (2013) in its Mental Health Action Plan 2013–2020, people with mental illness are still deprived of their right to receive evidence-based treatment as well as that of their basic human rights mostly in low-income countries with a large rural population.

Conclusion

People who live in rural areas deserve good-quality services for their mental health problems. Skilling frontline workers and primary care physicians in treating people with respect and dignity, diagnosing correctly their mental illness, and providing them with evidence-based treatments and care is the only way forward. This can be achieved by investing in mental health by developing fit-for-purpose strategies at government level, providing effective public mental health education, mobilizing health work force, and integrating mental health in general health. Modern technology and mobile- and computer-assisted tools and programs can further assist in this mission.

References

Behere PB, Behere AP (2008) A Farmers' suicide in Vidarbha region of Maharashtra state: a myth or reality? Indian J Psychiatry 50(2):124–127. https://doi.org/10.4103/0019-5545.42401

Behere P, Sharma V, Kumar V, Shah V (eds) (2017) Mental health training for health professional: global mental health assessment tool (GMHAT). Indian Psychiatric Society Publication, India. ISBN: 978-1-68419-385-1

Carpenter-Song E, Snell-Rood C (2016) The changing context of rural America: a call to examine the impact of social change on mental health and mental health care. Psychiatr Serv 68(5):503–6

Department of Health (2011) No Health without mental health: a cross-government mental health outcomes strategy for people of all ages. HM Government, London

Donaghy J (2012) Consulting rural mental health, stigma, services and support within the AWARD region. Research report

Epstein J (2017) Mental Healthcare in Rural Latin America. Práctica Familiar Rural 1(4). (Ver En Linea). https://doi.org/10.23936/pfr.v0i4.172.g227

Fekadu A, Hanlon C, Medhin G et al (2016) Development of a scalable mental healthcare plan for a rural district in Ethiopia. Br J Psychiatry 208(Suppl 56):s4–s12. https://doi.org/10.1192/bjp. bp.114.153676

Gamm L, Stone S, Pittman S (2010) Mental health and mental disorders – a rural challenge: a literature review. Rural Healthy People (1):97–114

Jack-Ide IO, Uys L (2013) Barriers to mental health services utilization in the Niger Delta region of Nigeria: service users' perspectives. Pan Afr Med J 14:159. https://doi.org/ 10.11604/pamj.2013.14.159.1970

Jacobson N, Greenley D (2001) What is recovery? A conceptual model and explication. Psychiatr Serv 52(4):482–485. https://doi.org/10.1176/appi.ps.52.4.482

Jenkins R, Kiima D, Njenga F, Okonji M, Kingora J, Kathuku D, Lock S (2010) Integration of mental health into primary care in Kenya. World Psychiatry 9:118–120

Jordans MJD, Luitel NP, Pokhrel P, Patel V (2016) Development and pilot testing of a mental healthcare plan in Nepal. Br J Psychiatry 208(Suppl 56):s21–s28. https://doi.org/10.1192/bjp. bp.114.153718

Kigozi FN, Kizza D, Nakku J et al (2016) Development of a district mental healthcare plan in Uganda. Br J Psychiatry 208(Suppl 56):s40–s46. https://doi.org/10.1192/bjp.bp.114.153742

Krishna M, Lepping P, Sharma VK, Copeland JRM, Lockwood L, Williams M (2009) Epidemiological and clinical use of GMHAT-PC (Global Mental Health Assessment Tool – primary care) in cardiac patients. Clin Pract Epidemiol Ment Health 13:5–7

Kumar A (2011) Mental health services in rural India: challenges and prospects. Health 3:757–761. https://doi.org/10.4236/health.2011.312126

Lund C, Tomlinson M, De Silva M et al (2012) PRIME: a programme to reduce the treatment gap for mental disorders in five low- and middle-income countries. PLoS Med 9:e1001359

Lund C, Tomlinson M, Patel V (2016) Integration of mental health into primary care in low- and middle-income countries: the PRIME mental healthcare plans. Br J Psychiatry 208(Suppl 56): s1–s3. https://doi.org/10.1192/bjp.bp.114.153668

Ma Z, Huang H, Chen Q, Chen F, Abdullah AS, Nie G, Feng Q, Wei B (2015) Mental health services in rural China: a qualitative study of primary health care providers. Biomed Res Int 2015:151053. https://doi.org/10.1155/2015/151053

Mental Health in Rural and remote Australia (2017) National Rural Health Alliance. http:// ruralhealth.org.au/sites/default/files/publications/nrha-mental-health-factsheet-2017.pdf

Mohit A (2001) Mental health and psychiatry in the Middle East: historical development. East Mediterr Health J 7:336–347

Monteiro NM, Ndiaye Y, Blanas D, Ba I (2014) Policy perspectives and attitudes towards mental health treatment in rural Senegal. Int J Ment Heal Syst 8:9. https://doi.org/10.1186/ 1752-4458-8-9

Ngui EM, Khasakhala L, Ndetei D, Roberts LW (2010) Mental disorders, health inequalities and ethics: a global perspective. Int Rev Psychiatry (Abingdon, England) 22(3):235–244. https://doi.org/10.3109/09540261.2010.485273

Pan American Health Organization (2016) Regional mental health atlas december 2015. PAHO Publication, Washington, DC

Patel V, Garrison P, de Jesus Mari J, Minas H, Prince M, Saxena S (2008) Advisory group of the Movement for Global Mental Health. The Lancet's series on global mental health: 1 year on. Lancet 372(9646):1354–1357. https://doi.org/10.1016/S0140-6736(08)61556-1. PMC2570037

Patel V, Boyce N, Collins PY, Saxena S, Horton R (2011) A renewed agenda for global mental health. Lancet 378(9801):1441–1442. https://doi.org/10.1016/S0140-6736(11)61385-8

Patel V, Saxena S, Frankish H, Boyce N (2016) Sustainable development and global mental health – a Lancet Commission. Lancet 387(10024):1143–1145. https://doi.org/10.1016/S0140-6736(16)00208-7

Petersen I, Fairall L, Bhana A, Kathree T, Selohilwe O, Brooke-Sumner C, Faris G, Breuer E, Sibanyoni N, Lund C, Patel V (2016) Integrating mental health into chronic care in South Africa: the development of a district mental healthcare plan. Br J Psychiatry 208(Suppl 56):s29–s39. https://doi.org/10.1192/bjp.bp.114.153726

Rathod S, Pinninti N, Irfan M, Gorczynski P, Rathod P, Gega L, Naeem F (2017) Mental health service provision in low- and middle-income countries. Health Serv Insights 10:1178632917694350. https://doi.org/10.1177/1178632917694350

Riva M, Bambra C, Curtis C, Gauvin L (2011) Collective resources or local social inequalities? Examining the social determinants of mental health in rural areas. Eur J Pub Health 21(2):197–203. https://doi.org/10.1093/eurpub/ckq064

Sharma VK (2015) Psychiatry in primary care- Indian perspectives. In: Malhotra S, Chakrabarti S (eds) Developments in psychiatry in India. Springer. ISBN: 978-81-322-1674-2

Sharma VK, Copeland JRM (2009) Detecting mental disorders in primary care. Ment Health Fam Med 6(1):11–13

Sharma VK, Lepping P, Cummins AG, Copeland JRM, Parhee R, Mottram P (2004) The Global Mental Health Assessment Tool: Primary Care Version (GMHAT/PC): development, reliability and validity. World Psychiatry 3(2):115–119

Sharma VK, Lepping P, Krishna M, Durrani S, Copeland JR, Mottram P et al (2008) Mental health diagnosis by nurses using the Global Mental Health Assessment Tool: a validity and feasibility study. Br J Gen Pract 58(551):411–416

Sharma VK, Jagawat S, Midha A, Jain A, Tambi A, Mangwani LK, Sharma B, Dubey P, Satija V, Copeland JRM, Lepping P, Lane S, Krishna M, Pangaria A (2010) The global mental health assessment tool-validation of GMHAT/PC in Hindi: a validity and feasibility study. Indian J Psychiatry 52(4):316–319

Sharma BB, Singh S, Sharma VK, Choudhary M, Singh V, Lane S, Lepping P, Krishna M, Copeland J (2013a) Psychiatric morbidity in chronic respiratory disorders in an Indian service using GMHAT/PC. Gen Hosp Psychiatry 35(1):39–44

Sharma VK, Durrani S, Sawa M, Copeland JR, Abou-Saleh MT, Lane S, Lepping P (2013b) Arabic version of the Global Mental Health Assessment Tool-Primary Care version (GMHAT/PC): a validity and feasibility study. East Mediterr Health J 19(11):905–908

Shidhaye R, Shrivastava S, Murhar V, Samudre S, Ahuja S, Ramaswamy R, Patel V (2016) Development and piloting of a plan for integrating mental health in primary care in Sehore district, Madhya Pradesh, India. Br J Psychiatry 208(Suppl 56):s13–s20. https://doi.org/10.1192/bjp.bp.114.153700

Simpson EL, House AO (2002) Involving users in the delivery and evaluation of mental health services: systematic review. BMJ 325:1265

Tejada P, Jaramillob LE, Garcíab J, Sharma V (2016) The global mental health assessment tool primary care and general health setting version (GMHAT/PC) – Spanish version: a validity and feasibility study. Eur J Psychiatry 30(3):195–204

Thornicroft G, Deb T, Henderson C (2016) Community mental health care worldwide: current status and further developments. World Psychiatry 15(3):276–286. https://doi.org/10.1002/wps.20349

Walker JF (2012) Mental health in the rural sector – a review, Farmsafe, New Zealand

Wang PS, Lane M, Olfson M, Pincus HA, Wells KB, Kessler RC (2005) Twelve-month use of mental health services in the United States. Results from the National Comorbidity Survey Replication. Arch Gen Psychiatry 62(6):629–640

WHO (2016) mhGAP intervention guide for mental, neurological and substance use disorders in non-specialized health settings: Mental Health Gap Action Programme (mhGAP): version 2.0. World Health Organization, Geneva

World Bank Population Data (2015) World Bank Group, Washington

World Health Organisation (2001) The world health report 2001- mental health: new understanding, New Hope. World Health Organization, Albany

World Health Organisation (2013) Mental health action plan 2013–2020. World Health Organization, Geneva

World Health Organisation (2015) Mental health atlas 2014. World Health Organization, Geneva

World Health Organization (2010) mhGAP intervention guide for mental, neurological and substance use disorders in non-specialized health settings: Mental Health Gap Action Programme (mhGAP) [Internet]. WHO Publication

Development of Rural Mental Health in Bangladesh

11

Muhammad Zillur Rahman Khan

Contents

Abstract

Most of the people of Bangladesh live in rural area, and it has good health indicator and economic growth. But the burden of mental disorders in rural areas is huge with limited resources and mental health professionals. A substantial proportion of persons suffering from mental disorders do not get medical treatment in time where availability of service is not the main reason. Bangladesh is renowned for its primary health-care network, and more than 13,000 community clinics are functioning (one clinic for 6000 rural population) at present. So, integration of mental health service into primary care with community-based approach and effective referral linkage with specialized and other tier-specific health services would be the key strategy for the development of effective rural mental health service. The current fourth Five Year Health Sector Program

M. Z. R. Khan (✉)
Department of Psychiatry, Patuakhali Medical College, Patuakhali, Bangladesh
e-mail: mzrkhan@gmail.com

© Springer Nature Singapore Pte Ltd. 2020
S. Chaturvedi (ed.), *Mental Health and Illness in the Rural World*, Mental Health and
Illness Worldwide, https://doi.org/10.1007/978-981-10-2345-3_29

(2016–2021) entitled "Health, Nutrition and Population Sector Strategic Investment Plan (HNPSIP)" with Essential Service Package (ESP) adopted by the Bangladesh government created a new hope for mental health including rural mental health. Newer mental health programs with provision of basic capacity for identification and management of mental disorders have been incorporated. But more effort should be put on preventive and promotive service on mental health. Besides, for expansion of service, new allocation of resources and capacity development are essential. Awareness and anti-stigma campaign is important to successful service development. The development of organized and effective rural mental health service should be prioritized for a better future of the nation.

Keywords
Bangladesh · Integration of mental health · Mental disorders · Mental health service · Rural mental health · Primary health care (PHC) · Treatment gap

Introduction

Bangladesh has recently stepped forward to a middle-income country group which has also steady annual economic growth rate facing limited resources and other obstacles. The burden of mental disorders is huge, and majority of people suffering from mental disorders live in rural areas like many other developing countries in the world. Globally urbanization increased day by day and Bangladesh is not an exception. Still majority of population live in rural areas which is nearly 65% (2016a) at present decreased from 95% in 1960. Burden of mental disorders is huge in rural areas with limited mental health professionals and resources. But still Bangladesh has better health indicator in comparison to other South Asian countries. The country has achieved Millennium Development Goals (MDGs) and received an award for MDG 4 by reducing infant mortality rate. The government is committed to achieve the Sustainable Development Goals (SDGs) like MDGs and prioritized mental health which creates a new hope in mental health service. The rural area has its own merits and demerits on causation of mental health conditions for its unique geographical location, environment, climate, population dynamics, culture, economy, and other attributes. Perception, attitude, and acceptance of mental health, mental disorders, and its treatment are also different from urban areas. Stigma and negative attitude toward mental health, disorders, and its treatment are one of the major hindrances to provide better mental health services rather than availability of services. But expansion of basic and specialized mental health service to rural and hard to reach areas is a timely need. Nearly more than half a century ago, the World Health Organization (WHO) defined health "as a state of completely, physical, mental and social well being," but still mental health is less prioritized. To providing basic health service including mental health service to all people of Bangladesh by the government is a constitutional mandate. In this perspective, the development of well-structured rural mental health is a priority developmental agenda and is also a challenge for Bangladesh.

Basic Health, Demographic, and Administrative Profile of Bangladesh

Bangladesh is a country of 160 million population. More than half (53.7%) of population are in 15–49 years age group which has a demographic dividend. Male and female ratio is nearly equal (100.3: 100), and life expectancy at birth has been increased, and at present it is 71.8 years (Bangladesh Bereau of Statistics, SID, Ministry of Planning 2016). The per capita gross domestic product (GDP) is 1384 USD, and the annual growth rate is 7.1% (2016b). Rural mental health should be prioritized if current economic growth and progress has to be maintained. It has 8 administrative divisions with only 4 metropolitan cities and 12 city corporations (2016b). Each division has several districts including 64 districts and sub-districts (Upazila), and there are a total number of 491 sub-districts. Each Upazila is divided into several unions. Each union is then subdivided to several wards and villages. The country has such 87,310 villages which clearly indicate rural predominance (2016b). So, providing mental health service to rural population will cover majority of the population in Bangladesh.

Brief History of Mental Health Services

There was only one mental hospital for the whole country before the independence in 1971. It was the Pabna Mental Hospital located in Pabna district established in 1958 and became a hospital of 200 beds in 1964. After independence, specialist mental health services were limited to outpatient department and started at Dhaka Medical College Hospital in 1974. The inpatient and outpatient service were extended to former Institute of Postgraduate Medical Research (IPGMR) which is the current Bangabandhu Sheikh Mujib Medical University (BSMMU). Then gradually psychiatry department was extended to eight old medical college hospitals outside Dhaka in 1980 (Karim et al. 2011). No formal mental health service was developed for rural population at that time, but rural people could get mental health services from a nearby psychiatry department of a medical college hospital located at the same or adjacent district. The Central Drug Addiction Treatment Centre was established in 1988 which is under the Ministry of Home Affairs for treatment and rehabilitation of persons suffering from substance abuse and dependence. Further expansion of psychiatry department in government medical colleges was under the process of development. In 2001, National Institute of Mental Health (NIMH), the only mental health institute of Bangladesh, was established in full swing in a separate location at the capital city Dhaka. It was a milestone in the history of mental health services in Bangladesh. At present it has 200 beds and consists of 6 separate departments with inpatient, outpatient, and 24 h and 365 days a year emergency department (Hamid 2016a). It has specialized clinics at outpatient department. NIMH is a teaching institute and hospital dedicated to mental health service, training of professionals, conducting research and academic courses, development of mental health legislation, and national and international collaboration and policy making.

Burden of Mental Disorders, Substance Abuse, and Its Correlates in Rural Population

Burden of mental disorders is huge in Bangladesh both in rural and urban settings. Evidence suggested that there is little difference in burden of mental disorders in rural and urban areas though dynamics and determinants of mental disorders are quite different in the two areas. Earlier study revealed lower prevalence of mental disorders such as 3.6% conducted by Chowdhury AK in 1981 (Chowdhury et al. 1981). But recent studies indicated the higher prevalence of mental disorders in rural population as evidenced by 16.5% in 2007 (Monawar Hosain et al. 2007; Hossain et al. 2014). Survey on mental health in adult population (n = 13,000) in whole Bangladesh revealed that overall prevalence of mental disorders was found 16.1% where proportion of mental disorders in rural and urban areas were nearly equal (49.5% vs. 50.5%) (Firoz et al. 2006). Common mental disorders among adult population were anxiety disorders (5.6%), depression (4.6%), somatoform disorders, obsessive compulsive disorder, schizophrenia, bipolar mood disorder, etc. Overall burden of mental disorders has been increased and found nearly doubled (32.8%) in a recently conducted community survey on mental health in adult population (Hamid 2016b). In the study, depression (12.8%), generalized anxiety disorder (7.0%), somatic symptom disorder (4.5%), obsessive compulsive disorder, schizophrenia, bipolar mood disorder, etc. were found common in rural population. Little difference was noticed in the prevalence of mental disorder in urban and rural areas (33.0% vs. 31.0%) in the study. Paradoxically, the prevalence of mental disorders was found higher among children and adolescents of rural areas (17.3%) than urban areas (14.3%) in a survey in 17 districts of Dhaka division (Rabbani et al. 2009). In the study common mental disorders among children and adolescent were mental retardation (3.8%), enuresis (3.0%), epilepsy (2.0%), anxiety disorder, somatoform disorder, depression, autism, etc. Similar finding was also observed for substance abuse. But rural areas have some protective factors for development of mental disorders like rural culture and joint family structure, etc.

Substance abuse is also common in rural Bangladesh. The prevalence of substance abuse in adult and children and adolescent was found 0.6% and 0.8% which was clearly an underestimation (Firoz et al. 2006; Rabbani et al. 2009). Recent community survey on mental health in adult population revealed that substance abuse was found lower (1.2%) in rural areas than urban (2.4%) one (Hamid 2016b). Factors related to substance abuse are easy availability, frustration, impulsive personality trait, peer pressure, unemployment, poverty, and underprivileged conditions which are nearly the same for rural and urban areas. Substance abuse is more common among young population (15–35 years), and a significant proportion of them are college or university students. It is evident from an expert opinion that about half of the college and university students are current substance abuser which is very alarming. But treatment facility for substance abuse for rural population is very scanty. Yaba (methamphetamine and caffeine) is now popular in both rural and urban youngsters. Besides, use of cannabis, benzodiazepine, opiates (heroin, phensidyl), and solvents is also observed in rural population.

Evidence suggested that there was little difference found in the presentation of mental disorders in rural and urban population. A substantial proportion of community population (both rural and urban) in Bangladesh presented with different somatic or physical complaints like headache, pain in different parts of the body, burning sensations, weakness, sleep disturbance, decreased appetite, palpitation, etc. was found in a recent community survey (Hamid 2016b). People suffering from mental disorders in rural areas usually presented to physicians with these somatic complaints which are often missed and are not diagnosed and treated properly which is a burden for both the treatment recipient and provider in terms of time, cost, and hassle.

Treatment Gap of Mental Disorders

Two-thirds of persons suffering from mental disorders in the low- and middle-income countries do not get appropriate medical treatment. Bangladesh is not an exception for it. Though there is no national data on treatment gap of mental disorders in Bangladesh, a large-scale community survey on decreasing the treatment gap of epilepsy in children in a rural community showed that 76.8% of children and adolescents (aged 5–15 years) did not get any medical treatment for epilepsy (Rabbani 2009). Stigma, negative attitude to mental disorders and its treatment, lack of public awareness, poor socioeconomic condition, and lack of service information are major causes of delay in early detection and treatment of mental disorders in rural areas. The lack of availability of mental health treatment facilities in rural areas is not the main reason, but underutilization of mental health service is the major problem.

Poverty, unemployment, substance abuse, perinatal birth injuries, poor maternal and infant care, overprotective child rearing, educational stress, and housing problems are the common factors contributing to mental disorders in rural population. Stigma and negative attitude to mental disorders and its treatment are still major challenges. Awareness on mental disorders and its treatment has improved significantly. But still a substantial proportion of the rural population seeks help from traditional healers (Kabiraj, Fakir, Ojha, Hekim, Pir, etc.) for complaints of mental disorders. Sometimes maltreatment causes increased morbidity, disability, disfigurement, and even death. Preference to nonmedical treatment for mental problem in rural (even urban) people is due to ignorance, poor communication skill of health professionals, cultural difference, easy availability, and acceptability. Poor knowledge on manifestations of mental disorder and lack of knowledge on information on availability of mental health service are important factors for determinants of help-seeking behavior in mental health condition of the rural population. But an optimistic picture was also observed. National survey on mental health in adult population in Bangladesh showed that 51.7% of the respondents did not believe that traditional healers were the appropriate treatment providers for mental patients and 54.2% of the responders were aware that mentally ill persons can be managed in primary health care (PHC) and other hospitals (Firoz et al. 2006).

Mental Health Service Structure, Manpower, and Governance

There is no separate mental health service in Bangladesh. Mental health service is mainly located under the umbrella of government health services under the Ministry of Health and Family Welfare (MOHFW) and Directorate General of Health Services (DGHS). There are also private mental health services available which is limited in nature. Since the independence in 1971 by the Liberation War, Bangladesh has a good primary health-care network. There are 482 government hospitals in Bangladesh at Upazila (sub-district) and union level. At grassroot level, there is a community clinic for every 6000 population, and more than 13,000 such clinics are at present functional which would be a potential area for providing mental health service for rural community people in the future (2016b).

There are three-tiered health services where mental health services are incorporated except specialized mental health services. The primary health-care system includes Upazila Health Complex (UHC) (30–50 bed hospital), Union Health and Family Welfare Centre (UHFWC), and union sub-centers and community clinics. Secondary care system includes a district hospital in 64 districts with inpatient, outpatient, and emergency services. There is no specialized mental health service in primary and secondary health-care levels. In tertiary care level, there are medical college hospitals where psychiatry departments are located in 36 districts (2016b). But more than half of the psychiatry departments of medical college hospitals do not have inpatient facilities, and some of them have very limited outpatient services. So, in district-level hospital, at least one post of consultant psychiatrist should be there like other consultants such as gynecology, medicine, surgery, etc. for better development of mental health service at district level. Without a consultant psychiatrist, it will be very difficult to coordinate mental health service in Upazila (sub-district), union, and village level.

Specialized hospitals are usually far away from rural areas and mostly located at capital city and in divisional cities. Besides NIMH, Dhaka, and psychiatry department of medical college hospitals, specialized mental health services are available at the Psychiatry Department of Bangabandhu Sheikh Mujib Medical University (BSMMU), Armed Forces Medical College, most of the Combined Military Hospitals (CMHs), Army Medical College hospital, and private medical college hospitals. Specialized mental health service center should act as a specialized treatment and referral center. Active referral linkage among primary, secondary, tertiary, and specialized hospitals is the key to develop sustainable rural mental health service. There are 68 private medical hospitals in Bangladesh, and most of them do not have any psychiatry department. So, they could not provide mental health service to public. In this regard, current medical legislation should be followed to have a separate psychiatry department with faculty members and other mental health professionals.

There is scarcity in mental health professionals in Bangladesh which is only 0.5 person per 100,000 population. Only 220 psychiatrists (0.07 psychiatrist/100000 population) are available for the whole country which indicate shortage of mental health professionals (Ministry of Health and family welfare 2009). Among them, nearly more than half of them provide services in capital city Dhaka. But a significant proportion of psychiatrists provide holiday private practice (usually once or twice a week) in different districts from their regular working place. This unrecognized service close to rural areas has a significant contribution in mental health service in rural areas. Other mental health professionals including clinical psychologist, psychologist, and occupational therapists are also small in number, and most of them work in capital city Dhaka and other districts in nearby public universities of Bangladesh. So, a specialist mental health team entitled "District Mental Health Team" should be incorporated in every district hospital of Bangladesh for better mental health service in rural areas. The team should include a psychiatrist, a senior doctor from health administration of the district hospital (civil surgeon), a clinical psychologist (or psychologist or allied professional), a nurse, a family member, a member from civil society, and supporting staffs. This team could coordinate and provide technical support to mental health service at rural level with support for National Institute of Mental Health (NIMH) and Non-communicable Disease Control (NCDC) of Directorate General Health Services (DGHS).

Though there is a shortage of specialist mental health professionals, the adequate number of trained health professionals in mental health remains in already existing health services. More than 10,000 doctors and 7246 health workers were trained in mental health (6 days basic mental health training) in the last several years from NIMH, Dhaka, with support from DGHS (NCDC and other departments). But the database of the training was not maintained, and the trained persons move from their workplace due to regular transfer and change of posting which is a drawback. This short-term training is a continuous process and very useful in case detection, referral, basic treatment of common mental disorders, and creation of awareness on mental health in rural areas. So, best utilization of existing health professionals is very important to expand the network of mental health services.

Integration of Mental Health Service into Primary Care

The Bangladesh government is committed to achieve SDGs and to ensure mental health, and well-being for all people is an important goal (Goal 3) of SDGs. But there is limited number of mental health professionals and resource. To provide mental health service for all rural population of Bangladesh, integration of mental health into existing primary care is the most feasible, cost-effective, and accessible strategy for proving mental health services for all people including rural population. Though this approach has some limitations, still it is the most effective strategy to provide comprehensive and quality mental health care to the grassroot pupation of the country which was evident from many low- and middle-income (LIMCs) countries and was also recommended by the World Health Organization (WHO). The integration increases the likelihood to positive outcomes for both mental and physical problems and will help to sustain mental health service for rural population.

Community-based mental health activities (CBAs) have been started since a decade ago as cost-effective and easily accessible means in Bangladesh. Community mental health services through integration into primary health care aim at providing cost-effective care to mentally ill patients in the communities. These services will reduce stigma among rural population and improves human rights of persons suffering from mental disorders. It will also ensure service from one-stop service center for both physical and mental disorders. Integration of mental health into primary care has been proved to be a successful and cost-effective strategy from a lesson learned from a pilot project in Sonargaon Upazila of Narayanganj District in 2008 (National Institute of Mental Health, Dhaka, and Directorate General of Health Services 2011). A medical team consists of two psychiatrists, two medical officers, a nurse, and an office staff that weekly visited the Upazila Health Complex (UHC), provided outpatient mental health service, and supplied essential psychotropic medication. Monthly awareness meeting and training of local doctors and health professionals were also conducted. A number of patients treated for mental disorders were increased significantly after 1 year in the outpatient department of the UHC which was a huge success. This model could be implemented in different districts to expand and strengthen rural mental health service in the community.

Newer Mental Health Program Under Government Sector Program

Mental health program in Bangladesh is mainly under government health sector program. In the previous Health, Nutrition, and Population Sector Development (HNPSD) Program (2011–2016), mental health was less prioritized. But the current fourth Five Year Health Sector Program (2016–2021) entitled "Health, Nutrition and Population Sector Strategic Investment Plan (HNPSIP)" (Ministry of Health and Family Welfare 2014) created new scope and hope in mental health service, and rural mental health has been emphasized. In the current sector program, many important mental health programs have been introduced, and the allocation to mental health budget has been increased which was only 0.4% of the total health budget in 2009 (Ministry of Health and Family Welfare 2009). But most of the mental health expenditure was allocated to mental hospitals (67%) (Ministry of Health and Family Welfare 2009), and limited resources were dedicated to promotive and preventive services which are very essential for mental health service development in rural and other community areas.

In the recent Health, Nutrition and Population Sector Strategic Investment Plan (HNPSIP 2016–2021), essential service package (ESP) was reorganized where a new mental health program was introduced under a new noncommunicable disease (NCD) component (Ministry of Health and Family Welfare 2016). The main objective of the ESP is to provide health services to all grass root and hard to reach people including rural population. The mental health program in ESP has been be supervised and monitored under the leadership of Non-Communicable Disease Control (NCDC) of Directorate General of Health Services (DGHS). In ESP screening, identification, awareness, and basic treatment of common mental disorders (schizophrenia, bipolar mood disorders, anxiety disorders, substance abuse and neurodevelopmental disorders, suicide prevention, counseling service, support for rehabilitation, provision of essential psychotropic medication, etc.) will be provided at primary health-care levels including community clinics, Union Health and Family Welfare Centre, and Upazila health complexes (Ministry of Health and Family Welfare 2016). Training of health professionals including doctors, nurses, and health workers is an essential component for the successful implementation of this program. This training of health professionals is a continuous process which is mainly conducted by the National Institute of Mental Health, Dhaka, with support and assistance from DGHS (mainly NCDC) and World Health Organization (WHO) county office. But sustainable capacity development for training of mental health professionals and decentralization of this program at district hospitals and medical college hospitals are vital for the successful implementation of the program. The manual for training of mental health professionals is already developed mostly by NIMH and DGHS supported by the WHO which needs further validation and upgradation. These activities will be monitored and coordinated at district level from psychiatry department of a medical college or district hospital. Unfortunately, there is no consultant psychiatrist placed at district level, and most of the psychiatry departments of medical college hospitals are not fully equipped in terms of facility, logistics, and manpower. Public awareness, Anti-stigma campaign, monitoring and supervision of the mental health services, mental health education, and referral should be the main focus of activities at this level. Essential psychotropic medications including haloperidol, risperidone, procyclidine, amitriptyline, fluoxetine, sodium valproate, carbamazepine, and diazepam were proposed for procurement and delivery to primary care from secondary care level such as district hospital under ESP coverage (Ministry of Health and Family Welfare 2016). Regular attendance of health professionals at this level is a big challenge for providing mental health service. Proper career planning, stoppage of frequent transfer of staffs, local recruitment, and backup support for housing, school, and other facilities should be ensured for sustainable services.

At secondary and tertiary level, a comprehensive treatment facility of mental disorders with inpatient, outpatient, and basic emergency service should be provided. At district hospital, no bed for mental patients is provided at present. A small unit consisting of at least 5 beds should be there, and all psychiatry departments in government medical college hospitals should have an inpatient unit of 10–15 bed facility and regular outpatient facility. At present tertiary and secondary care hospital have very little preventive and promotive mental health activities. To provide mental health service at rural and other grassroots level, they should come forward to coordinate and facilitate such activities for better rural mental health service in collaboration with NIMH and NCDC, DGHS. Community clinics in rural areas are very close to rural population. Community health-care providers (CHCP) with the help of other health workers (health inspector, family welfare visitor, etc.) and doctors could play a pivotal role in rural mental health service. There is an existing domiciliary service in the primary health-care system where health visitors visit village homes in a regular basis (at least once in a week) which has been proved effective in maternal, neonatal, and child health care (2009). They could take part in the awareness on mental health, mental health education, and basic follow-up (medication side effects, current status of illness, etc.) of persons suffering from mental disorders. This is not taking place at present in primary health-care system.

Mental Health Service for Vulnerable Population

A substantial proportion of the rural population are vulnerable to develop mental health conditions and disorder due to their geographic location (coastal, disaster-prone, and hard-to-reach areas), poor socioeconomic condition, and underprivileged conditions. Special mental health program needs to be incorporated to vulnerable or risk group of rural population such as children and adolescent, women, elderly, and young people at risk of suicide, self-harm, and substance abuse. To address the special needs of this vulnerable group, a mental health program aside from the existing health programs such as Maternal and Child Health program, Reproductive and Sexual Health program, and Injury prevention program should be incorporated. Under ESP such program has been chalked out, and the implementation of this program will ensure mental health for such population. Bangladesh is a disaster-prone country, so a program for psychosocial support of disaster-affected population is essential, and training manual for health professionals should be available. Screening, identification, and management of intellectual disability, autism, and other neurodevelopmental disorders of children and adolescent are also prioritized by government health authority (MOHFW and DGHS).

Evidence suggested that persons with severe and chronic mental illness are at greater risk of physical health morbidity and mortality and vice versa. There is no data available on the prevalence of physical comorbidity among persons suffering from severe and chronic mental illness. So, identification, screening, and management of physical health condition (diabetes, hypertension, etc.) should be an integral part of rural mental health services.

Policy, Legislation, and Collaboration of Care

Draft mental health act has been passed in the parliament recently and policy are not yet passed in Bangladesh. Mental health policy should specially emphasize rural mental health to provide better mental health service for rural and other grassroot population. The enactment of mental health act is necessary for protection of human rights of persons suffering from mental disorders for both rural and urban population. Policy should be based on equity and justice and provide quality mental health care to all people including rural population. Integration of mental health into primary care should come into action as soon as possible with community-based mental health activities. Rehabilitation service should be reorganized in collaboration with other health and non-health departments such as the Ministry of Social Welfare. There is no long-stay rehabilitation center for chronic mentally ill patients with vocational training facilities. Further community mental health facilities such as day treatment facility, half way home, foster home or care, occupational therapy, and rehabilitation center for substance abuse need to be developed in rural settings in the near future. Mental health matter is not only confided to medical department. Non-pharmacological management should be emphasized like pharmacological management for better mental health services of the rural population. Collaboration between different government ministries such as department of education, agriculture, social welfare, local government and rural development, and labor and youth development is necessary. Nongovernment organizations (NGOs) could play an important role in the development of rural mental health in collaboration with government agencies. National and international collaboration is needed in capacity development, training of mental health professionals, and research.

Promotion of Mental Health, Prevention of Disorders, and Anti-stigma Campaign

Mental health promotion and prevention of mental disorders remain less prioritized most often. Most of the mental health budget (two-thirds) and efforts are dedicated to curative or hospital service which is also the same for rural mental health. For best utilization of available health and mental health services by rural population public awareness, health education program and ant-stigma campaign are essential. Stigma, lack of awareness, negative attitude to mental health, and its treatment are major challenges for successful development of rural mental health in Bangladesh. Electronic and print media, and communication technology (mobile phone, Internet, etc.) could play a supportive role in developing public awareness among rural population. Celebration of important days (world mental health day, world health day, etc.) with colorful rally and road show, radio and television program (talk show, drama, etc.), and public and stakeholders awareness meeting have been proved very effective in creating public awareness in the last decade in Bangladesh. School mental health program and youth development program should be also emphasized.

Conclusion

Development of organized and effective rural mental service in Bangladesh is a big challenge for government and the people of the country. To achieve SDGs like MDGs and to maintain current economic growth and progress, mental health including rural mental health should be highly prioritized. Major hindrances to achieve this goal are not the resource constraints or the scarcity of specialist mental health professionals, but it is the effective policy and strategy and successful implementation of activities of mental health program. Reorganization of existing mental health service and integration of mental health into primary care will be the key strategy for successful development of rural mental health in Bangladesh. Effective referral linkage with different tiers of health and mental health service will make the rural mental service more effective and sustainable. In this perspective,

Bangladesh is on the right track. The new health sector program and essential service package adopted by the government created a new hope for mental health of rural population.

Policy makers usually prioritize health program in terms of mortality statistics which doesn't fit for mental disorders and mental health. The burden of mental disorders is huge which will cause huge morbidity, productivity loss, disability, and even death globally and also in Bangladesh. So, provision of more resource and budget is needed for capacity development, creation of new posts, development and training of mental health professionals, and research which is a timely need. It is high time to act together to make successful rural mental health services in Bangladesh a reality.

References

Bangladesh Bureau of Statistics, SID, Ministry of Planning G (2016) Report on Bangladesh Sample Vital Statistics 2015

Chowdhury AK, Alam MN, Ali S (1981) Dasherkandi project studies. Demography, morbidity and mortality in a rural community of Bangladesh. Bangladesh Med Res Counc Bull 7:22–39

Firoz AHM, Karim ME, Alam MF, Rahman AHM, ZM (2006) Prevalence, medical care, awareness and attitude towards mental illness in Bangladesh

Hamid M (2016a) Local Health Bulletin, NIMH 2016. In: Manag. Inf. Syst. (MIS), DGHS, Dhaka. http://app.dghs.gov.bd/localhealthBulletin2016/publish/publish.php?org=10000010&year=2016&lvl=5

Hamid M (2016b) Community Survey on Symptom Presentation of Common Mental Disorders

Hossain MD, Ahmed HU, Chowdhury WA et al (2014) Mental disorders in Bangladesh: a systematic review. BMC Psychiatry 14:216. https://doi.org/10.1186/s12888-014-0216-9

Karim M, Shaheed F, Paul S (2011) International perspectives on mental health. RCPsych Publication

Ministry of Health and family welfare B (2009) Mental Health System in Bangladesh

Ministry of Health and Family Welfare B (2014) Background Paper on Health Strategy for Preparation of 7th Five Year plan

Ministry of Health and Family Welfare B (2016) Bangladesh Essential Service Package (ESP)

Monawar Hosain GM, Chatterjee N, Ara NIT (2007) Prevalence, pattern and determinants of mental disorders in rural Bangladesh. Public Health 121:18–24

National Institute of Mental Health, Dhaka and Directorate General of health Services M (2011) Integration of Mental Health Services with Primary Health Care in Bangladesh

Rabbani MG (2009) Decreasing the treatment gap of Epilepsy in children in Bangladesh

Rabbani MG, Alam MF, Ahmed H, Sarkar M, Islam SAN et al (2009) Prevalence of mental disorders, mental retardation, epilepsy and substance abuse in children. Bangladesh J Psychiatry 23:11–52

(2016a) World Bank. https://data.worldbank.org/indicator/SP.RUR.TOTL.ZS?view=map

(2016b) Health Bulletin 2016

(2009) Early postpartum visits from community health workers reduce neonatal mortality in Bangladesh. Int Perspect Sex Reprod Health 35:208–209

Challenges of Research in Rural Mental Health in Sri Lanka

12

Athula Sumathipala, Dhanuja Mahesh, and Hiranya Wijesundara

Contents

A. Sumathipala (✉)
Research Institute for Primary Care and Health Sciences, Faculty of Medicine and Health Sciences, Keele University, Stoke on Trent, UK
e-mail: a.sumathipala@keele.ac.uk

D. Mahesh · H. Wijesundara
Ministry of Health, Nutrition and Indigenous Medicine, Colombo, Sri Lanka
e-mail: dhanuja76@gmail.com; hiranya.w@gmail.com

© Springer Nature Singapore Pte Ltd. 2020
S. Chaturvedi (ed.), *Mental Health and Illness in the Rural World*, Mental Health and
Illness Worldwide, https://doi.org/10.1007/978-981-10-2345-3_30

Abstract

This chapter discusses the challenges of conducting mental health research in rural Sri Lanka. The majority of the Sri Lankan population (77.4%) live rurally and the demand for mental health service remains high. Mental health service development in two districts, the Ampara District in the Eastern Province and Monaragala district from the Uva province of Sri Lanka, is detailed to describe the challenges faced by the mental health workforce. Real life experiences of two consultant psychiatrists who were leading these service development efforts bear witness to the deterrents and challenges in conducting mental health research.

Keywords

Rural mental health · Sri Lanka · Mental health research · Psychiatry

Introduction

Background and the Context

Sri Lanka has a land area of 65,610 km^2 (25,332 mi^2) and is situated in the Indian Ocean. According to the 2012 national census, the total population of the country was reported as 20,359,439, but the estimated population increased to 22.2 million in 2016, comprising Sinhalese (75%), Sri Lankan Tamil (11%), Indian Tamil (4%), Sri Lankan Moors (9%), and others 0.5% (Department of Census & Statistics Ministry of Policy Planning and Economic Affairs 2012). Anthropologically, Sri Lanka presents an interesting genetic diversity between five main population groups, resulting from ancestral and immigrant elements of European and Asian origin (Papiha et al. 1996).

For administrative purposes, Sri Lanka is divided into 9 provinces, further subdivided into 25 districts. More than half of the island's population is distributed in three provinces, the Western province (6,028,000), the Central province (2,690,000), and the Southern province (2,584,000), even though the land area of these three provinces consists of only 23% of the total land area of the country. The Northern Province reported the lowest population share of 5.2%, followed by 6.2% in each of the North-Central and Uva provinces, and 7.6% in Eastern province (Department of Census & Statistics Ministry of Policy Planning and Economic Affairs 2012).

The majority of the Sri Lankan population live in the rural sector (77.4%); the urban population share is 18.2%, while 4.4% live in large agricultural estates, considered a separate sector. Urbanization is relatively high in the Western province (38.8%) and very low in the North Central (4.0%) and North Western (4.1%) provinces. Rural population is therefore highest in the North Central province (96.0%) and the highest estate population is reported from Central province. Although Sri Lanka is categorized as a Low and Middle-Income Country (LMIC), social indicators, particularly on health and education, are more advanced than those of many other LMICS. The life expectancy at birth for females is 78.6 years and 72 years for males. Sri Lanka has a literacy rate of 96.9% for men and 94.6% for women, which compares very favorably with other countries in the region (Department of Census & Statistics Ministry of Policy Planning and Economic Affairs 2012).

Rural Society

Definition: Rural population refers to people living in rural areas as defined by national statistical offices. It is calculated as the difference between total population and urban population. The significant differences in rural and urban settings all over the world are equally applicable to Sri Lanka.

Urban Rural Disparity

Wide gaps have been identified between urban and rural populations in Sri Lanka with respect to income, savings, expenditure, level of education, language literacy, computer knowledge, etc. However, the health situation of the people in both sectors is reported to be at a satisfactory level (Anulawathie Menike 2015).

Health System in Sri Lanka

The Sri Lankan health system comprises different systems of medicine: Western, Ayurvedic, Unani, Sidha, Homeopathy, and Acupuncture. Among these, Western or Allopathic medicine is the main sector catering to the health needs of the population. Allopathic medical care is delivered through government and private providers. The government health care delivery system has two main arms: Curative care provided through the hospital system and Preventive care/Community health care delivered through primary health care teams under smaller community health service units called MOH (Medical Officer of Health) areas.

The Ministry of Health is the leading government agency providing stewardship to health service development and delivery in Sri Lanka. It also directly manages several large hospitals and specialized services including Provincial General Hospitals and selected District General Hospitals, while a decentralized system, i.e., nine provincial health authorities, manages the remainder of public services in the allopathic system. The Provincial Directorate of Health Service under the respective Provincial Council and regional Directorates of Health Services in each District administer the smaller curative care delivery units (some base hospitals, all divisional hospitals, all primary health care units as well as the community health service system through MOHs) (Ministry of Health, Sri Lanka 2017).

In 2015, government health expenditure as a percent of GNP Government was 1.66%. Health expenditure as a percent of total government expenditure was 5.65% (Ministry of Health, Sri Lanka 2017).

Health Indices

Sri Lankan health indices are comparable with that of resource rich countries. The crude death rate is 6.3 per 1000 population; maternal mortality rate is 33.7 per 10,000 live births; and the infant mortality rate is 16.0 per 1000 live births (Ministry of Health, Sri Lanka 2017). In the WHO's World Health Report GPE Discussion Paper Series: No. 30, 2012, Sri Lanka ranked 76th in overall performance in health compared to India at 112 and China at 144 (Tandon et al. 2002) (Tables 1–3).

Table 1 Total health workforce in Sri Lanka, 2015 (Ministry of Health, Sri Lanka 2017)

Category of healthcare professionals	Total
Medical officers	18,243
Registered/assistant medical officers	936
Nursing officers	42,420
Public health inspectors	1604
Public health midwives	6041
Hospital midwives	2765

Table 2 Mental health workforce in Sri Lanka, 2015 (Ministry of Health, Sri Lanka 2017; World Health Organization 2015)

Category of mental health professionals	Per 100,000 population
Psychiatrists	0.29
Medical officers not specialized in psychiatry	0.75
Psychologists	0.09
Psychiatric nurses	12.92
Psychiatric social workers	0.33
Occupational therapists	0.19

Table 3 Mental health workforce in Sri Lanka, 2017 (World Health Organization 2018)

Human resources	2017
Consultant psychiatrists	81
Consultant child psychiatrists	5
Consultant forensic psychiatrists	2
Medical officers mental health – focal point	20
Medical officers mental health (MOMH)	153
Medical officer mental health – diploma holder	59
Nurse (psychiatry)	192
Psychiatric social worker	53
Occupational therapist	35
Community psychiatry nurses	23
Speech therapists	8

Rural Health in Sri Lanka: The Divide

Distribution of facilities and human resources is unequal and specialized services are relatively less in rural areas. Access to health care facilities is more difficult due to greater distances and limited transport services (e.g., Monaragala, Nuwara Eliya districts). Poverty is a major problem and education levels are relatively lower in rural areas.

Rural Mental Health in Sri Lanka

It is mainly a hospital based system and there are no established community mental health services. The majority of the patients are seen by medical officers of Out-Patient Departments (OPD) who do not have specialized mental health training. Medical officers of mental health (MOMH) provide the service in some hospitals. In some areas, service has been decentralized to out-reach clinics, which the psychiatrists usually visit on a monthly basis.

Therefore, as a preamble to discussing the challenges of research in rural mental health in Sri Lanka, we will first present challenges faced by service providers in establishing and providing rural mental health services in Sri Lanka. We will present two narratives from the two co-authors from their clinical experience in two predominantly rural districts, one from the Eastern province – Ampara and the other from the Uva province – Monaragala.

Rural Mental Health in Sri Lanka: Experience in Ampara District in Eastern Province

Geographical Characteristics of the Ampara District

The Ampara District in the Eastern Province of Sri Lanka is a rural area in the dry zone of the country, with a surface area of 4415 km^2. It is a region mainly involved in agriculture, known as "the rice bowl of the country," and is the district with the largest population of rice-paddy cultivating farmers in Sri Lanka. The Ampara district has a significant number of water tanks, reservoirs and water canals, taking up an area of 193 km^2; a large area is covered with forests and wild life parks (Department of Census & Statistics Ministry of Policy Planning and Economic Affairs 2012).

Demographical Characteristics of the Ampara District

Ampara has a multiethnic and multicultural population of about 677,000, 3.2% of the Sri Lankan population, according to 2015 data from the Registrar General's Department. Population density is relatively low, 160 persons/km^2. Ethnicity wise, the majority are Sri Lankan Muslims (43.58%) followed by Sinhalese (38.73%) and Sri Lankan Tamils (17.39%) (Department of Census & Statistics Ministry of Policy Planning and Economic Affairs 2012). In line with national figures, 76.3% of the population in Ampara District live rurally.

Population Characteristics of Ampara RDHS Area

Unlike other districts of Sri Lanka where the health district is the same as the administrative district, Ampara District has two health administrative areas (Regional Directorates of Health Services – RDHS): Ampara and Kalmunai RDHS areas. Populations in the two RDHS areas differ significantly in ethnicity. While Kalmunai RDHS area follows the pattern similar to the district data, Ampara RDHS area has a majority of Sinhalese (99.08%), followed by a small minority of Sri Lankan Muslims and Sri Lankan Tamils. With the exception of the Ampara town, more than 90% of this region is a rural area (Department of Census & Statistics Ministry of Policy Planning and Economic Affairs 2012).

Government Health Administrative System in the Ampara RHDS Area

Similar to other districts of the country, the District General Hospital – Ampara (DGH Ampara), the only tertiary health care center in the Ampara District is administered by the Central government (Ministry of Health), whereas all other smaller local hospitals (Base and Divisional Hospitals) and preventive care services are managed by the Provincial Ministry of Health of the Eastern Province. These divisions in health administrative systems do bring significant challenges to working in rural areas.

Rural Mental Health Service Providers in Ampara RDHS Area

The co-author Wijesundara, the consultant psychiatrist at the DGH – Ampara since 2014, has been serving in Mental Health unit (MHU) of the hospital as well as engaging in community activities under RDHS – Ampara. According to her experience, rural mental health services are slowly developing in Sri Lanka with significant challenges with regard to sustainability and development. Rural mental health services are almost completely provided through the hospital (curative care) system in Sri Lanka, whereas community mental health services are virtually nonexistent.

At the Ampara District General Hospital, the consultant psychiatrist leads the mental health services, with a multidisciplinary team (MDT) including medical officers of mental health (MOMH), nursing officers, psychiatric social workers (PSWs), occupational therapists (OT), speech and language therapists (SLT), and assistant health staff members. This mental health team is mainly assigned for duties at DGH Ampara, but it caters to the entire population of around 268,000 people in the Ampara RDHS area.

In addition to the MDT team in the hospital, at the Ampara RDHS office, a medical officer (MOMH) works as the focal point to coordinate peripheral outreach curative services through clinics in other hospitals in the area, as well as to organize community mental health programs. A psycho-social worker is also available at the RDHS office to support these activities. In addition, one MOMH is available at the Base Hospital in Dehiattakandiya (a rural area), which is geographically isolated from the rest of the District and is farthest from the Ampara town.

There is no doubt that a mental health team is of benefit to the Ampara district; however, most staff categories in the team have limited specialist training in psychiatry. Some team members are not engaged full-time in mental health services due to other duties in the General Hospital setup. In addition, there is frequent staff turnover within the team due to internal and external transfers, which in turn leads to many issues in service delivery.

Type of Services Provided by Rural Mental Health Service in Ampara RDHS Area

System of Rural Mental Health Service: Human Resources and Types of Services

Mental Health Services provided by the General hospital psychiatric unit by MDT include inpatient psychiatric treatment, adult outpatient psychiatric clinics, child and adolescent outpatient clinic, consultation liaison psychiatric service, forensic psychiatric services, and perinatal psychiatric services. These services are offered at the mental Health Unit of DGH Ampara and cater to all hospitals in the Ampara RDHS area as well as adjacent areas of the Kalmunai RDHS area. Unlike in developed country settings, all services, encompassing the entire age range and all subspecialty services, are provided by a single mental health team to the entire population in the Ampara RDHS catchment area.

In addition to this centralized service at the main hospital, nine peripheral outreach clinics are held at the smaller hospitals and central dispensaries covering all the seven MOH (Medical Officer of Mental Health) areas under the Ampara Regional Directorate (RDHS).

Process of Service Development

Most improvements and service development have been collective efforts by the team. They used the "Monthly Quality Circle" of MHU and other staff meetings to discuss issues and to analyze the existing situation. Lack of data (records of past development activities, audits) and evidence from research on similar rural mental health services were barriers to finding suitable solutions. Therefore, the team developed ways to improve documentation and record keeping, as well as to assess the ground situation through various other methods before finalizing solutions for most issues. They worked closely with local health administrators, i.e., the Director of hospital and Regional Director of Health Services – Ampara as well as the central mental health administrators of the Mental Health Unit of the Ministry of health, to get their support to implement the planned activities.

Recent Development Projects and Challenges in Rural Mental Health Service of Ampara District, Sri Lanka

Hospital-Based Services

The basic components of the psychiatric unit including infrastructure (i.e., a well-designed building with an inpatient unit and a clinic complex) and human resources existed. Therefore, improving existing services, quantitatively and qualitatively, to provide an efficient service was the main objective. Limitations of skills, knowledge and negative attitudes of the team members, lack of support from health administrators, stigma regarding mental illness, and mental health services were significant challenges in developing hospital based services.

Improvements in Main Services Provided

(a) **In-patient unit** – The psychiatric in-patient unit is an open ward with male and female sides (Bed strength 10 on each side) on either side of the common nursing station. As the nursing station was a common open reception area with no clear separation or safety area, multiple issues were present.
   - Significant effort was put in to improve the safety and comfort level of patients in the ward setting. Steps taken include
     - Increasing bed strength (20–30)
     - Separation of the male and female sides
     - Identifying and blocking possible escaping points
     - Increasing the number of low-level and bar beds
   - Improving the safety of the environment of the reception, ward patient area, and seclusion rooms to reduce risk of harm to patients, staff, and property, e.g., removing glass cupboards and doors and removing possible hanging points.
   - Improving documentation of medical officers and nurses.
   - Re-establishing the direct telephone helpline to the Mental Health Unit.

- Establishing a computer laboratory with photocopy facilities, using a government donation to develop as a vocational training and income generating unit for patients and their family members.

(b) **Adult Clinics** – Over 1900 registered patients are followed up at adult clinics at the psychiatric unit of DGH Ampara. Clinics were held 3 days per week during the morning session. In order to improve service delivery efficiency to clinic patients, various strategies were implemented after experimenting. These include:

- Increasing staff
- Distributing the patient reviews equally across clinic dates
- Reducing door-to-doctor time (avoid delays in registration and retrieving records)
- Improving metabolic screening and management of physical comorbidities

(c) **Consultation Liaison service and forensic psychiatric service** – Initiation of a record system for documentation, report writing, and storage has been instrumental in improving care in this group of patients and persons. It has improved follow-up care and on subsequent contacts even after a significant time gap.

Establishing Systems and Protocols

Lack of consistent systems and protocols were a significant challenge to sustaining the quality improvement following implementation of new steps. In addition, lack of agreement among different categories of staff and lack of commitment to continue implementation were challenges within the team. Therefore, collective efforts were taken to develop consensus.

(a) **Ward protocols and guidelines** – Establishing guidelines and protocols specifically for the functioning of MHU such as admission and discharge protocols, managing a high-risk patient, etc. A booklet is available for reference for staff members.

(b) **Readiness for medical emergencies** – Following analysis of incidents, steps were taken to improve readiness to manage any medical emergencies including,

- Getting equipment for resuscitation and monitoring (defibrillator, oxygen saturation monitor, etc.)
- CPR training
- Emergency trolley readiness
- Daily checking of protocol

(c) **Orientation booklet for medical officers** – A document with basic guidelines for documentation and treatment in the MHU has been prepared to orient medical officers to clinical work, as changes of medical officers are frequent.

Service Expansion

(a) **Child mental health clinic**

After significant efforts made over 2014 and 2015, a fortnightly child mental health clinic (Saturday morning session) was opened in December 2015. The multi-disciplinary team has conducted 37 new-patient assessments and 28 periodic reviews in 2016. In 2017, 50 new-patient assessments and 69 periodic reviews have been performed in this fortnightly clinic.

(b) **Opening of electroconvulsive therapy (ECT) suite**

ECT was given in the surgical theatre previously leading to much hardship and stigma to patients and staff over the years. After repeated negotiations with the hospital administration since 2014, the team finally managed to furnish and open the ECT suite in January 2017.

(c) **Computerized patient database and patient reminder system**

A computerized database of all follow-up and new patients attending all the clinics was created in 2016. This has significantly improved retrieval of old clinic records and contact details of defaulters. In addition, it helps to summarize data for monthly returns and for audits. PSWs regularly update data to monitor clinic attendance and take remedial action to remind defaulters; some of them are given treatment following home visits.

(d) **Opening multisensory room for sensory modulation therapy**

Sensory integration abnormalities are commonly seen in children with developmental disorders, especially in autistic spectrum disorders. A multisensory room was developed and opened in July 2017, with funds from well-wishers, in order to provide sensory modulation therapy by occupational therapists and speech therapists for sensory abnormalities (e.g., hyper- and hypo- sensitivities of children).

At present, the team is working with another professional organization, the Sri Lanka Association for Child Development (SLACD) to establish an outdoor therapeutic sensory play area for similar children for above purposes, which would be the first of that kind in Sri Lanka.

Multidisciplinary Service Provision; Challenges

As mentioned earlier, most team members had minimal expertise and limited professional qualifications in the mental health field. Therefore, a significant level of continuous supervision and guidance was required from the consultant psychiatrist to develop new initiatives. In addition, high staff turnover due to transfers out of the unit or from the hospital leading to loss of expertise within the team is a basic barrier most rural mental health services have to face. This occasionally led to low morale in the entire team.

Some other achievements include:

(a) **Improvements in occupational therapy interventions**

Daily timetable and weekly activity groups, making clinic books and crafts for sale, use of horticulture unit and ward kitchen for development of Activities of Daily Living (ADLs) and vocational activities.

(b) **Interventions initiated under Psychiatric Social Workers' responsibility**

- High-risk patient register
- Family meetings and communicating with families of inpatients
- Home visits for social assessments and interventions
- Networking with local stakeholders (such as officials of Education, Social Services, Child protection)

- Developing horticulture unit under advice from the Provincial Agriculture Department
- Outreach medication program for four weekly intramuscular depot injections
- Arranging home visits for medical reviews (for patients who have not attended clinics themselves for more than 6 months)

Team Building and Continuous Education Activities for Staff

Some initiatives were taken to improve the team capacity and team spirit. This included improving communication and cooperation, developing mutual respect, sharing knowledge and learning skills from each other, and supporting continuous professional development programs.

(a) **Weekly educational forum**

A weekly educational forum is held in the unit with team members as resource persons. Other hospital staff members or external resource persons were also intermittently invited to present.

(b) **Team building**

Monthly quality circle meetings have become a space to discuss problems related to staff. Annual events organized by MHU such as the staff trip, running a free food counter one day in June ("Dansela" on the full-moon day in June), have become occasions to build team spirit. All staff categories were supported to attend special training courses in mental health at National Institute of Mental Health (NIMH) and to initiate higher studies such as Diplomas in relevant fields.

(c) **One to one supervision of staff members**

A system of monthly one to one supervision by the consultant is established for Medical officers, OT, and PSWs. This has provided a space to provide individual feedback, clarify doubts, and to discuss issues at a personal level.

Contribution for Academic Activities and Research

(a) **Annual scientific sessions of the clinical society of the hospital**

The Mental Health team carries out audits and contributes case presentations to the local clinical society sessions held annually. Though the level of research activities was minimal, in 2016, a successful session on "Psychiatric Rehabilitation" was conducted by the MDT at the "Pre-congress Workshop on Rehabilitation" for medical and nursing officers. In 2017, two oral presentations on descriptive studies on deliberate self-harm and self-poisoning were presented in the Free Paper session. Lack of interest in research among the Medical Officers of Mental Health (MOMHs) and other staff, lack of time due to heavy clinical workload, and limited knowledge on research methodology were the main hindrances identified. These will be discussed in more detail later in the chapter.

(b) **Clinical training**

The MHU functions as an academic unit providing teaching and clinical training for several categories of professionals including medical officers, undergraduate medical and psychology students. Although there is more than enough clinical material, the main hurdle is the medical team not having enough time to allocate for teaching.

Community-Based Services

Reviving a Community Organization with Special Interest in Mental Health ("Foundation for Promotion of Mental Health")

Engaging the community for mental health promotional activities is of paramount importance in rural settings. Lack of awareness regarding mental health and mental illness as well as limited knowledge regarding available mental health services and stigma are the significant issues preventing the community from reaching out for mental health services. The Foundation for Promotion of Mental Health was started in 2009 under a previous consultant psychiatrist, but was inactive since 2012. Since reviving this community organization in 2014, it has become a significant supportive force for all mental health promotional activities in the area.

Developing a Mental Health Volunteer Group

As a solution to the lack of resource persons to carry out mental health promotional activities, the mental health team, together with community organizations, decided to train a selected group of volunteers as a Mental Health workforce. This was organized as a training course over a period of 1 year including six (World Health Organization 2015) training workshops, three (Anulawathie Menike 2015) sessions of clinical training at MHU, individual and group-work assessments. Twenty two (22) successful volunteers are now working with us in the community. They contribute to awareness programs and are guided to carry out health promotional activities in their own communities/workplaces under guidance from the mental health team. Monthly meetings are used to continue this process. The team has identified a second group of volunteers to follow the first group as supporters and to learn from them.

Increasing and Improving Outreach Clinics

Increasing outreach clinics and thereby improving accessibility to services is a priority in order to counteract the current unavailability of community mental health services. A new monthly Mental Health clinic was started at another Divisional Hospital in May 2016, which has accomplished the task of covering all MOH areas in the Ampara RDHS area. Improving availability of medications (such as clozapine, lithium, and methylphenidate), record keeping, and follow-up of defaulters with home visits are some other measures taken to improve the quality of service provided to patients in very rural areas with minimal basic facilities.

Public Awareness Programs, Exhibitions and Workshops

Programs were organized to celebrate the World Mental Health Day every year since 2014. In 2016 and 2017 the team organized public exhibitions, which had high attendance. In addition, several programs on Child Mental Health have been organized with support from Consultant Child and Adolescent Psychiatrists and organizations with special interest in children such as the Sri Lanka Association for Child

Development (SLACD), to improve awareness in categories such as preschool, school, and primary school teachers, special education teachers and parents of children with special needs. Staff training sessions and workshops were held in parallel to these programs.

Multiple public education lectures and workshops have been conducted for target populations such as school children, preschool teachers, and professional groups. Lack of financial resources was a challenge the team managed to overcome with support from well-wishers. The lack of competent mental health professionals to engage in these community activities was compensated for by using support from community members and trained Mental Health volunteers.

Working with Other Organizations and Networking

There are a significant number of government officials (attached to central and provincial government authorities) working in fields which address psycho-social issues of individuals in general as well as that of those with mental or physical disability. Lack of adequate understanding and coordination among those service providers in health and other sectors such as education, social services, child and women's affairs have led to poor support for the patients in the community. The team has tried consistently to improve interactions and understanding to give the best possible psycho-social support to patients.

Developing Rehabilitation Facilities in RDHS-Ampara

Lack of psychiatric rehabilitation facilities (medium term and long stay) is a disadvantage in this rural area. Many patients with poor functional levels are losing out due to this. Difficulty in getting financial support and administrative backing as well as the lack of mental health cadre for these units are ongoing difficulties.

- **Reopening a medium stay residential rehabilitation center (Kedella)**
  A 6-bed unit at a small Divisional Hospital about 15 km away from Ampara town, which was previously used as a psychiatric rehabilitation unit but closed down in 2013, was renovated using funds from the central government (Mental Health Directorate of Ministry of Health). Despite delays, it was reopened in 2017.

With regard to future service development projects, the team is currently working to expand services of this rehabilitation center, open a central community day center in the Ampara town and a Drug and Alcohol residential rehabilitation facility in an existing building of another divisional hospital. As no separate staff cadre has been approved for these specialized mental health facilities by the Health ministry, these facilities are planned within existing hospitals or MOH office premises of the area, in order to run them using existing health staff.

It is evident from the above narrative description that extensive efforts have gone into developing clinical services, starting almost from scratch, and that service development can be done with commitment and leadership. In our opinion, two main factors contribute to the lack of research. Firstly, service provision is prioritized over research efforts, and secondly, research has not been identified as a policy priority by the health authorities in Sri Lanka, leading to the lack of an overarching research culture and required human skills.

Now we will compare and contrast it with the second rural district Monaragala, where mental health services development was led by the co-author, Dhanuja Mahesh.

Development of Rural Mental Health Services in Monaragala District, Sri Lanka: Challenges and Achievements

Monaragala is a second largest district of the country, situated in the south east region of the island. Health services are centered around one District General Hospital (DGH) and three Base Hospitals (BH). There are 11 MOH divisions. The population in the district is around 500,000 and the main occupation is farming (Department of Census and Statistics, 2016). Lack of transport facilities, poverty, drugs, and alcohol all contribute to the hardships faced by most inhabitants of the Monaragala district.

A permanent psychiatrist had not worked in the district before 2016, when co-author Dhanuja Mahesh was posted to the DGH Monaragala. He had to undertake a major task of developing services as a priority, with a small team of three medical officers and one nurse.

Hospital Based Development

When Dhanuja was posted to the DGH Monaragala in 2016, the hospital mental health staff consisted of three medical officers, one nurse, and one support staff worker, with only one small clinic room for the entire staff. Since the patient turnover was more than 100 per day, the physical structure and the surroundings were not appropriate to discuss the mental health problems of patients and it was not possible to ensure privacy. He was able to establish a new mental health clinic building with five rooms. Available human resources at the hospital clinic were not sufficient to provide an extended service to patients. Therefore, he increased the number of doctors in the mental health clinic to five by recruiting new doctors from the Ministry of health. After considerable efforts, he was also able to secure the appointments of a Psychiatry Social Worker (PSW) and an Occupational Therapist (OT).

Monaragala had no acute in-patient facilities, intermediate unit nor long-term units for psychiatric patients. Therefore, patients had to travel around 2 h to the nearest in-patient facility, which is not convenient for many of them. With the support of the Director of Mental Health in Sri Lanka, he was able to get approval for the following:

(a) New inpatient facility for DGH Monaragala

This new psychiatry ward will consist of 14 beds each for males and females, three interview rooms, ECT bay, doctors and nurses rest rooms, visitors area, and a garden. Rs. 5.3 million was allocated to initiate the building work and the hospital has received the money after completion of all the administration work.

(b) Intermediate unit at the Base Hospital, Siyambalanduwa

This will consist of 20 beds and will operate under the care of the District Psychiatrist, MOMH, and Physician Siyambalanduwa.

(c) Long term unit at Inginiyagala

This will be under the care of the District Psychiatrist, MOMH and Medical Officer-incharge, Inginiyagala.

(d) Alcohol rehabilitation facility at the District Hospital, Medagama

This will be a 15-bed in-patient facility with separate open garden area and space for indoor activities. This will be under the care of the District Psychiatrist and MOMH, Bibila.

(e) New vehicle for all mental health activities

A ten-seater air-conditioned van was received in December 2016, and it is used for travelling to peripheral clinics, home visits, and awareness programs, etc.

Mental health information services are rarely available in the Sri Lankan Health care system even though it is an important component of the service provision, particularly in a rural area such as Monaragala where there are limited public transport facilities and resources. A dedicated new direct telephone line was established at the mental health clinic – DGH Monaragala. It provides direct access for anyone to obtain information about mental health services available in the Monaragala district.

Patients have to spend time in long queues at the hospital pharmacy and at times miss the only available public transport to return home, with no choice but to sleep in the hospital corridors until the next morning. On one occasion, a father who is the only carer of his three sons with schizophrenia was attacked and killed by an elephant on his way back home in the night. Therefore, a separate medication dispensing counter was initiated in the hospital pharmacy of DGH Monaragala to minimize the difficulties to these patients and to provide medication without delay so that they can get back home before nightfall.

A new day activity unit, well equipped with six computers, photocopy machine, and a scanner, was set up under the supervision of IT unit, DGH Monaragala. A volunteer IT instructor comes in to teach younger recovered patients. This was planned to function as an income generating facility in future.

Community-Based Development

As it is the second largest district, patients in Monaragala had to travel far to get their monthly medication and for follow-up. Consequently, there is large number of dropouts and relapses. Therefore, decentralized services were planned to improve compliance.

1. Peripheral clinic development

Thirteen outreach mental health clinics were initiated in the district (namely, Bibila, Wellawaya, Siyambalanduwa, Medagma, Dambagalla, Ethimale, Buttala, Badalkumbura, Okkampitiya, Thanamalwila, Hambegamuwa, Kataragama, and Sevanagala). The consultant visits these clinics monthly. Initially managed by three MOMHs alone, six MOMHs are now serving in these 13 clinics.

A monthly review meeting was arranged at the RDHS office with all MOMHs, RDHS, Chief Pharmacists and Medical Superintendents, In-charge Nursing officers to discuss all matters related to mental health. Frequent communications were held with the RDHS and Chief Pharmacist to ensure continuous medication supply to the entire district.

2. Counselors

Eight development officers were recruited to work as counselors from the divisional secretariat offices. They work in mental health services at the DGH Monaragala and three Base Hospitals under the supervision of the District Psychiatrist and MOMHs.

3. Home visits and monthly injection program

Home visits and depot injections were arranged for necessary patients. Transport facility and required staff is now available for these home visits. A separate register was maintained for patients receiving the antipsychotic depot medication, to find default patients.

4. All MOH areas allocated to six MOMHs

Six MOMHs attend monthly MOH meetings and school medical inspections. This is to ensure good communication and to create awareness about the availability of mental health services at Monaragala.

5. Referral and follow-up systems for postpartum mental health issues.

A new referral system and a follow-up system were established for mothers who have postpartum mental health issues. In this system, there is direct communication between the MOH and the MOMH.

Staff and Public Education

1. All hospital staff (Medical, nursing, paramedical, administrative, support, and security staff) were sensitized and provided basic education about mental illnesses, available mental health facilities, and managing stress.
2. A clinical meeting was arranged monthly for all MOMHs to discuss problematic and challenging cases.
3. Conducted education and awareness programs for the following groups: school principals and teachers, school counselors, police department, armed forces, child probation service, SOS village Pharmacists in community pharmacies, school children, and government officers.
4. All the MOMHs were trained to continue the awareness programs in their attached hospitals, MOH offices, other government offices, schools, and public. This will provide a widespread and sustainable service in the area.

Other Special Activities/Services Provided

1. Prison Mental Health Service

A new prison mental health clinic was started at the Monaragala prison. This was conducted every Tuesday evening to provide court reports and to treat prisoners with mental health problems. This helps to enhance the mental wellbeing of prisoners.

2. Alcohol prevention program

Alcohol addiction is a severe problem in the district and most families suffer due to this. The alcohol prevention program was initiated with the setting up of the "District Committee for Alcohol Prevention" in collaboration with the RDHS, government administrative officers, police officers, and directors of the Education department. Eleven subcommittees were then formed in each MOH area, comprising the divisional secretariat, MOH, MOMH, OIC (officer in charge) of the police and education directors of the area.

3. Safe house

Sexual abuse of women and children is another serious problem in the area. Assuring the safety of the victims until legal procedures are completed is a challenge. With the support of the secretary of the Ministry of Women Affairs, a three-story building (safe house) with separate interview areas and a day center was approved for abused children and women. The building construction commenced in 2017.

Challenges Faced When Working as a Clinician in Rural Sri Lanka

Working in a remote area of rural Sri Lanka is full of challenges and requires personal sacrifices.

1. Away from the family

Residing far away from the clinician's home town and living without family, relatives and friends is stressful. In some areas there are no mobile connectivity or internet facilities. Monaragala is in the dry zone and for part of the year there is severe drought and it is hard to find even drinking water. To work in a rural area, a person needs to be flexible, tolerant, and able to adjust to the environment.

2. Lack of resources

It is very difficult to improve mental health services without adequate human and physical resources. There are many demands on the allocated health budget and therefore, persistence, frequent communication, and the ability to convince relevant authorities are required to obtain funding for service development plans.

3. Natural threats

Travelling in the evening is strictly prohibited due to the attacks of wild elephants and snakes. Vigilance is necessary at all times.

4. Resistance to change

Both the public and some staff members can be resistant to change due to many different factors.

From this narrative description too, it is once again evident that an extensive amount of effort has gone into developing clinical services, again starting almost from scratch. It is also evident that the leadership and commitment can lead to the development of clinical services, despite significant challenges.

Challenges Faced in Conducting Mental Health Research in Sri Lanka

Understandably, there is no systematic research carried out on the issue of the mental health research gap in Sri Lanka, and the description below is based on personal reflections and opinion. However, later in the chapter, we will discuss it in relation to the overall research gap in resource poor settings.

Reflections

Reflections of the Co-author Dhanuja Mahesh

As a postgraduate trainee, especially while working in the University Psychiatry Unit of the National Hospital of Sri Lanka, the co-author was enthusiastic in carrying out research under the supervision of seniors, however, that enthusiasm has started to vane since starting to work in the peripheral areas, due to many reasons.

Firstly, the heavy clinical work load, at the hospital as well as at outreach clinics (which involves a significant time spent travelling), community awareness programs, managerial tasks and responsibilities (planning and organizing), and service developmental initiatives are exhausting. Therefore, research slowly became a lower priority and settled at the bottom of the "to do list."

Furthermore, a strong research culture and external motivation was lacking within the health system until recently, especially in the peripheries. Although there is a clinical society which holds annual free paper sessions as part of scientific sessions to promote research, it is difficult to overcome the problems of poor motivation, inadequate knowledge, and limited support from juniors. In addition, seniors with expertise to support research or supervision are scarce. Poor knowledge of statistics is also a barrier in designing studies and data analysis.

Why Don't We Do Research?

The third author's reflections are as below.

1. Heavy work load – most clinics are overcrowded and sometimes the clinic duration extends 2–3 h over the scheduled time, leaving no quality time for research.
2. Less motivated to do research as it is not promoted as a part of the duties.
3. Limited facilities including human resources to conduct research.
4. No dedicated funding support.
5. Participants have a negative attitude towards research and have limited awareness about the importance of research.

The Gaps in Clinical Services Have an Impact on Research Being Difficult

The major issue is lack of resources, both human and material, including funding. Some districts do not have a psychiatrist or a proper follow up system to date. Many hospitals have no trained medical officers of mental health, nor trained nurses to provide clinical services, let alone staff for carrying out research.

Health Research in Sri Lanka

Health research in Sri Lanka can be discussed in the context of the global divide in research.

Less than 10% of research funds are spent on the diseases that account for 90% of the global disease burden. Though 93% of the world's burden of preventable mortality occurs in developing countries, too little research funding is dedicated to health problems in developing countries (Global Forum for Health Research 2000). There is also a significant publication divide in medical as well as in mental health research (Sumathipala et al. 2004; Patel and Sumathipala 2001), which could be indirect evidence of poor research capacity, whatever the reasons may be.

Mapping mental health research capacity and resources in LMICs has provided evidence for scarcity and unequal distribution of mental health research capacity in LMICs (Razzouk et al. 2010). The global survey mapped research and researchers from 114 LMIC countries around the world and found that researchers and publications were concentrated in 10% of the countries surveyed. The paper, based on a project by the Global Forum for Health Research, found among other things that Asian researchers were more likely to be based in private institutions than their colleagues from other continents. This finding poses the question whether the situation demands research infrastructure beyond traditional and conventional institutions such as universities. The paper further concludes that low publication rates from LMICs are due to a lack of human resources and cites access to journals and databases, research fellowships, and funding as main resources lacking in LMICs (Razzouk et al. 2010).

In a survey carried out by Sumathipala et al. (2004; Patel and Sumathipala 2001), the average contribution of countries outside Europe, North America, and Australia (Rest of the World, RoW) to overall research literature in five high-impact medical journals was 6.5%. An analysis of the authorship of 151 articles from RoW showed that 104 (68.9%) had co-authors from Europe, North America, or Australia. This study adds to the previous study (Patel and Sumathipala 2001), which revealed that over 90% of psychiatric research articles published in six leading journals originated from Europe and North America.

Being a Researcher from a Developing Country: Challenges, Barriers, and Rewards

Despite being a developing country in comparison to many others, Sri Lanka is well placed with regard to basic indices such as education and health. However, its research output remains among the lowest in the region. It is the lack of a research culture, inadequacies in research capabilities, the absence of a multidisciplinary approach to research, and the absence of incentives for research which prove to be major challenges to the advancement of research. However as the number of international collaborative projects carried out have increased, this at times has created new ethical challenges.

Case studies from our experience of research in Sri Lanka are presented here. They highlight hidden challenges such as bureaucratic hindrances, challenges posed by the new initiatives, resistance to change, unwillingness to learn, and professional politics. They also emphasize the lack of a critical mass of researchers and supervisors, inadequate funding and capacity in research methodology, lack of time earmarked for research, and other ethical issues in research practices.

In the case that these obstacles can be surmounted, gross under-representation of biomedical literature from developing countries in international journals contributes to poor dissemination of the work of developing-country researchers. Rewards for researchers in Sri Lanka are almost negligible. A mere increase in resource allocation or capacity building will not change the situation. What is needed is a radical reform of the research agenda of the scientific community.

The aim is to present insight into the spectrum of issues faced, rather than to quantify them or generate a consensus.

Research Culture in Sri Lanka

In 1995, the *Annual Health Bulletin* published by the Ministry of Health acknowledged the lack of a research culture. Inadequacies in research capabilities, absence of adequate incentives for research, and near absence of a multidisciplinary and intersectoral approach to research were identified as problems. However, significant remedial measures have been undertaken over the past 20 years.

Capacity building is essential to bridge the gap between the standards of health service provision and the research capabilities in Sri Lanka.

Sri Lanka invests only 0.19% of its GDP in research and development (R&D), compared to a world average of 1.4%, India at 0.9%, and developed countries at about 2% (The World Bank 2018). This has gone down to 0.1008% in 2013. However, the number of papers published in indexed journals (ISI citations) increased from 87 in 1997 to 561 in 2015 (National research Council Sri Lanka 2018). Of the 187 papers cited in 2001, 81 (43%) were related to the medical field, of which only five were on mental health. At present the limited health research in Sri Lanka is mostly carried out by academics attached to government universities and a few independent research organizations. There are postgraduate degrees with integrated research components in MSc and MD, but unfortunately very few academics present their findings at the international level. The research output arising through Health Ministry employees was limited until 2017 due to the lack of an overarching research culture within service provision, unlike in developed countries. In December 2017, the Health Ministry organized a national health research symposium where nearly 400 research projects were presented. This is a landmark in the improvement of health research, and popularizing research dissemination among health care workers in Sri Lanka. Numerous steps have been taken in the right direction, including providing incentives for those who conduct research.

In 2011, the Sri Lankan parliament budget proposals took a bold step to promote research by deciding to award a monthly research allowance of 25% of the basic salary to university academics and senior level officers engaged in research in public sector (Ministry of Higher Education 2011).

Experiences of Other Sri Lankan Researchers

The responses received from a small survey are categorized in Boxes 1–4.

Box 1 Hidden Challenges and Barriers

- Bureaucratic hindrance from people in higher managerial positions
- Envy against people who have successfully faced the challenges and barriers to research
- Authoritarianism and lack of flexibility expecting unquestioned conformity
- Rigid hierarchical system within academia
- Resistance to change and unwillingness to learn from new developments
- "Learned helplessness" – Demoralization secondary to various obstacles
- Research findings are not taken seriously for service development
- No protected time for research, against heavy clinical and teaching commitments.
- Publication hurdles, from drafts to publishing
- Some hospital administrators believe that priority should be given to clinical care and that there is "no role for research"
- Clinical care is perceived by doctors themselves as the only duty of a doctor
- Lack of incentives to promote research
- Lack of public engagement and understanding of the role of research in service development

Box 2 Challenges and Barriers: Research Supervision

- Inadequate critical mass of supervisors, particularly in mental-health-related research
- Inadequate opportunities for supervisors to keep abreast of the pace of new developments in their field
- Inadequate capacity in research methodology and statistics
- No mechanism to attract expatriate or foreign experts
- No concept of mentorship

Box 3 Inadequate Resources

Funding
Inadequate local funding, lack of information about existing external funding opportunities and inadequate capacity to compete for such funding
Equipment
Inadequate computers and information technology facilities, and laboratory equipment
Literature
Inadequate facilities to search for and acquire the relevant literature scarcity or difficulties in accessing journals and books

Box 4 Rewards for Research

- Self-satisfaction
- Sharing findings with interested academics
- Publication of results
- Helps to obtain promotions in academic jobs
- Recognition of work nationally and internationally
- Professional travel abroad
- Awards and material rewards (by National Research Council)

Mental Health Research in Sri Lanka

Due to the reasons described above and reflecting the global divide, health research is still a developing area, not happening on a larger scale. However, there is a growing trend in conducting health research, supported by the Ministry of health, local, and foreign universities and institutions. Understandably, the challenges of research in rural mental health in Sri Lanka are an invariable reflection of the lack of an overarching research culture, financial constraints, limited resources, high clinical work load, negative attitude in the health system.

However, in this backdrop, the first author founded and continues to provide strategic leadership to a research infrastructure; the Institute for Research and Development (IRD) which is now a sustainable research organization independent of government, industry, and universities. This infrastructure now complements its efforts by promoting capacity building through a cadre of high-caliber junior academics and clinicians.

The IRD is a not-for-profit research institution established in 2000 with full-time research staff, and many associates involved at various levels while employed in other academic or healthcare institutions in SL. This initiative was heavily supported by the Institute of Psychiatry, Kings College, London, UK, and there is an ongoing strong established partnership. It is also now supported by the Research Institute of Primary care at the University of Keele, UK. The IRD/Keele partnership is growing stronger, and Global health research is one of Keele University's strategic priorities.

The IRD has been a vehicle for developing a series of large-scale local and international research projects, particularly in mental health. To name a few out of a significant volume of mental health research done by the IRD: the impact of maternal migration on child and adolescents' health (Wickramage et al. 2015) vulnerability and resilience of displaced population (Siriwardhana et al. 2015a) and the impact of disasters and trauma on mental health (Siriwardhana et al. 2013). The Sri Lankan Ministry of Health (MoHSL) commissioned the IRD to conduct the first ever SL national mental health survey in 2007 ($n = 6000$ community survey and 4000 school survey) and another island wide survey (2011) on the impact of spouse migration on families left behind (Siriwardhana et al. 2015b), which contributed to the national migration policy in 2013. The TEENS Study, on the prevalence of Common Mental

Disorders and substance use among adolescents (including a qualitative component) in postconflict Vavuniya district (Northern SL), is underway.

Other pioneering work at IRD: two RCTs of Cognitive Behaviour Therapy for medically unexplained symptoms (MUS), which were identified in the Lancet Series on Global Mental Health (Prince et al. 2007) as among a handful of complex intervention trials from LMICs. The Inter-Agency Standing Committee guidelines incorporated this work as a front-line postdisaster intervention. The WHO-mhGAP training module on MUS is based on this work (World Health Organization 2010), and a pilot relevant to mhGAP was done among displaced and conflict-affected populations in SL. The WHO Colombo office commissioned the IRD to develop and evaluate a training program for primary care doctors in the Eastern province of Sri Lanka on identification, treatment and referral of epilepsy, psychosis, severe depression, Medically Unexplained Symptoms, and heavy alcohol use.

For the national mental health survey, a cross-sectional household survey was conducted to estimate the prevalence of mental health disorders. A multistage, cluster sampling design was used to collect data from 17 out of 25 administrative districts, 8 districts including the Ampara district were inaccessible due to escalated civil conflict during the study period. A structured interview including Patient Health Questionnaire, CIDI section K on posttraumatic stress disorder and war and Tsunami exposure questionnaires were used to ascertain mental disorder outcomes and socio-demographic correlates.

A total of 6120 participants were recruited. The study sample was representative of national figures for age and ethnicity, but females were over-represented and results were weighted for gender and district population variations. The overall prevalence of any mental disorder was 11. 7% (95% CI 10.7–12.8). Depressive disorders including major depression 2.4% (1.9–3.0) and other depression 6.7% (5.9–7.5) were prominent, followed by somatoform disorder 2.9% (2.4–3.4), PTSD 1.9% (1.5–2.5), and anxiety disorders 1.8% (1.3–2.2). Socio demographic correlations included older age, female gender, widowed, divorced, or separated marital status and a lower educational level. Puttalam district in North Western Province had the highest prevalence for depression, somatization, and anxiety, but significantly PTSD was of low prevalence, at the second lowest in the country. However, in Anuradhapura, Polonnaruwa, and Kurunegala districts (all of which were border conflict areas), PTSD prevalence remained high. The lowest overall CMD prevalence was reported from the Matara district in the Southern province. Significantly, this was a district severely affected by 2004 Tsunami and two leftist insurrections in 1971 and 1989.

There were variations in the prevalence of different categories of mental illness between districts. Major depression was highest in the Ratnapura district (4.0%, 95% CI 1.9–5.4), along with Anuradhapura (3.8%, 95%CI 2.2–6.6), Kalutara (3.3%, 95% CI 1.8–6.1), Puttalam (3.3%, 95%CI 1.9–5.4), and Gampaha (3.3%, 95%CI 1.7–6.3).

Overall prevalence of the category of other depression minor depressive, bipolar disorders, dysthymia, is 7.1% (95% CI 6.5–7.7). **Monaragala has the highest prevalence of 15.1% (95% CI 11.4–18.8),** followed by Puttalam 11.9% (95% CI 8.6–15.2) and Anuradhapura 10.4% (95% CI 7.2–13.6).

Somatoform disorder prevalence was highest in the Puttalam district (10.0%, 95%CI 7.0–14.1) followed by the Badulla district (7.3%, 95%CI 0.4–10.8). Anxiety disorder prevalence was highest in the Kalutara district (3.1%, 95%CI 1.5–6.5) followed by the Kurunegala district (2.7%, 95%CI 1.3–5.4). PTSD prevalence was highest in the Polonnaruwa district (5.2%, 95%CI 2.7–9.9) followed by the Anuradhapura district (4.2%, 95%CI 2.2–7.7).

Ingredients for Successful Research in Developing Countries

Will a good methodologist with a sound research question and a protocol, along with generous funding, succeed in completing a valuable piece of research in a developing country? Our experience suggests no. There are many other factors that contribute to and play a decisive role: learning to lead and to follow, propensity for teamwork and establishing networks, negotiating skills, managing difficult people, and supervising and training of others. North-South and South-South collaborations are also necessary to produce high quality research that can influence national policy and practice. Expatriates with active links in both developed and developing countries can play a significant role. If these ingredients are available, it will pave the way to producing high quality research in developing countries. However, providing reasonable opportunities for publication of research results and increasing the proportion of articles in journals on developing-country biomedical research are also critical. An increase in resource allocation or capacity building alone will not improve research capacity in a developing country. A radical reform of the agendas of the scientific community and of policy-makers must take place to create an environment conducive to research.

References

Anulawathie Menike HR (2015) Rural-urban disparity in Sri Lanka. IPASJ Int J Manag 3(3):1–12

Department of Census and Statistics (2016) Mid-year population estimates by district & sex, 2012–2016 [Internet], vol 2016. Population and Housing, Department of Census and Statistics, Colombo. Available from: http://www.statistics.gov.lk/PopHouSat/VitalStatistics/MidYearPopulation/Mid-year population by district.pdf

Department of Census and Statistics, Ministry of Policy Planning and Economic Affairs. Census of Population and Housing 2012

Global Forum for Health Research (2000) The 10/90 report on health research, 2000 [Internet]. Global Forum for Health Research, Geneva, 155 p. [cited 26 Dec 2017]. Available from: http://www.worldcat.org/title/1090-report-on-health-research-2000/oclc/44575649

Ministry of Health Sri Lanka (2017) Annual health bulletin 2015. Medical Statistics Unit, Ministry of Health, Nutrition and Indigenous Medicine, Colombo

Ministry of Higher Education (2011) Payment of research allowance in terms of Budget Proposals, 2011. MHE

National Research Council, Sri Lanka (2018) Statistics website [Internet]. [cited 5 Jun 2018]. Available from: http://www.nrc.gov.lk/index.php/statistics.html

Papiha SS, Mastana SS, Jayasekara R (1996) Genetic variation in Sri Lanka. Hum Biol 68(5):707–737

Patel V, Sumathipala A (2001) International representation in psychiatric literature: survey of six leading journals. Br J Psychiatry 178(May):406–409

Prince M, Patel V, Saxena S, Maj M, Maselko J, Phillips MR et al (2007) No health without mental health. Lancet 370:859–877

Razzouk D, Sharan P, Gallo C, Gureje O, Lamberte EE, de Jesus Mari J et al (2010) Scarcity and inequity of mental health research resources in low-and-middle income countries: a global survey. Health Policy [Internet] 94(3):211–220. [cited 5 Jun 2018]. Available from: http://www.ncbi.nlm.nih.gov/pubmed/19846235

Siriwardhana C, Pannala G, Siribaddana S, Sumathipala A, Stewart R (2013) Impact of exposure to conflict, tsunami and mental disorders on school absenteeism: findings from a national sample of Sri Lankan children aged 12–17 years. BMC Public Health [Internet] 13(1):560. Available from: https://www.ncbi.nlm.nih.gov/pmc/articles/PMC3698150/

Siriwardhana C, Abas M, Siribaddana S, Sumathipala A, Stewart R (2015a) Dynamics of resilience in forced migration: a 1-year follow-up study of longitudinal associations with mental health in a conflict-affected, ethnic Muslim population. BMJ Open 5(2):e006000

Siriwardhana C, Wickramage K, Jayaweera K, Adikari A, Weerawarna S, Van Bortel T et al (2015b) Impact of economic labour migration: a qualitative exploration of left-behind family member perspectives in Sri Lanka. J Immigr Minor Health 17(3):885–894

Sumathipala A, Siribaddana S, Patel V (2004) Under-representation of developing countries in the research literature: ethical issues arising from a survey of five leading medical journals. BMC Med Ethics [Internet] 5(1):5. [cited 3 Aug 2017]. Available from: http://www.biomedcentral.com/content/pdf/1472-6939-5-5.pdf

Tandon A, Murray CJ, Lauer JA, Evans DB (2002) Measuring health system performance for 191 countries. Eur J Health Econ [Internet] 3(3):145–148. Available from: http://www.pubmedcentral.nih.gov/articlerender.fcgi?artid=3605425&tool=pmcentrez&rendertype=abstract

The World Bank (2015) Research and development expenditure (% of GDP) | Data [Internet]. [cited 7 Jun 2018]. Available from: https://data.worldbank.org/indicator/GB.XPD.RSDV.GD.ZS?locations=LK

Wickramage K, Siriwardhana C, Vidanapathirana P, Weerawarna S, Jayasekara B, Pannala G et al (2015) Risk of mental health and nutritional problems for left-behind children of international labor migrants. BMC Psychiatry [Internet] 15(39):39. [cited 18 Mar 2015]. Available from: http://www.biomedcentral.com/1471-244X/15/39

World Health Organization (2010) mhGAP Intervention Guide for mental, neurological and substance use disorders in non-specialized health settings: mental health Gap Action Programme (mhGAP). mhGAP Interv Guid Ment Neurol Subst Use Disord Non-Specialized Heal Settings Ment Heal Gap. Action Program [Internet]:1–121. Available from: http://www.ncbi.nlm.nih.gov/pubmed/23741783

World Health Organization (2015) Mental health atlas 2014. Bull World Health Organ 93(8):516

World Health Organization. GHO | by country | Sri Lanka statistics summary (2002- present). WHO [internet]. [cited 2018 Jun 7]. Available from: http://apps.who.int/gho/data/node.country.country-LKA

Advantages and Disadvantages in the Management of Mental Illnesses in Rural Areas

13

Paola Tejada

Contents

Abstract

Inhabitants of rural areas live under specific conditions that make them more vulnerable to adverse economic, social, and natural situations. Besides this, the gaps in health care, particularly in mental health management, are bigger than the ones existing in urban areas. However, even though studies have mostly focused on the limitations of mental health management in rural areas, few of

P. Tejada (✉)
Universidad El Bosque, Bogotá, Colombia
e-mail: p.tejada@chester.ac.uk

© Springer Nature Singapore Pte Ltd. 2020   221
S. Chaturvedi (ed.), *Mental Health and Illness in the Rural World*, Mental Health and
Illness Worldwide, https://doi.org/10.1007/978-981-10-2345-3_16

them have addressed its strengths and advantages. These advantages include more and better paid jobs than in big cities, a less competitive environment, and a set of interesting and challenging mental health characteristics for the mental health professional who may dispose of various social, community, and religious sources.

On the other hand, the main disadvantages of managing mental problems in rural areas include stigma, isolation, limited resources for interventions, and limited opportunities for training and communication with other colleagues. Also, mental health professionals are unevenly distributed within urban and rural areas because trained psychiatrists work either in the main cities or in their nearest suburbs. Solutions proposed to solve these problems include training for rural service, telemedicine, dissemination of evidence-based treatments to rural clinics, and integration with the local ethnomedical system.

Keywords
Rural mental health · Advantages · Disadvantages · Telemedicine

Introduction

People in rural areas have the characteristics which describe them as a vulnerable population because they are more likely to live in poverty, lack health insurance, have a chronic health condition, and are unemployed (Jameson and Blank 2007).

As a population, rural inhabitants earn less income and include a higher proportion of the elderly (Jameson and Blank 2007). Households are mostly dependent on farm incomes or do not diversify their economic activities; small businesses and government employment are also other means of employment (Saavedra and Uchofen-Herrera 2016). They are more vulnerable to unexpected negative events such as economic impacts or natural disasters (Saavedra and Uchofen-Herrera 2016).

Additionally, some studies have highlighted that rural areas report poor health, higher inequality, and the biggest gaps in health care (Jameson and Blank 2007; Saavedra and Uchofen-Herrera 2016). Individuals in rural/remote areas are found to have higher rates of mental health disorders than those living in urban settings (Salgado-de Snyder et al. 2003). This could be explained by the unhealthy lifestyles related to poverty and the heavy workload associated with the rural environment.

On the other hand, other studies have shown better mental health indicators in small places than in big cities. A national study carried out in Colombia showed that mental disorders highly prevail in urban areas in comparison with rural areas with a ratio of around two to one both for life prevalence (10.0% vs. 6.1%) and for the last 12-month prevalence (4.4% compared to 2.7%) (Gómez-Restrepo et al. 2015).

The sequence of contacts that a person performs to solve health problems is known as the "pathways of care." Specifically, the concept of *paths of care* refers to "a structured pattern of interaction between social networks, informal systems and formal health care systems" (Salgado-de Snyder et al. 2003). The paths of care that rural people follow are long and complicated and rarely reach the professionals or institutions that offer specialized services (Salgado-de Snyder et al. 2003). Due to the scarce presence of specialized mental health services, rural inhabitants with mental symptoms often do not receive professional treatment (Salgado-de Snyder et al. 2003).

In the past decade, much attention has focused on rural mental health. Unfortunately, empirical research on mental health care in rural areas is poor, and existing research often highlights the disadvantages of available services (Jameson and Blank 2007). It remains unclear what the advantages are in mental illness management in rural areas. The purpose of this chapter is to give an overview of mental health-care problems in rural areas, point out the advantages and disadvantages in mental illness management in these areas, and offer suggestions on how to ameliorate the problems faced.

Advantages in Mental Illness Management in Rural Areas

The advantages could be divided into those for mental health workers (such as job and practice opportunities together with the benefits of a rural lifestyle) and those provided by the rural context which facilitate mental health work (personal resources to tackle mental problems in rural communities, religious resources, and social resources).

Job and Practice Opportunities

In contrast to urban areas, in which health professionals may need to specialize for getting a job, rural areas offer an opportunity to serve as a generalist, practicing across the life span (Hastings and Cohn 2013). In rural practice, there is a need to serve as a generalist in order to meet the needs of a heterogeneous group of patients. Because there are fewer referral options, mental health providers need to work with people presenting with issues across the life span, and this could be challenging and rewarding for clinicians (Hastings and Cohn 2013). In cities, competition is greater due to a large number of professionals who opt for the same job which forces people to increase their postgraduate studies if they want to be competitive.

Another advantage at a professional level is that vacancies in rural areas are not easily filled so opportunities to be hired are greater. Besides this, sometimes, the economic reward can be higher than in the city and, therefore, more attractive to staff.

In rural areas, it is common to work with members of the same family at the same time (Hastings and Cohn 2013). Working with multigenerational families provides a unique opportunity to understand the symptom or problem from multiple informants and may give a more balanced perspective from which to conceptualize the clinical situation (Hastings and Cohn 2013). The autonomy offered by rural clinical practice and the opportunity to work with a variety of diagnosis may be appealing to some mental health workers (Hastings and Cohn 2013).

Rural Lifestyle

Reasons such as a slower life pace, less violence and crimes compared to cities, and variability in client problems have been reported as factors that keep practitioners in rural settings (Hastings and Cohn 2013). In fact, some individuals find the values of rural life attractive (Hastings and Cohn 2013).

If mental health workers share the old-style values typically found in rural areas, it will be easier for them to adjust to the requests of the environment while they enjoy being away from some of the conditions of urban areas (Hastings and Cohn 2013). Rural areas typically feature tight communities with little crime, pollution, and traffic and provide abundant recreational activities (Hastings and Cohn 2013).

Personal and Religious Resources to Tackle Mental Problems in Rural Communities

A study on perceptions on health care in people with mental health problems in rural areas highlighted strength, self-care, self-control, willpower, or joy as internal personal resources that are positive health aspects which may serve as promotional elements in preventive mental health care (Saavedra and Uchofen-Herrera 2016; Salgado-de Snyder et al. 2003).

On the other hand, religion appears to be another important factor. Frequent attendance to religious services has been significantly associated with lower rates of suicide which suggests that religion and spirituality are resources that psychiatrists and clinicians may have been undervaluing (Saavedra and Uchofen-Herrera 2016). Instead, these resources could be used to promote participation in the social context (Saavedra and Uchofen-Herrera 2016). In order to recover health, rural people, especially in Latin America, often take part in religious rituals as they strongly believe in the healing power of God (Salgado-de Snyder et al. 2003). Participation in these ceremonies is socially promoted because it has a great transcendence in both the person and the community (Salgado-de Snyder et al. 2003). Religiosity and faith provide the individual with spiritual, personal, family, and social resources that can be easily mobilized for the solution of mental health problems (Salgado-de Snyder et al. 2003).

Social Resources

Usually, when there is not enough self-care to alleviate a mental symptom, help is sought from the social network, whose members offer their emotional support in the form of advice and guidance and their instrumental support through money, food, child care, etc. (Salgado-de Snyder et al. 2003). The resources offered by the network remain along the path of seeking help until healing (Salgado-de Snyder et al. 2003). Social resources and support are important in rural communities when their members are experiencing various social, economic, or health problems (Selamu et al. 2015). As a matter of fact, social leaders are highly respected, and their recommendations and advice are taken into account (Selamu et al. 2015). They have power to exclude a community member or mobilize the community in support of a member (Selamu et al. 2015).

Disadvantages

Difficulty in Recruiting and Retaining Qualified Staff

Nowadays, one of the most serious issues facing mental health care in rural areas is the difficulty to recruit and retain qualified personnel who can provide services to individuals in need (Jameson and Blank 2007). This can be explained by the great demands and efforts that this kind of work requires, not to mention the poor rewards (especially at an economic level) which produce high levels of job strain and job dissatisfaction.

Rural practice poses many special challenges for the clinician. Mental health workers face excessive workloads, limited availability of nongovernmental community support services, and limited access to crisis care which forces them to transfer the management of acute patients to metropolitan services (Buchanan et al. 2006). Inpatient facilities are virtually nonexistent in rural communities: only 13% of nonmetropolitan places had inpatient facilities, and none of the most rural settings had such services (Jameson and Blank 2007).

Some degree of professional isolation seems inevitable due to the shortage of mental health professionals in rural areas (Hastings and Cohn 2013). A shortage of mental health professionals means having fewer peers with whom to consult difficult cases and fewer referral options. Isolated clinicians may lack the professional and emotional support professional colleagues provide, and costs can be significant (Hastings and Cohn 2013). Rural clinicians who lack colleagues with whom to share interests and concerns, and who experience a deficiency of mutually nurturing relationships, were at higher risk for emotional exhaustion (Hastings and Cohn 2013). One study shows that, "Lack of sufficient guidance, reassurance of worth, social integration, and attachment were associated with the rural mental health counselors at high risk for burnout" (Hastings and Cohn 2013).

In addition to the shortage of specialty mental health professionals in rural areas, there is evidence that the workers who do practice in rural areas experience very high rates of burnout (Jameson and Blank 2007). Burnout was predicted by a lack of social integration with other professionals, a lack of guidance and advice from authoritative sources, and the absence of reliable support from others for assistance (Jameson and Blank 2007). This comes as no surprise, given the overall scarcity of professionals in these areas. The opportunities for support among co-workers in rural areas seem to be as rare as the providers themselves (Jameson and Blank 2007).

Job dissatisfaction and burnout due to professional isolation and lack of support threaten to prompt rural clinicians to leave the area, at a time when one of the most critical issues rural mental health care must face is recruiting and retaining staff to provide much-needed services (Hastings and Cohn 2013). As a matter of fact, one-third of the most rural places totally lack available health professionals to address mental health problems, and a much greater percentage of these places don't have specialty mental health services at all (Jameson and Blank 2007).

Besides the issues related to environment, nature of mental health work, and rural location, there are issues related to the organizations, management, and organizational structure that make recruitment and retaining of qualified staff difficult (Moore et al. 2010). Mental health care in rural areas is dominated by the public sector which may have difficulties in compensating professionals at the same competitive rate of urban service providers (Jameson and Blank 2007). Some studies suggest that in rural areas, recruitment is regarded by managers as a matter of greater urgency than retention or training and that organizations themselves are of a second order of significance in terms of perceived workforce difficulties (Moore et al. 2010).

In many countries, the pool of available and potential mental health staff is limited by global, national, and state shortages of suitable workers, competition with metropolitan and intra-regional organizations, inadequacy of specialist training, and the choice of qualified people to work in other places (Moore et al. 2010). Rurality means that most services are small, that workers need generalist skills to travel to clients, and that opportunities for career progression are limited (Moore et al. 2010). It therefore imposes extra demands on workers and costs per employee and, in turn, greater demand on the organization's capacity to recruit and train (Moore et al. 2010).

In addition, education and vocational training are difficult to provide in rural areas where funding for these activities may not be sufficient. Added to this, there isn't a multidisciplinary approach, and there are very few specific training programs for these professionals (Moore et al. 2010). The integration between primary care professionals (such as general practice physicians and nurses) and specialty mental health-care providers is often seen as low in rural areas (Jameson and Blank 2007).

Other reasons that make recruiting and retaining qualified staff difficult in rural areas are the cultural barriers and lack of respect for their professional judgment, thereby making it difficult to retain their services (Jameson and Blank 2007).

Individual, Social, and Geographical Factors as Barriers to Treatment

In general, rural residents face barriers to mental health care, including fewer local mental health providers, longer distance to specialty services, lack of insurance, social stigma, and a tendency to rely on family and other informal support (Buchanan et al. 2006).

Individuals in rural areas often do cite social stigma and lack of privacy as reasons not to seek help for mental distress (Jameson and Blank 2007). Some studies have found social stigma associated with mental illness to be higher in rural areas than in nonrural areas (Jameson and Blank 2007). Several studies have shown that people living in rural areas are under the influence of stigmatization when it comes to their attitude toward mentally ill patients as they are more likely to be rejected and stigmatized by society which is an important issue to consider (Gur and Kucuk 2016). Furthermore, the degree to which stigma was perceived predicted inclination to seek treatment for mental health problems. Individuals in rural areas also perceive a lack of privacy for primary care treatment of mental illness (Jameson and Blank 2007). Rural residents recognize each other by their vehicles and tend to know "everything about everybody" (Hastings and Cohn 2013). The stigma associated with seeking mental health treatment is exacerbated by the difficulty to remain discreet in small communities (Hastings and Cohn 2013). The lack of privacy is also associated with the fact that for mental health professionals, there is an increased likelihood of being engaged in multiple relationships with patients because of the reduced population density and the resulting likelihood of encountering one's patients outside the office (Hastings and Cohn 2013).

Two factors described as associated with stigma are the authoritarianism, which reflects a condescending view that the mentally ill is different from normal people, and the social restrictiveness, which is a manifestation of the view that mentally ill individuals need to be restricted during their hospitalization and later as well so that society can be protected from their actions and from their posing a threat to the community (Gur and Kucuk 2016). The studies that examined people's attitudes toward illnesses reported that, especially in mental diseases such as schizophrenia, negative attitudes prevail and people with sickness are rejected and social contact with them is avoided (Gur and Kucuk 2016). Particularly in rural areas, it is reported that attitudes may be even more negative and rejection may be even more evident (Gur and Kucuk 2016).

The stigma not only affects patients but also mental health professionals. Rural community values may make it difficult for a psychologist or psychiatrist to be accepted. Stigma regarding mental health practice and suspicion of outsiders are not uncommon aspects of rural social life (Hastings and Cohn 2013). Rural community values tend to be more conservative, with religion playing a central role in residents' lives. Yet mental health providers, as a group, generally endorse more liberal and less religious ideologies (Hastings and Cohn 2013). These cultural barriers and a lack of understanding regarding the mental health profession act like an obstacle for help-seeking behavior.

The stigma could be so strong that it also affects general practitioners. Primary care physicians often seem reluctant to diagnose mental disorders (Jameson and Blank 2007). One study found that approximately half of physicians in primary care rural settings deliberately misdiagnose depression (Jameson and Blank 2007). These physicians cited uncertainty about the diagnosis, problems with reimbursement for services if a diagnosis of depression is given, and fear that the patient may not be able to obtain health insurance in the future as the most common reasons for purposely misdiagnosing depression (Jameson and Blank 2007). Instead, they often give diagnoses of fatigue/malaise, insomnia, or headache to depressed patients (Jameson and Blank 2007).

In addition to the social stigma associated with mental illness, rural inhabitants often do not recognize the need for treatment (Jameson and Blank 2007). The denial of need for treatment may even be reinforced by social contacts in rural areas (Jameson and Blank 2007). This is perhaps one of the reasons why the mentally ill may choose to seek alternative methods of treatment. It is also known that especially individuals living in rural areas who have lower socioeconomic status do not seek psychiatric help for either themselves or their relatives. This avoidance seriously affects the ability of people to receive psychiatric treatment (Gur and Kucuk 2016).

When it comes to alleviating their problems, rural people do not usually consult a doctor. They would only come when the discomfort persists or when they have been repeatedly referred by a member of the local ethnomedical system or their social network (Salgado-de Snyder et al. 2003). Individuals residing in rural areas in need of mental health treatment often turn to informal sources of care like self-help, family, spouses, neighbors, friends, and religious organizations (Jameson and Blank 2007). For a long time, rural areas have been identified as having poorer access to health services than metropolitan areas due to a variety of different factors: poor geographical access to services, extended waiting periods for doctor appointments, and limited access to specialized services (Henderson et al. 2014). Thus, before going to the doctor, people consider the costs of transportation, consultation, medications, and subsequent consultations (Henderson et al. 2014). The cost of seeking medical care often becomes so high that patients repeatedly postpone the visit which contributes to their condition worsening (Salgado-de Snyder et al. 2003).

Rural people seeking mental health care face difficulties in accessing psychiatrists and rely upon doctors to provide primary mental health care (Henderson et al. 2014). Consultation with a mental health specialist, such as a psychiatrist or psychologist, is highly unlikely among rural people due to three factors: the difficulty of geographical access to these services, the cost of using them, and the cultural gap between the specialist and the patient which is even bigger than with the general practitioner (Salgado-de Snyder et al. 2003).

In case a rural inhabitant gets a psychiatric appointment, their commitment to the therapy is uncertain since it involves several trips for subsequent consultations and assuming the cost of medication (Salgado-de Snyder et al. 2003). Poverty, marginalization, greater levels of general pathology, and lower involvement in vocational activities in rural zones explain why it is difficult to access and adhere to mental health services (Jameson and Blank 2007). Although some people find temporary relief to their symptoms, these do not completely disappear or reappear, so a person can remain on the path of seeking help for a long time before actually finding a solution to their mental problems (Salgado-de Snyder et al. 2003). Additionally, significantly larger proportions of rural people lack adequate mental health coverage compared to urban groups (Hastings and Cohn 2013; Buchanan et al. 2006). Finally, rural inhabitants are more likely to report an inadequate number of mental health providers in their areas and are dissatisfied with the quality of the mental health care received (Buchanan et al. 2006).

Contributions to Help Alleviate Mental Health-Care Difficulties in Rural Areas

Training for Rural Service

Due to the need of successfully solving the shortage of staff in a particular rural area, it is important that universities adapt their contents to the needs of rural professionals (Jameson and Blank 2007). A number of scholars have asserted that graduate training provides inadequate preparation for rural mental health practice (Hastings and Cohn 2013). Academic programs have been described as adhering to an "urban training model" in which the boundaries between professionals and patients are clear and referral options are abundant (Hastings and Cohn 2013). A study conducted in Australia showed how a brief orientation program to rural mental health work affected participants' attitudes toward working and living in a rural area, two key factors known to influence recruitment and retention of rural practitioners (Sutton et al. 2016). This study examined impacts from an evidence-informed intervention designed to guide health and nursing students to mental health work and career opportunities in rural regions (Sutton et al. 2016). The study reveals that a targeted short-term intervention can positively change metropolitan tertiary-trained student attitudes toward living and working in a rural setting (Sutton et al. 2016).

It is equally important that psychiatrists and clinical psychologists participate in awareness-raising courses in the local culture, which would enable them to maintain an informed and flexible attitude toward the beliefs and expectations of treatment for mental illnesses (Salgado-de Snyder et al. 2003). In addition, they can learn some healing strategies of the local ethnomedical system that could be complementary to the therapeutic treatments they prescribed (Salgado-de Snyder et al. 2003). This approach would allow the necessary cultural congruence to facilitate the therapeutic adherence of the patient (Salgado-de Snyder et al. 2003).

The issue of recruiting and retaining mental health workers to work in rural areas is also worth examining. Incorporating specialized training for rural work is important, but has little value if trainees cannot be convinced to stand in these needy areas. Paramount to this effort is the need to make positions in rural areas as attractive as possible (Jameson and Blank 2007). There is a need to increase practice and internship opportunities in these places (Jameson and Blank 2007).

Many rural people who request medical attention in primary care services suffer psychiatric problems; therefore, the staff's mental health training in these centers is essential. The absence of professionals specialized in mental health in rural areas means that doctors, social workers, and nurses working in community health centers are integrated as interdisciplinary teams (Salgado-de Snyder et al. 2003). Their training requires incorporating the management of basic knowledge that allows them to appropriately channel cases, develop basic psychotherapeutic skills, and incorporate some healing tools recommended and used by members of the ethno-medical system (Salgado-de Snyder et al. 2003).

If health staff received mental health training to treat minor cases, psychiatric care would only be required by a small proportion of the population. In that case, mental health specialists would provide services to outpatient psychiatric patients on a regular basis at community health clinics, and clinic physicians would follow up (Salgado-de Snyder et al. 2003). It is necessary to point out that even when specialized psychiatric services were permanently available in small-town health centers, people would probably not come if they did not have the information that would enable them to identify the need for such services (Salgado-de Snyder et al. 2003).

Telehealth

Telehealth is the use of telecommunications and information technology to provide access to health screening, assessment, diagnosis, intervention, consultation, super-vision, education, and information across distance (Farrell and McKinnon 2003). The term telehealth is used to describe the use of communications technology in the educational, clinical, training, administrative, and technological aspects of health care (Jameson and Blank 2007). These technologies include the Internet in addition to telephone, television, video, and videophone (Farrell and McKinnon 2003). Telemedicine is used to describe the aspects of telehealth involved in patient care (Jameson and Blank 2007).

The use of new technologies such as broadband Internet and videoconferencing can potentially have a large impact on the delivery of services to rural areas (Jameson and Blank 2007). Telehealth has the potential to address a number of problems faced by rural caregivers. First, the use of telehealth gives patients in remote areas increased access to services (Jameson and Blank 2007). Assessment, psychotherapy, crisis, intervention, psychoeducation, medication consultations, and case manage-ment can be conducted from great distances, often through videoconferencing when economically feasible (Jameson and Blank 2007).

Research suggests that patients are generally happy using telehealth services (Jameson and Blank 2007). Researchers reported that patients not only responded positively to computer interviews but also gave honest answers (Farrell and McKinnon 2003). Subsequently, medical, marketing, staff, and social science researchers have explored the use of the computer as a means for reducing social desirability bias and obtaining more sensitive information from respondents than can be obtained by using traditional formats (Farrell and McKinnon 2003). The evidence that the computer encourages self-disclosure has led to the development of important applications such as computer interviews to detect risk conditions and behaviors (Farrell and McKinnon 2003).

An additional advantage of using telehealth systems is the potential increase in professional collaboration among rural mental health professionals (Jameson and

Blank 2007). Telehealth systems enable caregivers in isolated areas to interact with other professionals (Jameson and Blank 2007). Interactions include consultations, grand rounds, and supervision. These contacts could be important in keeping rural mental health professionals well-informed about developments and issues in the field, increasing the quality of care they are able to offer to their patients through consultations with specialists, and aiding the obtainment of continuing education credits required by most state licensing boards (Jameson and Blank 2007).

Telepsychiatry has been identified as being a cost-effective way of increasing treatment options in rural and remote locations and also allows people to be seen within their community (Henderson et al. 2014). Telepsychiatry services are usually well regarded by mental health teams, but in some communities, patients would rather prefer face-to-face services (Henderson et al. 2014). Despite this, it is acknowledged by consumers and caregivers that use of telepsychiatry services is better than not having a service (Henderson et al. 2014).

One of the first applications of the computer in mental health care was for assessment through computerized psychiatric interviews (Farrell and McKinnon 2003).

One example of a computer-based assessment tool is the Global Mental Health Assessment Tool (GMHAT). The GMHAT is a computer-assisted clinical interview to be used in routine clinical practice to detect and manage mental disorders in most settings (Sharma et al. 2010). The main purpose of developing GMHAT is to help people in bringing relief from the sufferings of their mental disorders. This will be even more relevant to places where there is, at present, no adequate service due to shortages of trained staff (Sharma et al. 2010). An epidemiological study of psychiatric disorders in rural populations shows that GMHAT can be used as a standardized diagnosing tool in primary health-care centers helping primary care workers to diagnose psychiatric cases in a short span of time and also in referring them to specialty centers (Sharma et al. 2010).

The Symptom-Driven Diagnostic System for Primary Care (SDDS-PC) is a computer-assisted program for identifying major depression, generalized anxiety disorder, panic disorder, obsessive compulsive disorder, substance abuse, dependence, and suicidal ideation (Broadhead et al. 1995).

The Case-finding and Help Assessment Tool (CHAT) detects the presence of alcohol use, psychoactive drugs use, gambling, depression, anxiety, stress, irritability, and eating behavior disorders in primary care (Goodyear-Smith et al. 2008). Patients self-administer eCHAT on an iPad in the waiting room and receive summarized results, including relevant scores and interpretations from a family physician on the website and in the electronic health record (EHR) at the point of care (Goodyear-Smith et al. 2013).

The advantages of using computer technology in the care of seriously mentally ill people living in rural areas include providing a means of education (Farrell and McKinnon 2003). Psychoeducation approaches have been used successfully in community mental health for years and are considered an evidenced-based approach (Farrell and McKinnon 2003).

When individuals with serious mental illnesses are discharged to the community, continuous care is highly desirable and necessary (Farrell and McKinnon 2003).

Lack of continuity places the individual at risk for becoming lost to services (Farrell and McKinnon 2003). For those in rural and geographically disperse areas, technology has the potential to reduce gaps in services, improve the connections between patient and provider, and give an alternative to traditional face-to-face contact (Farrell and McKinnon 2003). Early on, psychiatry was perceived as the ideal specialty for the application of telemedicine owing to the fact that assessment and treatment relies more on audiovisual information than use of laboratory tests and procedures (Farrell and McKinnon 2003).

The Dissemination of Evidence-Based Treatments to Rural Clinics

Rural areas suffer from a severe shortage of mental health-care professionals. Therefore, service providers in these areas must adopt strategies to administer effective treatments in a time-efficient manner if they expect to compensate for understaffing and insufficient funding (Jameson and Blank 2007). The use of evidence-based treatments in rural places alludes to interventions based directly on scientific evidence suggesting most significant contributors and risk factors for mental symptoms.

Evidence supports certain mental health treatments, including assertive community treatment, supported employment, family psychoeducation, illness management and recovery, and integrated dual diagnosis treatment for people with severe mental illnesses (Weaver et al. 2015). These evidence-based practices, generally combined with medication, have consistently been shown to improve symptomatology, functional status, and quality of life for people with mental problems (Weaver et al. 2015).

It is important to note that many mental health interventions have been created and tested in urban settings and therefore cannot be implemented in rural areas without taking into account the characteristics of this context (Weaver et al. 2015). For example, most interventions inherently assume an urban infrastructure (i.e., public transportation, adequate number of mental health professionals) (Weaver et al. 2015). Literature suggests that rural people's understanding of mental illness is not always congruent with the biomedical model, which informs most evidence-based treatments (Weaver et al. 2015). In order to facilitate the introduction of different types of interventions and to reduce stigma, it is important that the adaptations take into account the beliefs, practices, and attitudes of the rural area and that they integrate the local ethnomedical system (Weaver et al. 2015).

Research on the implementation of evidence-based practices for adults with mental problems living in rural places is poor (Weaver et al. 2015). For this reason, it is important to study the implementation of these practices in rural areas and pay special attention to identifying and testing adaptations. Implementation efforts are most effective when they address specific needs and concerns of target providers. Rural service settings face significant challenges when implementing evidence-based interventions, and adaptations are likely needed to address the unique aspects of rural service (Weaver et al. 2015).

In order to adapt and implement evidence-based interventions, work teams must be staffed by experts and health-care providers in the area. It is of particular importance that members of the institutions of the level to which the intervention will be addressed participate actively in all phases of the project. It is important that during the adaptation process, patients, policy makers, and other stakeholders are involved and provide information on the orientation of interventions.

Furthermore, the importance of providers' local knowledge cannot be understated. The limited availability of mental health professionals, low population density, geographic isolation, and stigma in rural areas may require the development of new implementation strategies and possibly new evidence-based practices that are effective, acceptable, and sustainable for communities (Weaver et al. 2015).

Integration with the Local Ethnomedical System

The local ethnomedical system in rural zones usually includes the priest, pharmacy manager, and traditional healers (Salgado-de Snyder et al. 2003). Practices such as dietary changes, use of home remedies or commonly used medications, self-control strategies, and faith in God's healing power have been reported as important sources of personal strength and self-help to deal with emotional problems. It is necessary to evaluate this type of actions as spontaneous responses of protection to the threat of loss of health (Salgado-de Snyder et al. 2003).

Therefore, it is essential not to devalue these types of practices when they are present, but to strengthen them and incorporate them as complementary actions to the treatment prescribed by the specialist or by the general practitioner (Salgado-de Snyder et al. 2003).

An important resource to maximize the effectiveness of the social network is to provide rural people clear and concise information through basic education on the identification of psychiatric symptoms and the evolution of mental illness. It also proposes the reaffirmation and promotion of the positive features already existing in the network, such as emotional support, advice, financial aid, instrumental support, and reciprocity (Salgado-de Snyder et al. 2003).

Psychoeducation among members of this system could be a more effective source of assistance, both by making interventions and by referring patients who so require it to the formal mental health services. It should be noted that this training is considered to be very useful since authorities and leaders of rural zones are recognized by the community and facilitate adherence to the therapeutic actions prescribed by the specialist (Salgado-de Snyder et al. 2003).

Because traditional healers are such an integral part of their communities, collaboration between them and medical practitioners would hold significant promise as a means to benefit patients. This partnership could improve access to care and alleviate the burden of mental illness experienced by patients and their communities (Schoonover et al. 2014).

Existing community resources can be used in order to enhance mental health care delivered at the primary care and community rural level (Selamu et al. 2015). In this context, community resources may be features of individuals, organizations, or the built environment that relate to health (Selamu et al. 2015). There are several examples of mental health interventions utilizing existing community resources to maximize their impact. Integrating mental health interventions to existing non-governmental organizations, in particular community-based rehabilitation programs, has been identified as a way to overcome problems when there is a lack of capacity and resources (Selamu et al. 2015).

Conclusion

The challenge of providing adequate mental health care for individuals in rural areas is a complex issue, and the solutions to the problems faced are not simple (Jameson and Blank 2007).

In this review, views of the benefits and limitations of rural mental health practice show many trends, primarily that, often, the very aspects of the area that make it most appealing can also lead to many challenges in providing mental health services (Hastings and Cohn 2013). For example, the advantage of the easygoing nature of rural life and a friendly context contrasts with the difficulty of accessing resources, isolation, and stigma.

Mental health professionals with emphasis on working within a developmental framework and focusing on prevention and psychoeducation may be especially compatible for rural practice (Hastings and Cohn 2013). Training programs could assist their students by incorporating more information about rural practice, especially regarding rural cultural norms and providing training experiences serving rural populations (Hastings and Cohn 2013). In addition, programs could help clinicians in training for the development of skills to assess the needs of small communities in order to identify any special areas of practice that would benefit those communities (Hastings and Cohn 2013).

Besides training for rural service, it is necessary to have detailed information derived from research in rural communities and the active participation of all human and institutional resources; from professionals and truly multidisciplinary researchers, educators and social, religious, and political leaders; and from members of the communities and local institutions (Hastings and Cohn 2013). The integration model of personal and community resources for the management of mental illnesses in rural areas requires establishing programs that sensitize, inform, and train community members as well as members of the ethnomedical system and the formal health system (Hastings and Cohn 2013).

Finally, it is important to point out that technology has the potential to decrease gaps in services and improve education, support, and care between patients and professionals in rural areas (Farrell and McKinnon 2003). As an alternative to traditional face-to-face contact, the Internet can potentially overcome the disparities in health-care access for rural mental health-care services (Farrell and McKinnon 2003).

Cross-References

▶ Mental Health Professionals in Rural Areas
▶ Organization of Mental Health Services in Rural Areas

References

Broadhead WE, Leon AC, Weissman MM, Barrett JE, Blacklow RS, Gilbert TT, . . . Higgins ES (1995) Development and validation of the SDDS-PC screen for multiple mental disorders in primary care. Arch Fam Med 4(3):211–219

Buchanan RJ, Schiffer R, Wang S, Stuifbergen A, Chakravorty B, Zhu L, . . . James W (2006) Satisfaction with mental health care among people with multiple sclerosis in urban and rural areas. Psychiatr Serv 57(8):1206–1209

Farrell SP, McKinnon CR (2003) Technology and rural mental health. Arch Psychiatr Nurs 17(1):20–26

Gómez-Restrepo, C., Escudero, C., Matallana, D., González, L., & Rodriguez, V. (2015). Encuesta Nacional de Salud Mental. Bogotá: Colciencias y Ministerio de Salud.

Goodyear-Smith F, Coupe NM, Arroll B, Elley CR, Sullivan S, McGill AT (2008) Case finding of lifestyle and mental health disorders in primary care: validation of the 'CHAT' tool. Br J Gen Pract 58(546):26–31. https://doi.org/10.3399/bjgp08X263785

Goodyear-Smith F, Warren J, Bojic M, Chong A (2013) eCHAT for lifestyle and mental health screening in primary care. Ann Fam Med 11(5):460–466

Gur K, Kucuk L (2016) Females' attitudes toward mental illness: a sample from rural Istanbul, Turkey. Iran Red Crescent Med J 18(5):e22267

Hastings SL, Cohn TJ (2013) Challenges and opportunities associated with rural mental health practice. J Rural Ment Health 37(1):37

Henderson J, Crotty MM, Fuller J, Martinez L (2014) Meeting unmet needs? The role of a rural mental health service for older people. Adv Ment Health 12(3):182–191

Jameson JP, Blank MB (2007) The role of clinical psychology in rural mental health services: defining problems and developing solutions. Clin Psychol Sci Pract 14(3):283–298

Moore T, Sutton K, Maybery D (2010) Rural mental health workforce difficulties: a management perspective. Rural Remote Health 10(3):1519

Saavedra JE, Uchofen-Herrera V (2016) Percepciones sobre la atención de salud en personas con problemas autoidentificados de salud mental en zonas rurales del Perú. Rev Peru Med Exp Salud Publica 33(4):785–793

Salgado-de Snyder VN, Díaz-Pérez MDJ, González-Vázquez T (2003) Modelo de integración de recursos para la atención de la salud mental en la población rural de México. Salud Publica Mex 45(1):19–26

Schoonover J, Lipkin S, Javid M, Rosen A, Solanki M, Shah S, Katz CL (2014) Perceptions of traditional healing for mental illness in rural Gujarat. Ann Glob Health 80(2):96–102

Selamu M, Asher L, Hanlon C, Medhin G, Hailemariam M, Patel V, . . . Fekadu A (2015) Beyond the biomedical: community resources for mental health care in rural Ethiopia. PLoS One 10(5): e0126666

Sharma VK, Jagawat S, Midha A, Jain A, Tambi A, Mangwani LK, . . . Lepping P (2010) The global mental health assessment tool-validation in Hindi: a validity and feasibility study. Indian J Psychiatry 52(4):316

Sutton K, Patrick K, Maybery D, Eaton K (2016) The immediate impact of a brief rural mental health workforce recruitment strategy. Rural Soc 25(2):87–103

Weaver A, Capobianco J, Ruffolo M (2015) Systematic review of EBPs for SMI in rural America. J Evid Inform Soc Work 12(2):155–165

Methodological Issues in Epidemiological Studies on Mental Health in Rural Populations

14

An Introduction

Vimal Kumar Sharma, Shiv Dutt Gupta, Nikhil Sharma, and Nutan Jain

Contents

V. K. Sharma (✉)
University of Chester, Chester, UK

Cheshire and Wirral Partnership NHS Foundation Trust, Chester, UK
e-mail: v.sharma@chester.ac.uk

S. D. Gupta · N. Jain
IIHMR University, Jaipur, India

N. Sharma
South London and Maudsley NHS Foundation Trust, London, UK

© Springer Nature Singapore Pte Ltd. 2020
S. Chaturvedi (ed.), *Mental Health and Illness in the Rural World*, Mental Health and
Illness Worldwide, https://doi.org/10.1007/978-981-10-2345-3_5

Abstract

Epidemiology is the study of distribution and determinants of health-related events (diseases) in populations as well as evaluating the impact of interventions on health in a population. A number of different study methods can be used, including case-control, cohort, and RCTs. Clear definitions regarding the subject group being studied must be applied. Studies looking at rural communities pose their own challenges. Most study tools are made for developed English speaking subjects, and therefore one has to be aware of local sentiments, language, and mindsets when implementing a study. There remains an urgent need to invest resources to carry out properly planned epidemiological studies in the villages.

Keywords

Epidemiology · Rural Mental Health · Research Methodology

Introduction

Epidemiology is essentially a study of distribution and determinants of health-related events (diseases) in each population as well as evaluating the impact of interventions on health in a population. It is one of the most important core competencies in public health. Its role has expanded from traditional investigations of outbreaks of diseases to measure magnitude and distribution of diseases in the community and identifying underlying causes to provide scientifically valid evidence for developing sound policies and strategies for prevention and control of various diseases, including mental health illnesses. It is increasingly employed in testing health interventions and evaluation of effectiveness of health programs and services. Methodological approaches in epidemiology include identifying disease and its magnitude using descriptive epidemiology; identifying causes and testing hypotheses using analytic epidemiological methods, namely, case-control studies and cohort studies; developing and designing preventive programs; and testing and evaluating health interventions and programs. A new era of implementation research has heralded in epidemiology that addresses the issues of implementation of successful interventions through health systems in real time.

The use of epidemiological studies on mental health and discipline of psychiatry lags far behind than its use in other disciplines. Most psychiatric studies are clinical in nature, hospital based, and urban oriented. Most epidemiological studies are descriptive in nature, aiming at measurement of prevalence of disorders and associated socio-demographic factors. Epidemiological studies have gained immense importance due to changing disease burden in the last two decades on account of currently undergoing health and demographic transition in low- and middle-income countries. A shift from acute infectious communicable diseases toward chronic and noncommunicable diseases has been taking place all over the world. While the mental health disorders are already in the top 10 disorders accounting for high disability-adjusted life years (DALY), they are expected to reach 2nd rank in the list in the next decade.

The myth that mental disorders are seen mostly in urban population is far from truth. These disorders are equally prevalent in the rural population. It is a well-established fact that the access and availability of affordable and high-quality mental healthcare is unavailable to the vast rural populations in LMICs. The frontline health workers are not trained in identifying and treating mental disorders; supply of medicines is severely limited; and the infrastructure to support mental healthcare is almost nonexistent. In view of the rising disease burden of mental health disorders, epidemiological studies on the mental health of rural populations are essential to learn about trends and patterns of mental disorders in these communities. Scrutiny of how sociocultural differences correlate as causative and protective factors for mental ill health is only possible through community-based studies in the population. Other areas such as perception of mental illness, help-seeking behavior, stigma, role of families, impact of poverty, etc. can only be understood through epidemiological studies. But such studies face a multitude of obstacles. This chapter addresses methodological challenges in planning and carrying out such studies.

Epidemiological Methods

Broadly, there are four categories or types of epidemiological study designs:

Descriptive Epidemiological Studies

Descriptive epidemiological studies include case identification studies, cross-sectional or longitudinal surveys, and epidemiological surveys. Case studies are usually carried out for in-depth analysis and descriptive reporting of a clinical case, experience, special situation, event, or phenomenon in clinical settings. The case studies provide in-depth information about disease phenomenon and develop new hypothesis.

Descriptive studies are mainly used to measure prevalence or incidence of diseases or disorders, patterns, and distribution according to person, place, and time characteristics and sociodemographic, economic, and environmental factors; examine association of these factors with the disease; and, finally, formulate hypothesis.

Analytical Epidemiological Studies

Analytical epidemiological studies are conducted mainly to test hypothesis and identify cause by evaluating association between the cause and effect (disease outcome). There are two types:

(i) Case-control or retrospective studies
(ii) Cohort or prospective studies

Case-control studies are also called as retrospective studies. The essential design of the study has cases of the disease under study and controls as those who do not have the disease under study.

The next step is to retrospectively measure frequency of exposure by the putative cause (risk factor) in a defined period. The risk of disease is measured by computing odds ratio (OR). The odds ratio is the ratio between frequency of exposure (risk factor) among cases and controls. If the odds ratio is equal to 1.0, it reflects that there was no casual association. An OR more than 1.0 suggests increased risk due to the exposure or risk factor, whereas OR less than 1.0 suggests reduced risk due to exposure or risk factor.

Case-control studies are useful and effective for studying rare diseases. The greatest advantage of these studies is that such studies can easily be done in hospital settings and are often less expensive.

Cohort studies are prospective and longitudinal in nature. The important feature of study design is that this includes group of people at risk without disease (study group) to begin with. The study population is divided into two groups: one that is exposed and the other that is not exposed to risk factors. The two groups are followed and defined over a period time. The frequency (incidence) of disease under study is measured and compared between these two groups. The causality is measured by computing relative risk (RR) which is a ratio between incidence rate among exposed and nonexposed. If the RR is equal to 1.0, then it suggests no difference in the risk of the disease. A RR more than 1.0 suggests increased risk due to the risk factor.

These studies are usually carried out on diseases which are commonly prevalent in the population. However, cohort studies are time-consuming and expensive. One of the biggest challenges of completing cohort studies is loss to follow-up (LFU) which may significantly influence the estimates of risk of disease. The advantage of these studies, on the other hand, is that they provide incidence rates of the disease. Other advantages of cohort studies are that bias (such as selection bias), information, and misclassification are much lower than case-control studies and determination of outcome is much more robust. Usually, cohort studies are planned after a series of case-control studies that have shown some consistent results.

Experimental Epidemiological Studies

Experimental epidemiological studies are conducted after case-control and cohort studies have firmly established the causative factor. These are called randomized controlled trials (RCT). RCTs could be done to test and evaluate the effectiveness of new drugs, therapies, and vaccines. RCTs could be categorized as clinical trials, therapeutic trials, or preventive trials. The basic design is similar to that of cohort studies, but the allocation of the subjects to treatment group and no treatment or placebo group is done through a process called randomization which rules out any selection bias. These studies may also be blinded (single or double blind) to remove any information bias. Strict protocols are used for recruitment of study subjects' inclusion and exclusion criteria, follow-up, and ascertainment methods of outcomes. The advantages of RCT are that it (a) ensures internal validity by removing possible common biases and (b) eliminates the effect of any confounding factors.

Intervention Studies

Intervention studies are undertaken to test effectiveness of intervention developed based on evidence produced by the analytical and experimental studies in real-time life settings. These interventions may be in the form of package of services, vaccine, drugs and therapy, nutrients, and behavior change/counselling.

Planning Epidemiological Studies

Detailed description of various types of epidemiological studies is beyond the scope this chapter. The focus of this chapter will be on planning for population-based community survey, using concepts of descriptive epidemiology.

Choice of Study Design

Nature of Research Question

Choice of the study design is dependent on the nature of research question asked in the study. If the research questions are about good understanding of disorders such as: "What is disease burden or prevalence mental health disorder (s)?" "What are the geographic patterns of distribution in the population?" "Who are affected most?" "When the disease occurs more?" then the natural choice of study design would be a cross-sectional survey using concepts of descriptive epidemiology.

Testing Hypothesis

Choice of study design is also dependent on the nature of hypothesis being tested. For example, if the hypothesis looks to answer questions such as "Why mental disorders occur in a particular population or in a community?" "Which community has higher risk compared to others?" then one needs to use analytical epidemiological studies, such as case-control, or a cohort study.

Resources

Most epidemiological research studies are carried out in high-resource countries. In low- to middle-income countries, studies are often limited to cities or surrounding areas. Maselko (2017), in her review of social epidemiology and global mental health (Maselko 2017), highlighted that only 10% of the health research conducted accounts for 90% of the world's population. A disparity arises as the majority of studies are undertaken in high-income countries. The results from these studies are then extrapolated to represent findings globally despite the majority of the world's population living in low- to middle-income countries.

Funding opportunities for epidemiological studies in rural communities are generally scarce, but this gets even worse when linked to mental health research. An additional problem is that studies in rural and remote locations are resource intensive due to the time taken in travel and arranging visits without the possibilities of making prior arrangements. Researchers who are culturally sensitive and fluent with local languages are often hard to find. The lack of "fit for purpose" researchers for studies in villages gives rise to significant challenges to gather good-quality data for epidemiological studies in rural population.

Population and Study Area: Issues in Defining the Rural Population

Having a clearly defined population is critical in planning epidemiological studies. For example is one studying rural or urban, men or women, specific age groups or an at risk population? Do you want to study the whole population or a subset/sample of population? For valid estimates, decide an appropriate size of sample, which is representative of the population.

The most pressing matter is in defining the rural community. It is important to establish clear criteria: should this be based on simply the number of people living in the area or should it be combined with other parameters such as a lack of facilities or the literacy level of the population. For example, the UK government classifies a rural area as one that is "outside settlements of more than 10,000 resident populations"(Bibby 2013). In the USA, "rural area" is defined as one which is not urban. The urban areas or clusters are based on population, i.e., over 50,000 or as urban clusters where there are over 1000 people per square mile. The national census (2011) (Census of India 2011) defined urban population in India as "statutory towns" (big cities) and "census towns" that have (a) a population over 5000, (b) at least 75% of male working population is involved with nonagricultural work, and (c) density of population over 400 per square kilometer. Any population that is nonurban is classified as rural. Similarly giving a definition of a "village" would be even more challenging as they are significantly different from one to the other. A number of factors influence villages and rural population in a modern and dynamic world.

Rapid Urbanization

In some low- to middle-income countries, the rate of development and urbanization is so rapid that the population data based on the previous census is no longer rural. Cities are encroaching rural areas at a very fast rate in China (Liu et al. 2017). Likewise, in India, towns are rapidly expanding at the expense of rural settlements (Gibson et al. 2017). These two examples are specifically important as, together, India and China make up over one-third of the world's population.

Migration

The younger population is moving to big towns and cities at a rapid rate all over the world and even more so in low-income countries. Thus, the rural communities' population is ever changing. Migration also affects demographic variables such as literacy and poverty.

With an improvement in transport links, there is now a category of people who commute daily from rural areas to work in urban places. While falling under the rural community, they are exposed to urban culture daily. If included in any community study, these people would not really represent true rural population.

Communications

All epidemiological mental health studies are dependent on interviews. A meaningful and proper communication between participants and researchers is important to establish accurate answers for the questions asked in a study. There are a number of reasons why a wide communication gap may develop during the process of studies. Language barriers could be one of them, as participants fail to grasp the meaning of questions asked to them. They may give short yes or no (more often no) answers to avoid further questioning. Researchers must also be aware of local dialects and able to discern whether the participants understand the questions put forward and are giving appropriate and adequate answers.

A final but important point regarding communication is the lack of trust of villagers toward researchers who are often city people. There could be a number of ill-informed rumors spread in the villages that the purpose of research could be to catch out villagers.

Threats to Validity and Reliability of Epidemiological Studies

The main epidemiological approach to study the distribution and determinants of diseases or health-related events is based on two ideas: asking questions and making comparisons. While planning and designing epidemiological studies in mental health, it must ensure that the results are valid and reliable. The main threat to validity arises from various forms of biases and confounding factors. Bias may occur because of inappropriate methods of selection of study subjects. Information bias results due to recall of information by subjects, methods of data collection, and the way questions are asked. Misclassification bias may result from inadequate diagnostic criteria and case definition. The other major threat to validity of study results arises from various confounding factors due to lack of comparability of the study population on its various characteristics or variables. Any factor which is associated with the disease and the causal factor may confound the study results.

In mental health epidemiology, the most important methodological challenge is the "case" identification method. Unlike physical disorders, identification of mental disorders is entirely based on clinical interview. Classification of mental disorders remains another challenge as both ICD (World Health Organization 1992) and DSM-IV (American Psychiatric Association 2013) are based on experts' consensus. This is based on a hierarchical model where the broad assumption is that more severe disorders may have symptoms of less severe disorders; therefore, "severe" disorders have priority on diagnosis over less severe disorders. For example, anxiety or depressive symptoms are ignored in arriving at a diagnosis if the person has symptoms of schizophrenia. The following section will highlight the problems of case identification in the context of the rural setting.

Case Identification Methods

There are a host of tools and schedules to identify cases of mental health problems. Some are broad screening tools such as GHQ (Goldberg et al. 1978); others are diagnostic tools such as CIDI (Composite International Diagnostic Interview (CIDI) 1993) and MINI (Sheehan et al. 1998). A number of case identification tools are for specific disorders such as depression, e.g., Center for Epidemiologic Studies-Depression Scale (CES-D), (Lewinsohn et al. 1997) or dementia such as Geriatric Mental State (GMS-AGECAT) (Copeland et al. 1986). The majority of the case identification tools are developed by academic centers in developed countries mostly in English. Researchers have made efforts in translating most of them into different languages and have carried out their reliability and validity studies in their respective countries. It is worth stating that reliability and validity studies are mostly carried out in a small sample, mostly at academic centers based at big cities. Therefore, when these tools are then taken to rural communities for a study, it raises questions on their proper applicability in this population. Ideally, any such case identification tool should be validated first before its use in a given rural community, but that becomes a separate study in itself. It is also worth noting that most tools are developed for research studies and rarely used in routine clinical practice. Their relevance to the identification of real-world clinical cases remains dubious. It is therefore important to consider using a method or tool that detects genuine clinical cases to detect prevalence or incidence of any mental disorders. One example of such tool is the Global Mental Health Assessment Tool (GMHAT/PC) (Sharma and Copeland 2009; Sharma et al. 2004, 2008, 2010, 2013; Tejada et al. 2016; Behere et al. 2017) that is primarily developed to identify mental disorders in primary care settings and used different populations.

In any case, using any standardized method of "case" identification, the researchers must be properly trained so that they reliably (by testing their periodic inter-rater reliability) identify "valid cases" (by ensuring that methods don't include a high number of false-positive cases as well as exclude genuine cases in the study. This refers to *specificity* and *sensitivity* of the tool used in the study). A proper scrutiny of reliability and validity of methods used in rural villages is often challenging for the reasons outlined above. Henderson (2000) in his review highlights the problems of reliability of case identification methods used in mental health epidemiology by lay researchers.

Comorbidity

Mental health comorbidity is much more common than we believe. Diagnostic classifications and practices, as outlined previously, ignore comorbid states. This gives rise to twofold problems: firstly, undermining the coexistent mental health problem and, secondly, that of double counting. There is a risk of this occurring if there is a prevalence study, whereby case identification tools that are disorder-specific end up identifying the two coexisting problems as separate entities thus impacting on results. Vella (Vella et al. 2000) and Van Loo (Van Loo and Romeijn 2015) describe in detail the complexities of comorbidity of mental disorders. A full understanding of mental disorders, their impact on people's lives, and their outcome in villages will be incomplete without taking account of mental comorbid states.

Incidence and Prevalence Studies

Case identification problems relevant to incidence and prevalence studies have already been outlined. But these studies are also problematic for other reasons in the rural setting. A representative sample necessary for such studies is difficult to identify as there is no accurate documental information readily available. Household surveys are more common methods applied in these settings, but this is very resource intensive. Follow-ups are necessary for incidence and outcome studies. Poor cooperation and long-distance travel for researchers are just two of the other obstacles in securing high inclusion rates of participants.

Experimental Studies

Studies that require examination of interventions give an additional challenge in villages. Most evidence-based interventions (e.g., investigating the impact of a package of care) are developed in the western world and may not be suitable for rural and remote communities of low- to middle-income countries.

Making a good assessment of morbidity using standardized methods, providing appropriate treatment and care, and evaluating its impact in rural areas remain a daunting task. Attempts have been made using WHO mhGAP (WHO 2016) guidelines in some rural areas (Fekadu et al. 2016; Shidhaye et al. 2016) but with only limited and questionable success.

The following is a brief outline of a study funded by the Indian Council of Medical Research in villages of the Alwar District in Rajasthan, India (Gupta et al.). This case study is used to give a real-life perspective on the challenge faced when conducting rural studies, as well as some solutions addressing them.

GMHAT Case Study

The case study is based on preliminary findings of an intervention study aiming to improve access and availability of mental healthcare services in line with the National Mental Health Programme (NMHP) (Directorate General of Health Services) of India, within the existing healthcare systems. The interventions included capacity building primary healthcare workers to enable them with the opportunity to improve their diagnostic skills in identifying common psychiatric problems at the health facility level. This is to empower them to manage the mental health problems independently. In instances where they do not feel suitably equipped to manage the case at their locality, referral support was provided by periodic visit of psychiatric specialists.

In addition, the intervention included increasing awareness among rural people through educating the community members to overcome their misconceptions about mental illness. This is expected to enhance community's participation in the mental healthcare program and to facilitate efforts toward identifying symptoms and behaviors of the mental illness. Community-level awareness will help in overcoming stigma attached to mental illness and in protecting their rights. In the long term, this may reduce treatment gaps at the local (village) levels.

The study covered population in six rural centers: three (Census of India 2011) were in the experimental group, and the other three were the control group. A rapid baseline survey, using qualitative and quantitative methods, was carried out to establish baseline indicators for perceptions and awareness of the community, healthcare providers of primary healthcare, access and availability of mental health, diagnostic and treatment skills, and disease burden due mental health illnesses.

For the experimental group, formal training and education based on the Hindi version of GMHAT/PC was organized for the Auxiliary Nurse Midwives (ANMs) – the frontline health workers in villages. Successful training was delivered in 5 days, following which the ANMs were provided with "tablet" computers to be used in villages for assessment of mental health. The tablet computers were also used to communicate with the project coordinators and specialist team at the district hospital and psychiatry center of the medical college.

The control group did not receive project interventions and continued to carry on as usual without receiving any formal training and support.

The study aimed to observe differences in detection of mental disorders, treatment provided to these people, and their outcome. The study also intended to evaluate the impact of raised awareness of mental health among professionals and public on their uptake of services. The project is running satisfactorily so far. The findings will be available in the next 12 months. Nevertheless, we see a great enthusiasm for learning diagnostic and treatment skills for mental illness. They are now fluent in using diagnostic GMHAT algorithm and interviewing skills. Since the start of the project, the community perception and awareness have increased and registered an increasing turnover at the health facility at the time field visit of team of specialists and researchers.

This project highlights that the assessment and intervention packages must be developed in conjunction with existing health system, clinicians, and frontline workers who understand the local issues well.

The use of technology, such as smartphone and tablets, is feasible in carrying out studies in villages. This not only assists in assessment and data collection but equally helps in maintaining timely communication with experts.

It was identified early on in this study that ongoing and timely support from experts keeps the project's momentum at right speed and that can be provided by using communication technology.

Of note, this study used the participation of the routine workers (ANMs), unlike in other studies where research assistants are funded for a limited period. By instilling the training in the staff who works at the local facilities, it is hoped to make this project sustainable in the local rural population and serve as scalable and replicable model for adoption in the mental health program enhancing access and availability of mental health services in the vast rural areas.

Qualitative Evaluation

Qualitative methods of research are essential in epidemiological studies to have a deeper understanding of the area examined in the study. To understand any phenomenon, qualitative methods help in answering "why" and "how" questions about the area chosen for the study. This helps in understanding peoples' beliefs, values, and behavior. The methods used in rural and remote places, such as observations, in-depth interviews, and focus groups, are not easy for a number of reasons. The interviewers or observers have to be well aware of local knowledge, social structure, and language. They should have a good grasp of local language and communication skills needed to explore the areas of the study. Translators can sometimes be used for qualitative research interviews, but they are hard to find, and their reporting has a risk of misinterpretation. Long distances, time demands for qualitative research, and cooperation from rural people add further obstacles to carry out qualitative research in villages.

Ethical Issues

Obtaining informed consent of rural participants in any study is as important but equally difficult. Researchers have to make sure that they not only explain the purpose of a study but also ensure that participants fully understand how the study will run. They also understand the confidentiality issues and their ability to withdraw from the study any time they wish. These issues put pressure on researchers who are hard pressed for time. There is risk that researchers may at time cut corners and fail to follow all necessary ethical considerations.

Where a researcher finds a subject living with psychiatric morbidity, they should be able to aid or direct them toward help, and not just be content with gathering information. This is easier said than done, with limited psychiatric services available in rural areas. This can potentially leave researchers in the unenviable position of revealing significant mental illness present in the rural community but being unable to help them adequately.

Conclusion

Nearly half of the world's population still lives in rural areas and almost three-fourth in the LMICs. Strategy and planning of health and social care services for this population has to be evidence based. So far, few attempts have been made to carry out properly planned epidemiological studies to understand the needs and problems of this population. Carrying out such studies is not easy for a multitude of reasons as outlined in this chapter. There is an urgent need to invest resources to carry out properly planned epidemiological studies in the villages.

References

American Psychiatric Association (2013) Diagnostic and statistical manual of mental disorders, 5th edn, Washington, DC, https://dsm.psychiatryonline.org

Behere P, Sharma V, Kumar V, Shah V (eds) (2017) Mental health training for health professional: global mental health assessment tool (GMHAT). Indian Psychiatric Society Publication. ISBN: 978-1-68419-385-1. http://www.indianpsychiatricsociety.org

Bibby P (2013) Urban and rural area definitions for policy purposes in England and Wales. Department for the Environment, Food and Rural Affairs; Gov.UK 2013. https://assets.publishing.service.gov.uk/government/uploads/system/uploads/attachment_data/file/239477/RUC11 methodologypaperaug_28_Aug.pdf

Census of India (2011) Rural urban distribution of population. Ministry of Home Affairs. http://censusindia.gov.in/2011-prov-results/paper2/data_files/india/Rural_Urban_2011.pdf

Composite International Diagnostic Interview (CIDI) (1993) Core version 1.1. World Health Organization, Geneva

Copeland JRM, Dewey ME, Griffiths-Jones HM (1986) Computerised psychiatric diagnostic system and case nomenclature for elderly subjects: GMS and AGECAT. Psychol Med 16:89–99

Directorate General of Health Services. The National Mental Health Program. Ministry of Health & Family Welfare. Government of India. https://dghs.gov.in/content/1350_3_NationalMental HealthProgramme.aspx

Fekadu A, Hanlon C, Medhin G et al (2016) Development of a scalable mental healthcare plan for a rural district in Ethiopia. Br J Psychiatry 208(Suppl 56):s4–s12. https://doi.org/10.1192/bjp.bp.114.153676

Gibson J, Datt G, Murgai R, Ravallion M (2017) For India's rural poor, growing towns matter more than growing cities. Policy research working paper; no. 7994. World Bank Group, Washington, DC. http://documents.worldbank.org/curated/en/903311488807371325/For-Indias-rural-poor-growing-towns-matter-more-than-growing-cities

Goldberg DP et al (1978) Manual of the general health questionnaire. England NFER Publishing, Windsor

Gupta SD, Jain N, Sharma P, Sharma VK. Assessing and managing mental health problems through frontline health workers: a pilot study- funded by Indian Medical Research Council - unpublished

Henderson S (2000) Epidemiology of mental disorders: the current agenda. Epidemiol Rev 22(1):24–28

Lewinsohn PM, Seeley JR, Roberts RE, Allen NB (1997) Center for Epidemiological Studies-Depression Scale (CES-D) as a screening instrument for depression among community-residing older adults. Psychol Aging 12:277–287

Liu Z, Liu S, Jin H, Qi W (2017) Rural population change in China: spatial differences, driving forces and policy implications. J Rural Stud 51:189–197. https://doi.org/10.1016/j.jrurstud.2017.02.006

Maselko J (2017) Curr Epidemiol Rep 4:166. https://doi.org/10.1007/s40471-017-0107-y

Sharma VK, Copeland JRM (2009) Detecting mental disorders in primary care. Ment Health Fam Med 6(1):11–13

Sharma VK, Lepping P, Cummins AG, Copeland JR, Parhee R, Mottram P (2004) The global mental health assessment tool-primary care version (GMHAT/PC). Development, reliability and validity. World Psychiatry 3(2):115–119

Sharma VK, Lepping P, Krishna M, Durrani S, Copeland JR, Mottram P, Parhee R, Quinn B, Lane S, Cummins A (2008) Mental health diagnosis by nurses using the global mental health assessment tool: a validity and feasibility study. Br J Gen Pract 58(551):411–416

Sharma VK, Jagawat S, Midha A, Jain A, Tambi A, Mangwani LK, Sharma B, Dubey P, Satija V, Copeland JR, Lepping P, Lane S, Krishna M, Pangaria A (2010 Oct) The global mental health assessment tool-validation in Hindi: a validity and feasibility study. Indian J Psychiatry 52(4):316–319

Sharma VK, Durrani S, Sawa M, Copeland JR, Abou-Saleh MT, Lane S, Lepping P (2013) Arabic version of the global mental health assessment tool-primary care version (GMHAT/ PC): a validity and feasibility study. East Mediterr Health J 19(11):905–908

Sheehan DV, Lecrubier Y, Sheehan KH, Amorim P, Janavs J, Weiller E, Hergueta T, Baker R, Dunbar GC (1998) The mini-international neuropsychiatric interview (M.I.N.I.): the development and validation of a structured diagnostic psychiatric interview for DSM-IV and ICD-10. J Clin Psychiatry 59(Suppl 20):22–33

Shidhaye R, Shrivastava S, Murhar V, Samudre S, Ahuja S, Ramaswamy R, Patel V (2016) Development and piloting of a plan for integrating mental health in primary care in Sehore district, Madhya Pradesh, India. Br J Psychiatry 208(Suppl 56):s13–s20. https://doi.org/ 10.1192/bjp.bp.114.153700

Tejada P, Jaramillob LE, Garcíab J, Sharma V (2016) The global mental health assessment tool primary care and general health setting version (GMHAT/PC) – Spanish version: a validity and feasibility study. Eur J Psychiat 30(3):195–204

Van Loo HM, Romeijn JW (2015) Psychiatric comorbidity: fact or artifact? Theor Med Bioeth 36(1):41–60. https://doi.org/10.1007/s11017-015-9321-0

Vella G, Aragona M, Alliani D (2000) The complexity of psychiatric comorbidity: a conceptual and methodological discussion. Psychopathology 33(1):25–30

WHO (2016) mhGAP intervention guide for mental, neurological and substance use disorders in non-specialized health settings: mental health gap action Programme (mhGAP): version 2.0. World Health Organization, Geneva

World Health Organization (1992) The ICD-10 classification of mental and behavioural disorders: clinical descriptions and diagnostic guidelines. World Health Organization, Geneva

Vishal Bhavsar and Dinesh Bhugra

Contents

Abstract

Stigma against people with mental illness is rife. Stigma can be defined as the assignation to a person of a label that marks them out, tarnishes them, and leads to their exclusion by society. Consequently funding for research and services is very strongly affected, and that feeds into ongoing stigma. Most of the community psychiatric services are in urban areas, whereas asylums and old-fashioned institutions are on the verge of urban conurbations or in rural settings, thereby increasing isolation and perhaps contributing further to stigma. Reasons for the persistence of this stigma are many and include lack of knowledge as well as

V. Bhavsar (✉)
Section of Women's Mental Health, Department of Health Services and Population Research, Institute of Psychiatry, Psychology and Neurosciences, De Crespigny Park, London, UK
e-mail: vishal.2.bhavsar@kcl.ac.uk

D. Bhugra
Mental Health and Cultural Psychiatry, Institute of Psychiatry (KCL), London, UK

Department of Psychosis Studies, King's College London, London, UK

© Springer Nature Singapore Pte Ltd. 2020
S. Chaturvedi (ed.), *Mental Health and Illness in the Rural World*, Mental Health and Illness Worldwide, https://doi.org/10.1007/978-981-10-2345-3_31

creation of the other which confirms individual's own identity. There are some differences in attitudes toward people with mental illness between rural and urban populations. In this chapter these reasons are explored.

Keywords
Stigma · Rural settings · Urban areas · Differences · Education

Introduction

Efforts toward conceptualizing and intervening on stigma toward people with mental ill health have proven to be a rich avenue for theoretical, empirical, and policy work (Wahl 1999). These efforts offer the promise of achieving societal improvement and, ultimately, health equity between those with mental ill health and the rest of society, including those affected by physical illness (Millard and Wessely 2014). The understanding and application of the concept of stigma (Goffman 2009) to mental ill health has seen rich conceptualization in clinical samples, empirical research to characterize self-stigma experienced by people with mental illness, public stigma, and its material consequences for people affected by mental illnesses, their carers as well as their families, and interventions that have aimed at improving attitudes toward mental illness among the general public. Public campaigns to reduce stigma such as Time to Change in the UK have had some impact on attitudes toward mental illness in the short term, but there are concerns that the effects of anti-stigma campaigns on stigmatizing attitudes and behaviors toward people with mental illness may not persist (Henderson et al. 2016). One possible complementary strategy is to focus interventions on areas where stigma is more persistent; moreover, particular circumstances and social conditions in which people live, work, and study may require refined stigma intervention strategies. One such example is rural settings.

In this chapter, we consider rural attitudes to mental health. Negative, labeling, exclusionary, and discriminatory attitudes toward people with mental health problems are usually characterized as mental health stigma. Here, we compare the occurrence of these attitudes among rural dwellers with those living in urban areas. After formally defining mental health stigma, we provide an overview of research – briefly overviewing theoretical developments in how stigma has been conceptualized, defined, and measured, before describing current knowledge into which factors influence a person's attitudes to mental illness, and intervention efforts aimed at improving attitudes, including in rural settings where this data is available. We extend this description by also considering mental health stigma research that has adopted a more structural ("top-down") approach and attempt to integrate these explanations with current literature on rural mental health, focusing in particular on the role of rural infrastructures, availability of services, as well as cultural and other differences between rural and urban populations.

Definition and Conceptualization of Mental Health Stigma

Stigma is now a widely used term to describe how society excludes certain people. So powerful has the idea of stigma been that it has been argued that it is in fact a concept that has been overused and under-conceptualized (Manzo 2004). At the same time, stigma is clearly a broadly useful conceptual model to explain exclusion experienced by people everywhere – stigma is attached to gender, sexual orientation, racial background, caste, physical or psychiatric disability, as well as health status. Therefore, there is a tension between the breadth with which the concept of stigma is used (in line with its perceived utility) and its pervasiveness.

Stigma has been defined as the assignation to a person of a label that marks them out, tarnishes them, and leads to their exclusion by society. In this regard therefore, conceptualizing stigma experienced by an individual also involves a conceptualization of community, how that community works together, and how that community is experienced phenomenologically. Another challenge with defining stigma is to disentangle stigma from its consequences – if a person with schizophrenia who is qualified for an employment role is denied work because of mental illness, is this economic exclusion a *result* of stigma or a *reflection* of the stigma itself? Moreover, given the importance of the wider community for determining what is stigmatized and by whom, there are clear problems with definitions of stigma that locate it as an issue of individual attitudes. Indeed, stigma may play out at higher levels – for mental health, families with members affected by mental illness may find themselves stigmatized by the community, above and beyond stigma experienced by the person themselves. Furthermore, the intersection of mental health stigma with other exclusions is also important – communities with high prevalence of mental ill residents, for example, in institutions, are often characterized by other forms of marginalized groups, ethnic, religious, and other cultural minorities in multiple number of ways (Hartwell 2004). In some instances, these areas might themselves be stigmatized ("geographic stigma"). Rural populations are often stigmatized on a number of levels from educational to financial variations (Li et al. 2006). Stigma might also serve useful purposes for society as a whole – it is closely tied to the notion of "the other," which may provide communities with the opportunity to locate social and moral problems away from the main group. This purpose may make it very difficult for stigma to be eliminated.

For mental health professionals working in the field, as well as for patients with ill mental health, stigma feels and is indeed important because it affects investment in service development, research, and infrastructure allowing access to services. In this way, mental health stigma is also an important component of public mental health. It is likely that a person who experiences stigma because of mental illness has their health impacted negatively. Reintegration into society is impaired, and recovery is therefore difficult. On the other hand, if structures for reintegration do not exist in the first place, due to problematic infrastructure, lack of services, or lack of professional assistance, then no end of interpersonal positivity toward people with mental illness will assist in positive recovery.

Rurality, Rural Life, and Its Implications for Stigma

The definition of rurality is under tension from the increasing recognition given to place in health research and changing economic and political landscapes which are rapidly altering the layout of more remote communities. One major change is asymmetric environmental degradation which, it is supposed, will have far-reaching effects on the ability of rural dwellers to live, work, and study. While we are familiar with the concept of ever-expanding urbanization, as previously remote parts of the world are becoming increasingly developed and densely populated, there is increasing attention paid a complementary process, where specific communities are being denuded as economies, industries, and modes of production change. For example, Fraser and others examine the mental health impact of "desertification," which is a ruralification process manifest in reducing population density, decline in the agrarian sector, and demographic instability, finding that living in an area affected by reducing population density is associated with worse mental health. This highlights the importance of understanding rural residential settings as dynamic systems (Fraser et al. 2005).

There are many reasons why more and more agencies are paying attention to rural mental health. It is possibly a response to the great emphasis laid on urban health since the 1990s, a reflection of greater devolution of political and health policy-making influence to regions (which usually vary in population density and degree of urbanization), a consequence of increasing national development and economic growth among poorer countries, or a recognition that where there is health stability and attention to the psychosocial determinants of health, there is economic empowerment and political stability. Rural mental health is increasing in the spotlight, but falls into broader context of rural health and, in turn, rural policy. Nevertheless around the world, rural settings systematically receive less funding, and there is lower policy-maker and public understanding of specific rural problems, which in turn tend to receive less attention than urban health problems. Rural-urban distinctions are, in themselves, difficult to make – most studies employ definitions based on population density. However, rural settings are also typified by lower density of infrastructures, including transport links, health services, employment opportunities, educational centers, and systems for public order and safety, such as police and fire stations. These factors may in themselves be considered manifestations of policies and institutional processes which unwittingly discriminate against individuals, including people with mental illness, residing in rural communities.

Rural Environments as "Hotbeds" of Mental Health Stigma

Most previous models of mental health stigma emphasize individual characteristics of either people with mental illness or the general public (Evans-Lacko et al. 2013). There are now many critiques of the individualized model, citing evidence from social sciences that aspects of the social world may be better understood as products of structural as well as individual circumstances (Krieger 2012). This theoretical approach has been complemented by quantitative methods that allow the separate assessment of contextual and individual effects, most prominently in the form of multilevel modeling (Diez Roux 2001). What is sometimes overlooked is that stigma, according to traditional models, is framed as a problem of, by, and against individuals, rather than a problem of society as a whole. This becomes less tenable in the consideration of rural mental health stigma, given the relative difficulty in sustaining the claim that rural individuals are systematically different from urban dwellers in the same country. Thus, structural understandings of stigma, which emphasize the behavior, pronouncements, and restrictions applied to institutions and organizations within a community, may be of particular use in understanding rural mental health stigma. Indeed, mental health stigma research has been informed by aspects of research into other forms of stigma. There are likely to be similarities between prejudice experienced because of one's mental ill health and that experienced due to specific ethnic groups or religion (especially if they are in the minority) or sexual orientation. These identities are likely to co-occur in individuals, and over time in particular interactions, spaces, and life phases, particular identities will be most relevant. In this regard, the concept of micro-identities attempts to capture the variegated forms in which identity is enacted in human life and provides a model for understanding how one's identity as someone with mental illness is affected by other identities – such as a mother and a provider (Wachter et al. 2015). In this context, dwelling in a rural area might itself be considered a reason for stigma, as is seen in rural-urban economic migrants in China (Li et al. 2006).

Physical, structural, and microenvironments are important in establishing identity, through the support and input of family, school setting, and social peer groups. Rural settings also have implications for the possible effectiveness of stigma interventions and their take-up ("intention to treat"). We suggest that the relationship between stigma and rural mental health should be considered as dynamic. For example, although the recent trajectory of urbanization has suggested a worsening of rural mental health stigma, technology as applied to mental healthcare through teletherapy and e-mental health might in the future make a positive difference for rural populations (Velasquez et al. 2012). An increased emphasis on online interfaces for the assessment and evaluation of psychiatric symptoms and the increasing role of technology in the provision and monitoring of treatments could address current rural-urban inequalities in mental health treatment, with subsequent implications for reducing mental health stigma in the longer term, assuming equity of access to technological improvements between rural and urban dwelling people.

Farming is a key driver to the rural economic, social, and political structure.

At the same time, the productivity of farming is exposed to random and uncontrollable factors, which may have important implications for explanation of untoward and adverse events. Rural environments and their possible lack of alternative economic and occupational opportunities may present insurmountable challenges for farmers who fail, without obvious avenues for alternative employment. There is complex relation between mental health stigma and agrarian suicide. Merriott (2016), for example, carried out literature review into Indian famer suicides, finding that socioeconomic factors (indebtedness, illiteracy) were more strongly implicated in farmer suicides than mental health problems per se. There is now well-established evidence that farmers are more susceptible to the development of mental illness and attempted and completed suicide (Malmberg et al. 2018) and some suggestion that famers attempt suicide through more violent methods. This raises the possibility that public health interventions to reduce suicide, for example, through restricting the availability to violent methods of self-destruction and the distribution of economic resources in society, are closely connected. Access to pesticides is both an economic and occupational necessity for farming, and a potential gateway to suicide in the right context, and economic circumstances. Although specific policies to reduce farmer suicides may bring benefits, these approaches have not been evaluated, as far as we are aware. Farmers should be considered a group at high risk of mental illnesses because of economic pressures they are under, their social isolation, and the discrete set of global pressures, such as automation, outsourcing, and falling global food prices, which could directly result, in some, in suicidal behaviors (Malmberg et al. 2018). It is entirely possible that globalization may have left many rural communities behind and furthermore a lack of access to cheap migrant labor may further add to both stress and stigma. As urban communities grow, rural communities and cultural identities may become excluded and overlooked in terms of policy and economic development, and rural ways of life increasingly considered outmoded. This may lead to greater stigmatization of rural cultural groups themselves, as well as interacting with mental illness stigma to create a toxic mix of negative attitudes toward mental illness and toward the cultural group that is affected by them. Educational and socioeconomic differences may enhance the apparent greater flexibility of urban laborers in comparison to rural workers and a sense that rural workers are an increasing irrelevance to the economic and cultural productivity of nations.

Although rural mental health is rightly a concern for clinicians and policy-makers, it is worth reflecting on the overwhelming seriousness of suicide – would stigma interventions reduce this problem? Or do farmer suicides spring from deeper-seated cultural forces excluding rural populations from economic benefits of development? Gregoire asserts that farmer suicides are attributable mainly to changes in farming practice and economic factors, but aggravated by geographic isolation, inaccessibility, and more negative attitudes toward mental illness in rural dwellers (Gregoire 2002). Rural attitudes toward mental illness are likely location-specific and linked to geographically and historically located processes of exclusion. For example, Mohanty (2005) conceptualizes agrarian suicide in India as a product of a process of entrapment between aspirational economic liberalism, land reform, and decolonization on the one hand and rising debt, declining income, and falling living standards on the other. In population-based data, Vijayakumar and others (Patel et al. 2012) found that around half of suicides in India 2001–2003 were due to poisoning, predominantly with ingestion of pesticides, implying that means restriction might affect suicide rates in farmers, but also have broader public health impact. Improving mental health-related attitudes in particular occupational and geographically defined groups could have material impact on negative outcomes – reducing stigma might, for example, allow farming workers to access help and support earlier and prevent suicide at a population level. These mechanisms need urgent research.

Mental Health and Other Stigma: Microidentities and Intersections

An important factor influencing the experience of rural mental health is the intersection with physical illness, whose occurrence, course, and prognosis may also be influenced by rural infrastructure and cultural factors. Infrastructure differences in rural settings have clear implications for mental health stigma, above and beyond the possible reduced availability of health centers and treatments (and professionals to provide them). Top-down investment in infrastructure is also affected by the ability of scientific literature to accurately reflect and assess the mental health of rural dwellers – clearly there remains a large mismatch in academic interest in urban centers, compared to rural areas, because of this discrepancy in data availability. Thus, a vicious circle is set up involving limited understanding of rural lives and rural mental health, limited funding allocated to the development of rural infrastructures for mental health and awareness, a lack of research to characterize rural problems and attitudes, limited public awareness of rural difficulties, and continued limited resourcing of services and infrastructure to help rural communities and their health. Rural residence has other downstream implications – countries with low levels of urbanization tend to be less resilient to environmental degradation and disaster (Johnston 1995; Lora-Wainwright 2010), which are both important in determining population stress. Empirically, greater detail is necessary on the distinct social, historical, and cultural environments in which mental health stigma is formed, sustained, and enacted in everyday life. Literacy may also be a problem, given that rural literacy rates consistently fall below urban environments (Zhang 2006; Hannum 1999). Moreover, where literacy does exist, rural settings may have greater linguistic diversity or present other communication challenges.

Compositional and Contextual Influences on Rural Mental Health Stigma

Compositional influences on rural mental health stigma are those individual-level factors which are likely to influence stigma and which are also more common in rural environments (Zhang 2006). In this regard, the most frequently offered explanation for greater mental health stigma in rural dwellers is that rural people are more poorly educated. Moreover, food insecurity is an established risk factor for mental illness and social stress and likely to be more concentrated in rural settings (Lund et al. 2010). Living standards in rural environments may be more variable. Job opportunities may be more limited, enhancing the deleterious impact of unemployment on mental health and limit the acceptability of marginalized individuals to rural communities (Brockman and D'Arcy 1978; Clark and Binks 1966). In particular settings, rural environments may contain groups which are ethno-culturally diverse, and cultural differences are noted to be relevant for understanding stigma, for example, in UK South Asians (Gilbert et al. 2004). Although an individual's attitudes to mental illness might be shaped by cultural background, the cultural context may be conceptualized as a structural influence on stigma, as discussed below.

Jackson and others (2007) review help seeking for mental health problems in rural communities, taking a narrative review approach. Access to medical care for mental health problems is shaped by the availability of different options for care, geographic accessibility, the possible specific lack of specialized input (e.g., in more remote settings), pressures on time and staffing in rural environments, and self-recognition, awareness, and appraisal of mental health problems, their implications, and treatment. Consonant with greater stigma and exclusion of people with mental health problems in rural areas, levels of help seeking and service use for mental illness in rural settings appear to be significantly lower than urban comparators (Caldwell et al. 2004). Caldwell and others found similar prevalence of mental illness in young men in urban and rural dwellings, based on Australian survey data, and however found significantly lower help seeking in rural males. Parslow and Jorm report similar levels of help seeking behavior for mental health problems in rural residents for primary care, but lower levels of service use for both psychological and psychiatric services (Parslow and Jorm 2000). General practitioners in Australia appeared to treat fewer individuals for psychological problems, with reducing population density (Caldwell et al. 2004). Barry et al. (2000) found rural men to have lower awareness and more negative mental health-related attitudes, compared to rural females. Fox et al. (1999) found low levels of help seeking for mental illness in a cross-sectional study on a community sample. In one of the few studies directly comparing stigma and attitudes between urban and rural dwellers, Gunnell and Martin reported 30% lower rates of consultation for mental illness in rural men compared to urban men and 16% for rural women compared to urban women (Gunnell and Martin 2004), reporting evidence for an interaction between gender and rurality in the effect on help seeking (although they did not measure stigma or attitudes per se). Hoyt et al. (1997) reported greater depressive symptomatology, greater mental health stigma, and reduced willingness to seek help in a longitudinal study of American adults. In a cross-sectional study of Canadian residents (Lin et al. 1996), urban participants were more than three times more likely to use services for mental illness compared to urban participants, with evidence of interaction between urban-rural setting and use of public healthcare on the risk of help seeking for mental health problems. Significant negative attitudes to mental illness and its treatment have also been reported in a range of other studies located in rural settings (Smith et al. 2004; Wrigley et al. 2005).

It may be that in certain rural settings, there are ingrained cultural differences with respect to the urban population, which may be relevant for attitudes toward mental illness. Fuller and colleagues (2000) examine mental health-related attitudes in Australia through a questionnaire, finding that attitudes toward mental illness were shaped by (a) reluctance to acknowledge illness, (b) stigmatization of mental illness, and (c) the influence of rural and remote circumstances. Rural communities in Australia face great economic hardship, which has implications for mental health. Interviewees emphasis economic circumstances, and the sense of "collective depression" driven by failure of local/regional economic productivity, due to the seasons, or due to random variation in economic productivity over time. Rural Australians reported a style or self-image of self-reliance, invulnerability, pride, and conservatism. The investigators propose that, in the context of these factors, mental symptoms may not be identified, seen as in keeping with local eccentricities, or affected by fear of disclosure or loss of anonymity. Policies may be considered influences on mental health stigma (Corrigan et al. 2005). Corrigan's "two square model" presents models for the maintenance of psychiatric stigma (a) as interplay between attitudes and behaviors/opportunities and (b) as interplay between structural discriminations (with and without intent) and attitudes/behaviors (Corrigan et al. 2004). This although useful in many respects, but for rural communities may overlook the interlocking systems of prejudice discrimination and attitudinal differences that may exist in relation to rural settings and rural residents per se. Systematic differences in policy may exist in relation to rural environments compared to urban environments, from rules on land use making life more challenging for farmers to the unequal impact of policies on automation and healthcare provision.

Conclusions

Stigma has proven to be a useful concept to explain the life worlds of both rural dwellers and people with mental illness, but may also play an important role in identity formation of the individual who is holding the stigmatizing negative views. Under these circumstances it is important to differentiate between stigma, prejudice, and discrimination. Discrimination can be eliminated using legal frameworks for equity and equality. Reducing mental health stigma in rural settings will require coherent conceptualization of how mental health-related attitudes are shaped, including the role of systems, agencies, campaigns, and infrastructure. Stigma toward people with mental illness appears more marked in rural environments; however the explanations are unclear, but clearly have to do with more than simple differences in education, socioeconomic status, and material resources.

Future conceptualizations of rural mental health stigma should incorporate structural factors that influence attitudes toward mental illness and those factors which, deliberately or not, negatively discriminate against people with mental illness and which constitute "structural stigma." At the same time, such theories must return to the complex intersection of marginal identities, where stigma is experienced because of mental illness, rurality, and a host of other factors. Established research and practice does not fully reflect the intersectionality of mental health stigma, but future coherent understanding of mental health stigma in rural settings will require this.

Cross-References

▶ Common Mental Disorders and Folk Mental Illnesses
▶ Methodological Issues in Epidemiological Studies on Mental Health in Rural Populations
▶ Organization of Mental Health Services in Rural Areas

References

Barry MM, Doherty A, Hope A, Sixsmith J, Kelleher CC (2000) A community needs assessment for rural mental health promotion. Health Educ Res 15:293–304

Brockman J, D'Arcy C (1978) Correlates of attitudinal social distance toward the mentally ill: a review and re-survey. Soc Psychiatry 13:69–77

Caldwell TM, Jorm AF, Dear KB (2004) Suicide and mental health in rural, remote and metropolitan areas in Australia. Med J Aust 181:S10

Clark A, Binks NM (1966) Relation of age and education to attitudes toward mental illness. Psychol Rep 19:649–650

Corrigan PW, Markowitz FE, Watson AC (2004) Structural levels of mental illness stigma and discrimination. Schizophr Bull 30:481–491

Corrigan PW, Watson AC, Heyrman ML et al (2005) Structural stigma in state legislation. Psychiatr Serv 56:557–563

Diez Roux AV (2001) Investigating neighborhood and area effects on health. Am J Public Health 91:1783–1789

Evans-Lacko S, Henderson C, Thornicroft G (2013) Public knowledge, attitudes and behaviour regarding people with mental illness in England 2009–2012. The British Journal of Psychiatry, 202(s55), s51–s57.

Fox JC, Blank M, Berman J, Rovnyak VG (1999) Mental disorders and help seeking in a rural impoverished population. Int J Psychiatry Med 29:181–195

Fraser C, Jackson H, Judd F et al (2005) Changing places: the impact of rural restructuring on mental health in Australia. Health Place 11:157–171

Fuller J, Edwards J, Procter N, Moss J (2000) How definition of mental health problems can influence help seeking in rural and remote communities. Aust J Rural Health 8:148–153

Gilbert P, Gilbert J, Sanghera J (2004) A focus group exploration of the impact of izzat, shame, subordination and entrapment on mental health and service use in South Asian women living in Derby. Ment Health Relig Cult 7:109–130

Goffman E (2009) Stigma: notes on the management of spoiled identity. Simon and Schuster. New York

Gregoire A (2002) The mental health of farmers. Occup Med 52:471–476

Gunnell D, Martin R (2004) Patterns of general practitioner consultation for mental illness by young people in rural areas. A cross-sectional study. Health Stat Q 21:30–33

Hannum E (1999) Political change and the urban-rural gap in basic education in China, 1949–1990. Comp Educ Rev 43:193–211

Hartwell S (2004) Triple stigma: persons with mental illness and substance abuse problems in the criminal justice system. Crim Justice Policy Rev 15:84–99

Henderson C, Stuart H, Hansson L (2016) Lessons from the results of three national antistigma programmes. Acta Psychiatr Scand 134:3–5

Hoyt DR, Conger RD, Valde JG, Weihs K (1997) Psychological distress and help seeking in rural America. Am J Community Psychol 25:449–470

Jackson H, Judd F, Komiti A et al (2007) Mental health problems in rural contexts: what are the barriers to seeking help from professional providers? Aust Psychol 42:147–160

Johnston BR (1995) Human rights and the environment. Hum Ecol 23:111–123

Krieger N (2012) Methods for the scientific study of discrimination and health: an ecosocial approach. Am J Public Health 102:936–944

Li X, Stanton B, Fang X, Lin D (2006) Social stigma and mental health among rural-to-urban migrants in China: a conceptual framework and future research needs. World Health Popul 8:14

Lin E, Goering P, Offord DR, Campbell D, Boyle MH (1996) The use of mental health services in Ontario: epidemiologic findings. Can J Psychiatry 41:572–577

Lora-Wainwright A (2010) An anthropology of 'Cancer Villages': villagers' perspectives and the politics of responsibility. J Contemp China 19:79–99

Lund C, Breen A, Flisher AJ et al (2010) Poverty and common mental disorders in low and middle income countries: a systematic review. Soc Sci Med 71:517–528

Malmberg A, Simkin S, Hawton K (2018) Suicide in farmers. Br J Psychiatry 175:103–105

Manzo JF (2004) On the sociology and social organization of stigma: some ethnomethodological insights. Hum Stud 27:401–416

Merriott D (2016) Factors associated with the farmer suicide crisis in India. J Epidemiol Glob Health 6:217–227

Millard C, Wessely S (2014) Parity of esteem between mental and physical health. BMJ 349:g6821

Mohanty BB (2005) 'We are like the living dead': farmer suicides in Maharashtra, Western India. J Peasant Stud 32:243–276

Parslow RA, Jorm AF (2000) Who uses mental health services in Australia? An analysis of data from the National Survey of Mental Health and Wellbeing. Aust N Z J Psychiatry 34:997–1008

Patel V, Ramasundarahettige C, Vijayakumar L et al (2012) Suicide mortality in India: a nationally representative survey. Lancet 379:2343–2351

Smith LD, McGovern RJ, Peck PL (2004) Factors contributing to the utilization of mental health services in a rural setting. Psychol Rep 95:435–442

Velasquez SE, Duncan AB, Nelson E-L (2012) Technological innovations in rural mental health service delivery. In: Smalley, K. Bryant. Rural mental health: Issues, policies, and best practices. Springer Publishing Company, 2012. New York

Wachter M, Ventriglio A, Bhugra D (2015) Micro-identities, adjustment and stigma. Sage, London

Wahl OF (1999) Mental health consumers' experience of stigma. Schizophr Bull 25:467

Wrigley S, Jackson H, Judd F, Komiti A (2005) Role of stigma and attitudes toward help-seeking from a general practitioner for mental health problems in a rural town. Aust N Z J Psychiatry 39:514–521

Zhang Y (2006) Urban-rural literacy gaps in Sub-Saharan Africa: the roles of socioeconomic status and school quality. Comp Educ Rev 50:581–602

Mental Health Professionals in Rural Areas

16

Prabha S. Chandra

Contents

Abstract

Most mental health professionals all over the world prefer working in urban areas. As a result rural and remote area psychiatry is often neglected. While the situation is concerning in high-income countries, it is even worse in low- and middle-income countries. Rural psychiatry can be challenging because of poor resources, professional isolation, lack of a team, and perceived lower opportunities for professional growth.

Many countries are thinking of creative and innovative ways of recruiting and retaining psychiatrists for rural mental health services. These include ongoing training, linking them to an academic center, enhancing incentives, and encouragement to develop newer services that are context specific. In addition, a major thrust on training postgraduate trainees in rural psychiatry through immersive methods, such as rural postings, has also shown promise.

P. S. Chandra (✉)
Department of Psychiatry, National Institute of Mental Health and Neurosciences, Bangalore, Karnataka, India
e-mail: prabhasch@gmail.com; chandra@nimhans.ac.in

© Springer Nature Singapore Pte Ltd. 2020
S. Chaturvedi (ed.), *Mental Health and Illness in the Rural World*, Mental Health and Illness Worldwide, https://doi.org/10.1007/978-981-10-2345-3_26

Keywords
Rural · Psychiatry · Training · Equity · Mental health · Psychiatric services

Introduction

World over there is a dearth of mental health professionals (MHPs) working in rural areas. Even in countries where rural and remote area mental health is an established specialty, recruitment and retention in rural areas has remained a challenge. Norman Sartorius (2009) in his book *Pathways of Medicine* discusses several important issues regarding psychiatrists and the nature of care they provide worldwide. First is the lack of equity in mental healthcare in remote and rural areas with most professionals concentrating in urban centers, and second is the decrease in public mental health services and the burgeoning of private care. The third aspect that he describes is the changing values among health professionals with career choices being driven more by prestige or money rather than by a sense of purpose and altruism (with few exceptions). This is probably true in most countries except where there is limited private care and most mental health is provided by a public health service.

However, providing mental healthcare to people who live in rural areas is necessary, and no government can shy away from it. There will need to be efforts to ensure that there are enough professionals who live and work in rural areas and are able to meet the unique needs of the mental health of people living there. For this to happen, we need to understand the challenges that mental health professionals face in such situations, learn from solutions developed across the world to enhance recruitment and retention of MHPs in rural areas, and also focus on psychiatric training. We also need to consider what different mental health professionals can offer to rural mental health instead of focusing only on psychiatrists.

What Are the Challenges of Working in Rural Areas?

The challenges to working in rural areas could be professional, social, or economic. There are unfortunately not too many studies addressing this problem. A qualitative study among psychiatrists working in rural New Zealand identified the following ten barriers – feelings of isolation, the lack of a multidisciplinary team, frequently being on emergency cover or on call, the lack of academic opportunities, donning the role of both a psychiatrist and a general physician, negative attitudes toward psychiatry and psychiatrists in the rural communities, poor funding and inadequate resources, poor retention of staff, and absence of psychiatry trainees (Lau et al. 2002).

Another problem reported was the absence of peers and opportunities for second opinions which in turn contributed to the feeling of professional isolation. The rural psychiatrists interviewed in the study also reported missing the opportunities to be teachers and to be around trainees. Trainees help by sharing the workload and also provide intellectual stimulation and the opportunity for a rural doctor to teach.

Social factors included an inability to integrate into rural society, lesser opportunities for education for children, infrequent avenues for leisure and recreation, and the career of the partner being compromised.

Economic factors included lack of incentives and less opportunities for promotion. Even though the cost of living in rural areas might be cheaper, professionals reported spending money on travel to visit family who might live in urban areas and also bemoaned the lack of health facilities.

Many of these factors appear to contribute to psychiatrist's preference for urban centers and a worldwide dearth of psychiatrists in rural areas.

What Are the Charms of Working in a Rural Area?

However, all doesn't seem to be lost as there is some attraction of practicing psychiatry in rural areas. The study, among New Zealand psychiatrists described above, also asked about what was positive about working in rural areas (Lau et al. 2002). Some of the factors reported included a chance to include nonmental health caregivers in the team, lower cost of living, being able to see patients at their homes, the variety of cases seen in practice, and also being able to practice general medicine alongside psychiatry. Some other aspects included an opportunity to develop a bond with both patients and the community, flexibility in work schedules, and the chance to be close to nature. Psychiatrists also reported low stress levels and the absence of politics to be the nonmaterial perks of working in rural areas.

What Is the Situation in Low- and Middle-Income Countries?

In a recently conducted situation analysis in South Africa, all 163 public rural primary healthcare facilities (district hospitals, primary care clinics, and community health centers) were approached and asked the following questions:

- Do you have psychiatrists employed at your health facility?
- Do you have medical officers dedicated to mental health?
- Does your facility receive outreach psychiatrist services and at what frequency?
- How many sessions per week does your service dedicate to mental health?

The results revealed a glaring absence of psychiatrists in rural areas. Only seven psychiatrists (2% of the total number of SA's public sector psychiatrists) were found to be employed in rural primary care health facilities that serve a population of 17,143,872 rural South Africans. Psychiatrists were represented at a rate of 0.03 per 100,000 population, and 99 (62%) of the primary healthcare facilities did not receive monthly psychiatric outreach by a psychiatrist. This was much below the recommended psychiatrists for 100,000 populations recommended for high–middle-income countries (De Kock and Pillay 2016).

In India, the National Mental Health Survey has shown the psychiatrist/population ratio to be 0.3 per 100,000 population as against the recommendation of 1 per 100,000 population, with wide variability between regions (Gururaj et al. 2016).

The number of psychiatrists ranged from a low of 0.1–0.15 to a high of 0.56–1.2 per 100,000 population. The survey however did not focus on rural urban differences. Medical doctors trained in mental health ranged from a low of 0.05 to a high of 2.74 per 100,000. However, the district mental health program which aims mainly at rural mental health shows coverage that varies from a low of 13.7% in some states to a high of 60–199% in other states.

The situation of mental health professionals in rural areas in other countries of Africa and South Asia is even more dismal.

How Can We Enhance the Recruitment and Retention of Mental Health Professionals in Rural Areas?

Is there something we can learn from countries where rural psychiatry has long been a specialty? What have been some of the innovative models?

Most countries like the USA, Australia, and New Zealand where rural psychiatry has been established for long have tried out several ways of recruiting and retaining psychiatrists for rural areas. Most such efforts describe the main ingredient of a successful program to be one where psychiatrists and indeed all mental health professionals see a value of their work (Wilks et al. 2008; Sutton et al. 2015a; Cosgrave et al. 2016). Knowing that working in rural areas is encouraged and rewarded is important for maintaining staff morale and retention. Describing their experience in rural California where they had a mix of successful and unsuccessful programs (Ng et al. 2013) made several relevant observations. Among the successful programs was a linked outpatient and consultation liaison service for patients who had been admitted to rural general hospitals. The other was an adult day treatment center for the elderly and a clinical trials division that ran multiple research projects in collaboration with the local hospitals. Psychiatrists enjoyed working closely with trained community clinic health teams to identify risk and triage and follow up patients at home. However, services that did not work included support groups and an electroconvulsive therapy service. ECT was associated with stigma because of lack of knowledge about the treatment and fears. Support groups were not successful because of issues related to confidentiality in close-knit communities. It is therefore important to plan programs well when they are being launched and take the view of the community seriously. If a program fails, psychiatrists may get demotivated to stay on in such areas.

Another important factor that motivates rural psychiatrists to stay on in remote areas appears to be a connection with academic centers to foster collaboration, participate in teaching, and be part of research collaborations. Links with academic centers decrease the sense of isolation, help in an objective evaluation regarding the kind of work that is being done, and encourage development of new services (Bonham et al. 2014).

Table 1 How can rural psychiatry be made more attractive to practicing psychiatrists?

Decrease social isolation	Online teaching, tele-psychiatry, peer supervision, links with academic centers
Funding	Extra incentives, promotional opportunities, chances to attend conferences
Enhance professional satisfaction	Posting trainees so that rural psychiatrists get chances to teach, encourage publishing of unique experiences in rural psychiatry
Sharing load with other health professionals	General practitioners joining on-call duties, using non-psychiatrists such as psychologists and nurses to triage, developing a team

The authors emphasize the role of academic centers in building academic bridges and creating mentorship as well as teaching opportunities. In addition to encouraging psychiatrists in rural areas, it also spurs interest among young trainees toward rural psychiatry. Needless to say, the role of psychiatric leadership in fostering these partnerships is vital.

Burnout and frustrations are known to be a problem among rural psychiatrists. One way to overcome this is through providing new online resources on self-care, providing tele-psychiatry for getting connected with peers during case discussions, and continuing medical education through online or tele-psychiatry courses (Watanabe-Galloway et al. 2015) (Table 1).

How Can Training in Rural Psychiatry Be Enhanced?

It is now clear that if one wants psychiatrists to work in rural areas, then the topic should be introduced in their training itself. There need to be adequate opportunities for training and exposure to a range of rural situations. Sometimes, what is defined as rural may just be a semi-urban posting some distance from an academic center. The definition of what is a rural posting may vary from geographical location and cultural issues to even a way of practice. The trainees need to be exposed to different geographical locations and to different cultural milieus.

A curriculum that is unique to psychiatric practice in rural areas must be developed which focuses on the specific mental health problems that might be more common in rural areas, the ways in which they present, the need to be both a general physician and a psychiatrist, and ways of handling issues of stigma, boundaries, and confidentiality and of conducting home visits (Bennet-Levy and Perry 2009). The curriculum should also focus on community consultation and participation, networking with other physicians and in many countries with local healers and faith-based healing (Berntson et al. 2005; Nelson et al. 2007).

Imparting these skills involves using innovative methods such as creating a rural psychiatry fellowship and clinical and contextual teaching with rural medical mentors. It has been suggested that rather than visits to rural areas or once a month postings, a longer rural posting where trainees live in the area and get to understand the community and the working patterns is more useful.

There also needs to be involvement of leadership in academic centers to ensure funding for transport and housing for sending trainees to rural areas. Finding a way for the academic center to stay in touch and have teaching through video conferencing when trainees are in rural areas is also important. Using medical informatics to coordinate rural curricula, research, or continuing education have also been successful methods.

The important thing to remember is that exposure to rural psychiatry should start early on in one's career. Once a physician becomes settled in a certain way of life in an urban area, it is more difficult to relocate. Some other ways of increasing recruitment include admission policies that prioritize students from rural areas, taking in students with an aptitude for rural practice, maximizing the quality of the educational experience, encouraging doctors from rural areas to go back to their roots, and starting rural medical colleges which take students from that area who are more likely to give back to their own communities.

In the department of psychiatry at the National Institute of Mental Health and Neurosciences, we have developed a 2-week rural psychiatry training program as part of the community psychiatry training. Residents in their second year of postgraduate training are posted to a rural district mental health program where they live in the rural area and work with the district mental health team. During this period they conduct clinics at primary health centers, are involved in awareness programs in rural schools and colleges, and make home visits. Preparation for the posting involved several visits and meetings from the team at NIMHANS to the district, formal permissions, collaborating actively with the local district team leader, and pre-posting briefing of the trainees. While the program has just begun and is showing acceptance from the trainees as well as positive feedback from both the trainees and the trainers, its long-term feasibility and gains achieved will need to be evaluated systematically (Fig. 1).

What About Other Mental Health Professionals?

Compared to studies among psychiatrists, the literature on psychologists and mental health nurses working in rural areas is quite inadequate. Most studies among psychologists are from Australia and focus on competencies that clinical psychologists may require to work in rural areas.

Using qualitative research methods, competencies that rural psychologists identified included being able to handle professional isolation, manage rural life, work without too much supervision, deal with a lack of support services, become used to issues such as visibility in the community, and handle concerns of confidentiality. Psychologists also identified resilience, autonomy, and confidence as important personality attributes of being able to work in rural life (Conomos et al. 2013; Sutherland and Chur-Hansen 2014).

Psychologists who placed less value on academic prestige and more value on service were thought to be more successful in rural areas. These findings are important in recruitment.

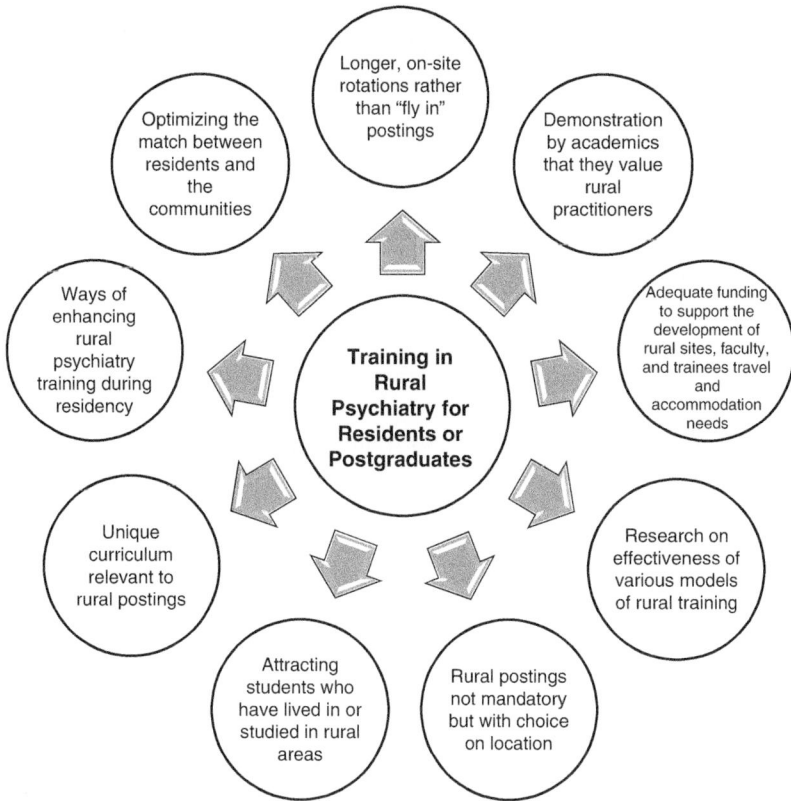

Fig. 1 Ways of running a successful rural mental health training for postgraduates in psychiatry

There is even less information on mental health nurses in rural areas (Sutton et al. 2015b). Efforts to integrate mental healthcare into primary care rely heavily in countries like South Africa on nonpsychotic mental health professionals such as mental health nurses.

Similar to their study on psychiatrists working in rural areas in South Africa, De Kock and Pillay (2016) did a situation analysis of mental health nurses in the 160 rural South African primary care centers. They describe an acute shortage of mental health nurses in rural South Africa's public health facilities. Only 62 (38.7%) of the 160 facilities employ mental health nurses. These nurses served an estimated population of more than 17 million people, suggesting that they were employed at a rate of 0.68 per 100,000 population in South Africa's public health rural facilities. Compared to the national average of 9.7 mental health nurses per 100,000 population, this is again inadequate (De Kock and Pillay 2017).

The World Health Organization (WHO) has recommended the use of task shifting and task sharing as a way to deal effectively and rapidly with mental health workforce shortages. Task shifting involves shifting some tasks such as routine medical examination (BP, weight, pulse) and basic prescribing of certain medications to cadres of nonmedical health professionals who have been specifically trained for the purpose. In many parts of the world, pharmacists and psychologists have been trained to prescribe basic psychotropic medications to overcome shortage of mental health professionals in areas of shortage. However, there is a need to develop proper systems of training and ongoing supervision when task shifting or task sharing is used.

Conclusions

It is quite clear that rural mental health should be a priority in most countries both because of need and the lack of professionals. The leadership needs to pay attention to this problem, find innovative ways of recruitment and retention, start discussing the option of working in rural areas early in a psychiatry student's postgraduate career, and make rural mental health a part of clinical postings during training. In addition, there is also a need to pay attention to other mental health professionals so that a team approach is used which may handle the professional isolation often felt by psychiatrists working in rural areas. The appropriate use of technology to link them to academic centers and for ongoing training is another important strategy to reduce burnout.

Cross-References

▶ Common Mental Disorders and Folk Mental Illnesses

References

Bennett-Levy J, Perry H (2009) The promise of online cognitive behavioural therapy training for rural and remote mental health professionals. Australas Psychiatry 17(1_suppl):S121–S124. https://doi.org/10.1080/10398560902948126
Berntson A, Goldner E, Leverette J, Moss P (2005) Psychiatric training in rural and remote areas: increasing skills and building partnerships. Can J Psychiatr 50(9):1
Bonham C, Salvador M, Altschul D, Silverblatt H (2014) Training psychiatrists for rural practice: a 20-year follow-up. Acad Psychiatry 38(5):623–626
Conomos AM, Griffin B, Baunin N (2013) Attracting psychologists to practice in rural Australia: the role of work values and perceptions of the rural work environment. Aust J Rural health 21(2):105–111
Cosgrave C, Maple M, Hussain R (2016) Factors affecting job satisfaction of Aboriginal mental health workers working in community mental health in rural and remote New South Wales. Aust Health Review. https://doi.org/10.1071/ah16128
De Kock JH, Pillay BJ (2016) Mental health nurses in South Africa's public rural primary care settings: a human resource crisis. Rural Remote Health 16(3):3865
De Kock J, Pillay B (2017) A situation analysis of psychiatrists in South Africa's rural primary healthcare settings. Afr J Prim Health Care Fam Med 9(1). https://doi.org/10.4102/phcfm. v9i1.1335

Gururaj G, Varghese M, Benegal V, Rao GN, Pathak K, Singh LK (2016) National mental health survey of India, 2015–2016: prevalence, patterns and outcomes. National Institute of Mental Health and Neuro Sciences, Bengaluru

Lau T, Kumar S, Thomas D (2002) Practising psychiatry in New Zealand's rural areas: incentives, problems and solutions. Australas Psychiatr 10(1):33–38. https://doi.org/10.1046/j.1440-1665.2002.0389a.x

Nelson W, Pomerantz A, Schwartz J (2007) Putting "Rural" into psychiatry residency training programs. Acad Psychiatry 31(6):423–429. https://doi.org/10.1176/appi.ap.31.6.423

Ng B, Camacho A, Dimsdale J (2013) Rural psychiatrists creating value for academic Institutions. Psychiatr Serv 64(11):1177–1178. https://doi.org/10.1176/appi.ps.641110

Sartorius N (2009) Pathways of medicine. MedicinskaNaklada, Zagreb; 185p, (ISBN 978-953-176-454-4)

Sutherland C, Chur-Hansen A (2014) Knowledge, skills and attitudes of rural and remote psychologists. Aust J Rural Health 22(6):273–279. https://doi.org/10.1111/ajr.12152

Sutton K, Maybery D, Patrick K (2015a) The longer term impact of a novel rural mental health recruitment strategy: a quasi-experimental study. Asia Pac Psychiatry 7(4):391–397. https://doi.org/10.1111/appy.12183

Sutton K, Patrick K, Maybery D, Eaton K (2015b) Increasing interest in rural mental health work: the impact of a short term program to orientate allied health and nursing students to employment and career opportunities in a rural setting. Rural Remote Health 15:3344

Watanabe-Galloway S, Madison L, Watkins KL, Nguyen AT, Chen L (2015) Recruitment and retention of mental health care providers in rural Nebraska: perceptions of providers and administrators. Rural Remote Health 15:3392

Wilks CM, Oakley Browne M, Jenner BL (2008) Attracting psychiatrists to a rural area-10 years on. Rural Remote Health 8(1):824

Mental Health Literacy in Rural India

17

Journey Forward

Meena Kolar Sridara Murthy, Madhuporna Dasgupta, and
Santosh Kumar Chaturvedi

Contents

Abstract

Poor knowledge about mental health is one of the main barriers for seeking
treatment in our country, especially among rural areas. It also contributes significantly toward stigma and discrimination that persons with mental disorders face.
Early recognition and intervention for persons with mental health problems have
positive outcomes. Mental health literacy is being seen as one of the key
indicators in predicting successful outcomes in persons with mental disorders.
People with mental illness are subjected to discrimination and are treated with
contempt due to the deep-rooted stigma present in the rural society. Due to this,
there is a need to understand the local and cultural contexts of stigma to develop
effective and innovative methods to change such attitude. With newer research,

M. K. S. Murthy (✉) · M. Dasgupta
Department of Mental Health Education, National Institute of Mental Health and Neurosciences,
Bangalore, Karnataka, India
e-mail: meenaksiyer@gmail.com; madhupornadg@gmail.com

S. K. Chaturvedi
Psychiatric Rehabilitation Services, Department of Psychiatry, National Institute of Mental Health
and Neurosciences, Bangalore, Karnataka, India

Department of Mental Health Education, National Institute of Mental Health and Neurosciences,
Bangalore, Karnataka, India
e-mail: skchatur@hotmail.com

© Springer Nature Singapore Pte Ltd. 2020
S. Chaturvedi (ed.), *Mental Health and Illness in the Rural World*, Mental Health and
Illness Worldwide, https://doi.org/10.1007/978-981-10-2345-3_34

today, various new techniques have been implemented to increase the mental health literacy among the rural societies in India. This chapter will look into the diverse methods of increasing mental health literacy among the rural population in India.

Keywords
Mental health literacy · Stigma · Knowledge · Help seeking · Rural India

Introduction

Knowledge and information about mental health and mental illness are important not only for detection and treatment of mental illnesses but also for mental health promotion and prevention of mental illnesses. Adequate and correct knowledge about mental health and mental illnesses is needed for minimizing of stigma toward mental illness and towards those suffering from them. Mental health literacy (MHL) is a concept, the importance of which has grown lately in the domain of health literacy (HL). Mental health literacy refers to the "knowledge and beliefs about mental disorders which aid their recognition, management or prevention" (GlobeNewswire 2018). Mental health literacy can be explained as the awareness and knowledge about mental health and its related problems that help in early recognition, management, as well as prevention of the same to lead an overall healthy life. In other words, mental health literacy is the very foundation of the promotion of positive mental health so that best available care can be provided to a person in need (Kutcher et al. 2016). Mental health literacy is a part of health literacy that shapes understanding about our own health which in turn aids us to become responsible individuals living a successful and healthy life (Kutcher et al. 2016). The importance of mental health literacy cannot be underestimated as knowledge about mental health at large help us to take good care of ourselves, our friends and families, as well as people in our community. Additionally, it makes us empowered individuals living in a better society.

Mental Health Literacy in Rural India

Despite its importance, MHL still seems to be quite a neglected area of health literacy, especially in the developing countries. According to the National Mental Health Survey, 2015–2016, the Indian community is burdened with mental disorders. Statistically speaking, around 11% of people in India above 18 years suffer from mental disorders do not receive proper treatment (Pradeep et al. 2018). One of the reasons of such poor treatment seeking is because of the limited knowledge people have with respect to mental health disorders. For instance, in a study assessing the mental health literacy among adolescents of rural South India, the percentage of mental health literacy among the participants was quite low

(Ogorchukwu et al. 2016). Studies in rural Maharashtra show that people in the community tend to keep a social distance from a person having mental health disorder, perceiving it to be dangerous (Kermode et al. 2009). Caregivers of people with mental health disorders also express feelings of shame about a family member having mental illness (Poreddi et al. 2015).

Mental health literacy in rural India is expected to be relatively poorer than in urban areas. Despite all the cultural challenges, there are quite a few numbers of researches showing that people in the community are aware of the concept of mental health to some extent. A study assessing the community attitude toward mentally ill among students from medical and arts college in Gujarat showed that the students had a positive attitude toward mentally ill (Lakdawala and Vankar 2016). Education having made a positive impact on the literacy level of people is proven through researches done from time to time. One study assessed the impact of mental health training program for community health workers on their knowledge and attitude toward mental health. The study was conducted for 3 months, to assess the efficacy of the training program so conducted. The results of this study conducted in a rural district of Bengaluru, India, showed that the training improved the mental health literacy among the community health workers and also played a role in reducing the stigmatizing attitude of the participants toward mental health (Armstrong et al. 2011).

Nevertheless, the success stories of mental health literacy are still in its early stage. One of the reasons for the same may be poor treatment turn-up. Thus, the question of the hour is what can lead to such poor treatment turn-up? The society continues to view a person with mental health problems to be dangerous and harmful. People in the community tend to stay away from persons going through a mental health crisis. Stigma and lack of support prevent a person with mental health problems from seeking professional help. This delays the treatment seeking and increases the chances of disability in general. Ideally, every person in the community should be aware about mental health. In order to catalyze the procedure, the community health workers and social health activists should be educated with respect to the same, as they are the first contact in the community; and they in turn can increase the knowledge about mental health among the public.

Researches in India as well as in other developing countries show that mental health literacy is quite limited. One such study exploring the perception of the social health activists on depression in urban Bengaluru showed that although the participants identified it to be a form of *manoroga* (mental illness), the level of their knowledge with respect to the specific signs and symptoms and the need for treatment were quite limited (Kapanee et al. 2018). Limited knowledge and high stigma were also seen among health providers in Gujarat with respect to depression (Almanzar et al. 2014). One method to overcome this deficit is by providing proper training to the health activists and community health workers about various key aspects of mental health. A number of researches assessing the attitude of the community health workers before and after a short training program have shown positive results. An interesting study carried out by Maulik et al. demonstrated that the use of mobile technology has increased the use of mental health services in rural India (Maulik et al. 2017). The research mentions about training of the village health workers and primary care doctors in screening people with common mental health disorders of the community, with the help of an electronic decision support system (Maulik et al. 2017). Thus, practical training of the community health workers gives them a clear understanding of mental health that in turn can help provide proper information in the community.

Continuing on the theme of researches on the mental health literacy among the Indian population, it is seen that the scenario does not differ much among the rural and urban population. A study on the perception and attitude toward mental illness in South Delhi showed that the attitudes of the participants toward mental illness were stereotyping and pessimistic (Salve et al. 2013). Venkatraman et al. assessed the stigma toward mental illness among higher secondary school teachers in Puducherry in South India. It showed that around 70% of the participants experienced stigma toward mental illness (Venkataraman et al. 2019). Thus, the paucity of mental health literacy among the Indian population is aggravating stigma, myths, and misconceptions related to mental illness. Hence, stigma gets all the more strengthened due to lack of proper education methods and improper information dissemination.

Proper information dissemination through mass media among the community can also aid mental health literacy. However, the contrary is seen to occur in the Indian context. Portrayal of mental illness in a derogatory and stigmatizing manner by the media is quite a common phenomenon (Gururaj et al. 2016). *Pagal* (mad) is a common term used in movies to characterize a person having a slight behavioral change (Gururaj et al. 2016). In other times, a character with mental illness is shown as felon or violent in the movies in India (Gururaj et al. 2016). Such depiction of mental illness shown in movies and other sources of media stands far from the actual clinical practice. The portrayal of electroconvulsive therapy in the Indian cinemas as well as in Hollywood is mostly inaccurate, fictitious, and scientifically inaccurate. Thus, such depictions are a source of misinformation among the general public (Sharma and Malik 2013).

Mental health literacy among the people in the community has been an area of interest among number of researchers worldwide. The deep-rooted stigma, the lack of support, and the ignorance, being some of the factors, among the community have made it crucial to assess the level of mental health literacy of the public so that proper steps can be taken to increase mental health literacy in order to ensure better treatment seeking. For instance, a study done in China with respect to understanding the level of mental health literacy among the public showed that effort to increase mental health literacy in rural areas of China is the need of the hour, which can resolve the problems of treatment gap in the country (Huang et al. 2019). In yet another study among rural women of south central Appalachia (United States), it was identified that few of the barriers of mental health screening among primary care clinic are stigma, lack of support, and also lack of education (Hill et al. 2016). Thus, lack of education leading to poor mental health literacy needs to be continually addressed in today's world. A study in the rural communities of Australia identified that people tend to avoid availing of mental health services or even to acknowledge the fact that they are going through a mental health crisis. This is because of the high degree of stigma and more importantly the fear they associate with mental health problems. People who avail mental health services are labelled as "weirdos," and thus, the last thing the common people would like to do in the community is to seek help (Fuller et al. 2000). This throws a considerable amount of light on the mental health literacy in rural and remote Australia. The perspective of mental health among the rural African-Americans is again somewhat similar (Sullivan et al. 2017). Even though the community members showed concern about the widespread stigma attached to mental illness, they were unable to identify the mental health problems and when one should seek help (Sullivan et al. 2017). Results of the investigations done on the level of community mental health literacy in sub-Saharan Africa show that most of the participants were unable to identify psychiatric symptoms. It also showed that the belief in supernatural and ultra-human views was quite common and various forms of alternative mental health services were preferred (Atilola 2015).

Similar findings were seen in an investigation on mental health literacy carried out in Pakistan. Statistical analysis done in the study showed that the type of disorder, education, and area of living significantly contributed one's ability to diagnose the presence of mental illness. The study further showed that a considerable number of participants thought that hakims, homeopaths, and religious or magical healers are the most appropriate people to contact in case of a mental health crisis (Suhail 2005). Lack of knowledge is related to decrease in mental health literacy. A study done to assess the attitude of Sri Lankan medical undergraduates and doctors toward mental illness revealed that the medical undergraduates showed compara- tively more negative attitude toward mental illness compared to the doctors. In general, a high level of stigma was identified toward patients with depression and alcohol and drug addiction. There was a consistent high degree of blaming attitude toward persons with mental illness (Fernando et al. 2010). Thus, we see that the stigma toward mental illness is so deep rooted that even medical professionals are found to have the same. Such negative attitude among medical professionals can negatively affect the treatment procedures of patients with mental illness. The same study also mentions one way of reducing such stigma. If ensured that medical students have contact with recovered patients in the community psychiatric settings, it may decrease the stigmatizing attitude to a great extent (Fernando et al. 2010).

Another cross-cultural study was carried out to investigate the ability to identify different psychiatric problems among British, Hong Kong, and Malaysian participants. It showed that the British participants could identify most of the case vignettes, followed by Hong Kong Chinese and Malaysian. What is interesting to see in this study is that when it was investigated about the type of help the participants would like, it was seen that a higher percentage of British participants recommended professional help com- pared to Hong Kong and Malaysian participants (Loo et al. 2012).

To address this issue, attempts have been made by some countries to conduct programs at the community level for providing psychological first aid for people having mental illnesses. The results shared by researchers across the world have shown positive results in this context. A typical psychological first aid program consists of sessions for identifying symptoms of mental health disorders like acute suicidal thoughts; depressive, anxiety, or psychotic behavior; their possible risk factors; and how to get professional help. One such program which deserves a mention is the Young Mental Health First Aid Program, conducted in Australia with an aim to help young adults learn skills required to recognize early signs of mental illness of adolescents and provide help as and when required (Kelly et al. 2011). This study done among the Australians assessed their knowledge and attitude on mental health and also showed that the participants expressed interest in knowing about mental health disorders so that they can be empowered enough to identify the warning signs of a mental health crisis (Kelly et al. 2011). These Mental Health First Aid (MHFA) training programs for identifying mental disorders after gaining such positive results have begun to roll out in other countries as they are being accepted worldwide as an excellent method of empowering people at a community level as well as increasing mental health literacy. Based on the positive feedback on these training programs, similar training programs were conducted for the Chinese as well as Vietnamese communities in Melbourne, Australia, to help members of these communities to identify persons with mental illness (depression and schizophrenia) and help in seeking treatment (Lam et al. 2010; Minas et al. 2009). As culture plays an important part when it comes to disseminating mental health literacy, such training programs were conducted, specifically for various cultural groups within Australia, like for the Australian Aboriginals and Torres Strait Islander people (Kanowski et al. 2009). The Mental Health First Aid Program was also conducted among the Bhutanese refugees in the United States which again had a positive impact. The pre- and post-assessments of the study showed that the program had significantly helped the participants to identify the symptoms of psychiatric illness (Subedi et al. 2015). Thus, programs like this need to be encouraged in the community for facilitation of increased mental health literacy.

However, people in the community should also be empowered to know the day-to-day mental health crisis which one can go through and those which need attention. Efforts should be made in identifying the symptoms of the mental health crisis, as they can act as a predisposing factor to a forthcoming mental illness in an individual. It is thus essential to empower general public with knowledge and skills to provide support to people with mental health problems. Efforts toward this have been made from 2016 by the National Institute of Mental Health and Neuro-Sciences [NIMHANS], Bengaluru, Karnataka, India, to create such community empowerment programs to educate people about the day-to-day mental health problems which one can go through. Such a program includes understanding of the psychological impact of a range of mental health crisis like loss of job to bereavement of a loved one and how to deal with the same. These training programs are designed to help common people in the community with early identification of crisis among others, to provide preliminary support to people facing a crisis, and also to refer them to appropriate medium for an effective treatment early recovery. This initiative is quite different from the other first aid training programs where mental illness is emphasized in particular. The objective of the program is to assess and increase knowledge of mental health problems and build the skills needed to intervene, assist, and refer individuals experiencing a mental health issue. The 3-h workshop program conducted by NIMHANS has been a success in the community as it has helped to sensitize people about mental health problems and also taught them skills in providing the preliminary help required by someone with mental health problem. It is an excellent method to make people aware about the mental health problems which one can encounter during their lifetime due to a number of causes and situations. Such a program can also help in sensitizing people in the community so that they can be empathetic to the people who are facing a mental health crisis. They can help the people facing mental health problems with the support and care which are very crucial in any kind of mental health problem.

Art therapies, like dance, theater, folklore, and street plays apart from providing entertainment, have their origins in India from decades and are one of the means of imparting health education in rural India. Loganathan et al. (Loganathan and Varghese 2015) in 2015 used street play to enhance knowledge about mental illness in a rural population near Bangalore. A script for a street play with a local theater group on schizophrenia was devised that enhanced knowledge and shift attitudes and beliefs about mental illness in a manner familiar to them. The researchers were able to successfully conduct street plays on mental illness in a rural setup in Bangalore, in a feasible manner, keeping in the cultural aspects of the community.

In another study conducted by John et al. (2015) in 2015 in Bengaluru, psycho-drama was used to target and enhance knowledge and attitude of community members toward mental illness. The drama communicated the common symptoms of mental illness, myths and misconceptions about mental illness, need for proper treatment, and long-term medication and possibility of recovery. The community-based intervention brought about changes in the stigma, reduced discrimination, and increased social acceptance and social support of the family members of persons diagnosed with schizophrenia.

The Schizophrenia Research Foundation [SCARF] in Chennai, India, is another organization, which has been organizing film festival called the "Frame of Mind," which features several films portraying mental illness and an international competition for short films on mental health and stigma. This method has been a huge success and has had three editions so far. Similar festivals have since been held in other cities like Kolkata. Many NGOs use short films to spread awareness about their work/cause, but the efficacy of such interventions is not evaluated.

Culturally specific media of communications are accepted and powerful tools in the rural communities. They can be used for enhancing mental health literacy in rural communities successfully. In an article titled "Destigmatizing day-to-day practices: What developed countries can learn from developing countries," the author mentions how societal changes like being more inclusive and re-integrating people with mental illness in the society can make a huge difference. This includes finding socially useful and culturally valued works of people with mental illness (Rosen 2006). Many newer studies have been initiated looking at their efficacy with specific focus on theater and drama. Use of media can be an apt method of increasing mental health literacy if used in the right direction. Use of community radio for correct and proper information dissemination can help increase mental health literacy as it has the capability to reach out to a large number of people in no time. Sharing the success stories of people suffering from mental health problems and how treatment helps them through print media can inculcate positive mental health outlook among the people in the community. Media professionals should work in close contact with the mental health professionals, thereby gaining proper information for the masses. More movies should concentrate on promoting positive mental health so that the community is educated positively about mental health (Tables 1 and 2).

Table 1 Major researches on mental health literacy in the Indian context (2009–2019)

Authors and place	Main findings
Venkataraman et al. (2019) Puducherry	Most of the participants showed perceived stigma toward case vignettes of mental illness
Kapanee et al. (2018) Bengaluru	Inadequate knowledge about the signs and symptoms of depression among the participants
Maulik et al. (2017) Rural Andhra Pradesh	Rural health activists were empowered with the mobile technology to identify persons with mental illness to reduce treatment gap
Lakdawala and Vankar (2016) Gujarat	Significantly negative attitude toward persons with mental illness Mentally ill people were considered to be violent
Ogorchukwu et al. (2016) Udupi, Karnataka	The percentage of mental health literacy among respondents was quite low Prevalence of stigmatizing attitude toward mental illness
John et al. (2015) Bengaluru	Community-based intervention brought a positive change, reduced the discrimination, and increased the social acceptance of persons with mental illness
Loganathan and Varghese (2015) Rural Bengaluru	Use of street play had a positive response as a form of awareness method to increase mental health literacy
Poreddi et al. (2015) Bengaluru	61.5% of respondents opined that people with mental health problems blame for their condition Participants also opined that if they or their family members had some mental illness, they would be ashamed and would not want to tell other people
Almanzar et al. (2014) Gujarat	Considerable stigma and misinformation about depression were identified
Kumar (2013) Punjab	There is a need to improve the mental health literacy of the community people in Punjab
Salve et al. (2013) South Delhi	Respondents showed negative attitude toward persons with mental illness in the form of stereotyping, restrictiveness, and pessimistic prediction Few respondents preferred treating mental illness with the help of tantric/ojha
Sharma and Malik (2013) Bollywood movies	Portrayal of ECT in Indian cinemas is misleading, unscientific, and fictional
Armstrong et al. (2011) Rural Bengaluru	Mental health literacy among the participants was poor before the training Training program improved the ability of participants to identify the disorders
Kermode et al. (2009) Rural Maharashtra	False beliefs and negative attitudes toward mental illness were prominent Persons with mental illness were considered dangerous

Table 2 Major researches on mental health literacy in the international context (2000–2019)

Author and place	Main findings
Huang et al. (2019) China	Recognition of mental health problems is low by the people in the community
Sullivan et al. (2017) Rural African-American community	Respondents were concerned about the widespread stigma toward mental illness However, respondents thought that the people in the community are not capable of identifying the disorders
Hill et al. (2016) Rural Appalachia	Factors like stigma, lack of support, and lack of education act as hindrances for mental health screening and treatment
Subedi et al. (2015) Bhutanese community in Australia	The Mental Health First Aid Program was a promising approach to improve the mental health literacy among the Bhutanese refugees in Australia
Atilola (2015) Sub-Saharan Africa	Respondents were largely unable to identify the common mental disorders Supernatural and ultra-human views were quite famous Alternative mental health services were preferred
Loo et al. (2012) Hong Kong, Malaysia, Britain	Hong Kong and Malaysian populations were poor in identifying the mental illness compared to the British population
Kelly et al. (2011) Australia	The Mental Health First Aid Program improved the mental health literacy among the sample. They could recognize the disorders better
Lam et al. (2010) Chinese community in Melbourne	The Mental Health First Aid training program significantly improved participants ability to identify the mental illness
Fernando et al. (2010) Sri Lanka	Study revealed high levels of stigma toward people with depression and alcohol and drug addiction Blaming the people with mental illness was common Medical students showed negative attitude toward persons with mental illness
Anstiss and Ziaian (2010) Afghanistan, Bosnian, Iran, Iraq, Liberian, Serbian, and Sudanese community in Australia	The study found that participants were reluctant to seek help beyond their friendship network in case of mental health problems
Kanowski et al. (2009) Torres Strait Island	The culturally appropriate training program empowered the participants to identify mental health issues
Minas et al. (2009) Vietnamese community in Melbourne	Participants could recognize mental health problems better after the training program Negative attitude toward persons with mental illness significantly declined the post-training program

(continued)

Table 2 (continued)

Author and place	Main findings
Suhail (2005) Pakistan	Depression was identified better than psychosis Few respondents preferred going to hakims and homeopath for treatment, while few others preferred religious and magical healers to be appropriate
Fuller et al. (2000) Australia	The following themes influence help-seeking behavior in rural communities: Reluctance to acknowledge mental health problems Avoidance of appropriate help Stigma and avoidance of mental health services

Implication

The key adverse effect of low mental health literacy is the wide treatment gap. It reinforces the stigma related to mental illness which makes it hard for people to seek treatment. Even if some people do turn up for treatment, they do not adhere to advice and treatment. The popularity of witchcraft and beliefs on quacks in rural parts of India makes it difficult for mental health experts to imbibe concepts related to mental health literacy among the rural population (John et al. 2015).

Future Directions

There is a lot of scope to bring about awareness among the Indian community. Government focus on mental health literacy is something that India needs. The reach of the accredited social health activists (ASHA workers) seems to be promising as they are the first contact person in the community. They work as the intermediary between mental health professionals and the common people in the community. Experts are also joining hands with the common people in rural areas to increase mental health literacy through indigenous methods. Inclusion of user-friendly health applications for healthcare workers in rural areas may empower them to help people in diagnosing mental health problems and educate them about treatment plans and the advantages for the same. Inclusion of mental health education program by the government for proper awareness will not only help the health workers but in turn increase the mental health literacy among people in the community at large.

Conclusion

Promotion of mental health literacy is an important part of health service to the community. To ensure this, people must become aware of their mental health needs and have adequate mental health literacy. There are a number of ways in which this can be done. Campaigns, indigenous mediums of mass communication, government training programs, inclusion of mental health literacy in school curriculum, and proper dissemination of information through the media promoting positive mental health are the need of the hour.

References

Almanzar S, Shah N, Vithalani S et al (2014) Knowledge of and attitudes toward clinical depression among health providers in Gujarat, India. Ann Glob Health 80(2):89–95. https://doi.org/10.1016/j.aogh.2014.04.001

Armstrong G, Kermode M, Raja S et al (2011) A mental health training program for community health workers in India: impact on knowledge and attitudes. Int J Ment Heal Syst 5(1):17

Atilola O (2015) Level of community mental health literacy in sub-Saharan Africa: current studies are limited in number, scope, spread, and cognizance of cultural nuances. Nord J Psychiatry 69(2):93–101

De Anstiss H, Ziaian T (2010) Mental health help-seeking and refugee adolescents: qualitative findings from a mixed-methods investigation. Aust Psychol 45(1):29–37

Fernando SM, Deane FP, McLeod HJ (2010) Sri Lankan doctors' and medical undergraduates' attitudes towards mental illness. Soc Psychiatry Psychiatr Epidemiol 45(7):733–739

Fuller J, Edwards J, Procter N, Moss J (2000) How definition of mental health problems can influence help seeking in rural and remote communities. Aust J Rural Health 8(3):148–153. https://doi.org/10.1046/j.1440-1584.2000.00303.x

GlobeNewswire (2018) For immediate release: the importance of mental health literacy among all of the awareness. Available via: https://www.globenewswire.com/news-release/2018/01/30/131 4605/0/en/Forimmediaterelease-The-importance-of-mental-health-literacy-among-all-of-the-a wareness.html. Last accessed 22 July 2019

Gururaj G, Varghese M et al (2016) National Mental Health Survey of India, 2015–16: prevalence, patterns and outcomes. NIMHANS Publication No. 129. National Institute of Mental Health and Neuro Sciences, Bengaluru

Hill SK, Cantrell P, Edwards J, Dalton W (2016) Factors influencing mental health screening and treatment among women in a rural south central Appalachian primary care clinic. J Rural Health 32(1):82–91. https://doi.org/10.1111/jrh.12134

Huang D, Yang LH, Pescosolido BA (2019) Understanding the public's profile of mental health literacy in China: a nationwide study. BMC Psychiatry 19(1):20

John S, Muralidhar R, Raman KJ, Gangadhar BN (2015) Addressing stigma and discrimination towards mental illness: a community-based intervention programme from India. J Psychosoc Rehabil Ment Health 2(1):79–85

Kanowski LG et al (2009) A mental health first aid training program for Australian Aboriginal and Torres Strait Islander peoples: description and initial evaluation. Int J Ment Heal Syst 3 (1):10

Kapanee ARM, Meena KS, Nattala P, Manjunatha N, Sudhir PM (2018) Perceptions of accredited social health activists on depression: a qualitative study from Karnataka, India. Indian J Psychol Med 40(1):11. https://doi.org/10.4103/IJPSYM.IJPSYM_114_17

Kelly CM, Mithen JM, Fischer JA et al (2011) Youth mental health first aid: a description of the program and an initial evaluation. Int J Ment Heal Syst 5(1):4

Kermode M, Bowen K, Arole S, Pathare S, Jorm AF (2009) Attitudes to people with mental disorders: a mental health literacy survey in a rural area of Maharashtra, India. Soc Psychiatry Psychiatr Epidemiol 44(12):1087–1096. https://doi.org/10.1007/s00127-009-0031-7

Kumar R (2013) Attitude to people with mental illness: a mental health literacy survey from Punjab State. Int J Health Sci Res 3:135–145

Kutcher S, Wei Y, Coniglio C (2016) Mental health literacy. Can J Psychiatry 61:154–158. https://doi.org/10.1177/0706743715616609

Lakdawala B, Vankar GK (2016) Mental health literacy amongst college students: a community based study. Indian J Ment Health 3(3):342

Lam AY et al (2010) Mental health first aid training for the Chinese community in Melbourne, Australia: effects on knowledge about and attitudes toward people with mental illness. Int J Ment Heal Syst 4(1):18

Loganathan S, Varghese M (2015) Play it street smart: a street play on creating awareness about mental illness. World Cult Psychiatry Res Rev 10:189–200

Loo PW, Wong S, Furnham A (2012) Mental health literacy: a cross-cultural study from Britain, Hong Kong and Malaysia. Asia Pac Psychiatry 4(2):113–125

Maulik PK, Kallakuri S, Devarapalli S et al (2017) Increasing use of mental health services in remote areas using mobile technology: a pre–post evaluation of the SMART Mental Health project in rural India. J Glob Health 7(1). https://doi.org/10.7189/jogh.07.010408

Minas H et al (2009) Evaluation of mental health first Aid training with members of the Vietnamese community in Melbourne, Australia. Int J Ment Heal Syst 3(1):19

Ogorchukwu JM, Sekaran VC, Sreekumaran Nair LA (2016) Mental health literacy among late adolescents in South India: what they know and what attitudes drive them. Indian J Psychol Med 38(3):234. https://doi.org/10.4103/0253-7176.183092

Poreddi V, Birudu R, Thimmaiah R, Math SB (2015) Mental health literacy among caregivers of persons with mental illness: a descriptive survey. J Neurosci Rural Pract 6(3):355. https://doi.org/10.4103/0976-3147.154571

Pradeep BS, Gururaj G, Varghese M, Benegal V, Rao GN, Sukumar GM (2018) National Mental Health Survey of India, 2016 – rationale, design and methods. PLoS One 13(10):e0205096. https://doi.org/10.1371/journal.pone.0205096

Rosen A (2006) Destigmatizing day-to-day practices: what developed countries can learn from developing countries. World Psychiatry 5(1):21

Salve H, Goswami K, Sagar R et al (2013) Perception and attitude towards mental illness in an urban community in South Delhi–A community based study. Indian J Psychol Med 35(2):154. https://doi.org/10.4103/0253-7176.116244

Sharma B, Malik M (2013) Bollywood madness and shock therapy: a qualitative and comparative analysis of depiction of electroconvulsive therapy in Indian cinema and Hollywood. Int J Cult Ment Health 6(2):130–140. https://doi.org/10.1080/17542863.2012.669769

Subedi P, Li C, Gurung A et al (2015) Mental health first aid training for the Bhutanese refugee community in the United States. Int J Ment Heal Syst 9(1):20

Suhail K (2005) A study investigating mental health literacy in Pakistan. J Ment Health 14(2):167–181

Sullivan G, Cheney A, Olson M et al (2017) Rural African Americans' perspectives on mental health: comparing focus groups and deliberative democracy forums. J Health Care Poor Underserved 28(1):548–565. https://doi.org/10.1353/hpu.2017.0039

Venkataraman S, Patil R, Balasundaram S (2019) Stigma toward mental illness among higher secondary school teachers in Puducherry, South India. J Family Med Primary Care 8(4):1401. https://doi.org/10.4103/jfmpc.jfmpc_203_19

Culture and Mental Health Care in Asia-Pacific Rural Communities

18

Ee Heok Kua

Contents

Abstract

This chapter focuses on mental health care of rural communities in the vast Asia-Pacific region. The myriad cultures of this colorful part of the world are a boon to the tourism industry, but beyond the opulent holiday resorts are meagre health services and deficient mental health care. The primary health professionals like family doctors or nurses provide care for the mentally ill. In these rural districts, the traditional healers have a pivotal role from time immemorial, and psychiatric patients seek their help because they share the same sociocultural beliefs of health and illness.

There is a dearth of data on mental disorders of rural communities in Asia-Pacific countries. Provision of mental health care in these communities is challenging, and the care of the mentally ill is now more critical with the migration of young people to the cities and overseas for better wages. The family with aged parents is the main caregiver.

E. H. Kua (✉)
Department of Psychological Medicine, National University of Singapore, Singapore, Singapore
e-mail: pcmkeh@nus.edu.sg

© Springer Nature Singapore Pte Ltd. 2020
S. Chaturvedi (ed.), *Mental Health and Illness in the Rural World*, Mental Health and Illness Worldwide, https://doi.org/10.1007/978-981-10-2345-3_33

Educational programs through e-learning can upgrade skills of the primary health team in mental health care, and the digital technology can be a platform for preventive psychiatry in these rural communities.

Keywords

Culture · Mental health · Services · Rural · Communities · Asia-Pacific

Introduction

A century ago the mental asylum, often situated in the countryside, was the focus of mental health care in most countries around the world. Mental patients were perceived as dangerous and should be sequestered far away from the cities. The same provision of care of mentally ill people was replicated in European colonies in Asia, Africa, and Latin America. Unfortunately, with the passage of time, villages where these asylums were built have become stigmatized; for example, in Malaysia, Tanjung Rambutan village and, in Singapore, Yio Chu Kang village are associated with mental illness.

In the past 40 years, most European countries, including the United Kingdom, have begun to close the mental asylums. But in the post-colonial era, the provision of mental health care of independent states in Asia, Africa, and Latin America is still largely in these mental institutions – albeit the emphasis of care in all countries today is general hospital and community psychiatry in the cities. This transition of care in Asian countries like Singapore and Malaysia is well documented by Loh et al. (2017).

Reports on mental health care of countries are derived primarily from research in cities or towns, and there is a paucity of data from rural communities. This is evident in many Asia-Pacific countries where mental health resource deficiencies are pressing problems (Tasman et al. 2009). In the low- and middle-income countries (LMICs) around the Asia-Pacific rim, where infrastructure and health services are meagre, people living in rural districts are even more deprived. The mental health-care centers in these communities are poorly equipped and often have difficulties in recruiting health professionals like doctors and nurses. In recent years, there is a diaspora of youths to cities to seek employment, and globalization has attracted young people from LMICs to migrate overseas for better opportunities. In rural communities, the family is the sole caregiver, and caring for people with mental illness can be demanding especially if parents are aged (Tseng et al. 2001).

From the mountains of the Andes to the Pacific island of Fiji and rainforest of Borneo, there are many health-care centers providing services for the mentally ill with penurious facilities. Despite the paucity of resources and immense demands, the primary health professionals press on with the arduous tasks. These professionals always radiate a sense of humility and quiet pride – the inhabitants of rural communities admire and respect them for their commitment and dedication to people with mental illness (Kua 2010a).

Paradox of the Idyllic Countryside

The countryside of Asia, Africa, and Latin America are often depicted in tourist brochures as idyllic resorts for city slickers to recuperate from the stress of urban living. In fact many inhabitants from these tourist enclaves eke out a living as guides and benefit from the tourism industry. Unfortunately, away from the opulent holiday resorts, rural communities encounter social economic difficulties which may precipitate or exacerbate their mental health problems. Poor facilities, lack of mental health professionals, and dwindling family caregivers are pressing issues besides financial hardship.

In many Asian societies, the emphasis on respect for the elderly and family support may be crucial in elevating the status of old people and minimizing stress in old age. Filial piety is more often observed in rural communities. The sociocultural influence on the perception of elderly people is a protective factor that may account for the low rate of depression (Kua and Mahendran 2017). However, in recent times, with the migration of young people to the cities for employment and a lack of family support, the mental health of elderly people may be seriously affected not just by financial reason but by loneliness.

In China, the suicide rate of women from 1995 to 1999 was 25% higher than men, and there were a large number of suicides among young rural women (Phillips et al. 2002). Yip et al. (2005) reported a higher suicide rate among elderly in rural areas and explained that easy accessibility of potent pesticides and lack of mental health services as well as suicide prevention programs were possible factors. The shift of young people from rural areas to cities has affected the family-centered rural communities (Chiu et al. 2012; Ho et al. 2012).

In Taiwan, Liu et al. (1998) studying a rural community with a sample size of 2055 people 65 years or more reported the prevalence of dementia as 2.5% using the *Diagnostic and Statistical Manual of Mental Disorders* (DSM IIIR) criteria – this is not too different from reports in other urban centers (Kua and Mahendran 2017). The selectively lower risks of dementia among population groups such as Chinese and rural Indians are intriguing and suggest that some possibly unique elements of Asian lifestyles including traditional dietary patterns and vitamins intake (Feng et al. 2009).

In a hill resort in Malaysia, a study of suicidal behavior in this rural region of the Cameron Highlands showed that young Indian women were the most vulnerable and pesticide was the commonest method of suicide (Maniam 1988).

The community program for the identification and management of psychoses in rural communities described in the paper by Li et al. (2012) is unique. Studies coordinated by the World Health Organization in the 1980s found that patients with schizophrenia from low-income countries had a more benign course than those from high-income countries (Leff et al. 1992), a finding that is confirmed by a more recent worldwide outcomes study of schizophrenia (Haro et al. 2011). Li and colleagues reported that treatment adherence, level of education, and short duration of illness were the best predictors of better outcomes. Another important factor not listed is family support. Though not stated in the paper, one expects that in rural Sichuan almost all of the patients are living with family members. A long-term follow-up study of schizophrenia in Singapore – in which 90% of the patients were Chinese – found that all the patients with good outcomes were living with their families (Tsoi and Kua 1992; Kua et al. 2003). For patients with chronic disabling conditions like schizophrenia, the family provides the basic necessities of shelter, food, and finance support; but, perhaps even more importantly, it also provides emotional support and encouragement in treatment adherence. Besides the family, there is probably less stigmatization in villages where psychiatric patients are often gainfully employed in the farms and hence they have a role which is important for self-esteem.

Aspirations of old age are also influenced by cultural beliefs. For more than a thousand years, Chinese families worship in their homes three deities who personify longevity, happiness, and wealth (Kua 2017). They represent their aspirations – to live to a ripe old age, have sufficient finances, and enjoy good health. Embedded in the Asian tradition of filial piety is the expectation that children should take care of their aged parents and provide financial, social, and emotional support. Such tradition is still present in many rural communities, but with the rapid global economic changes and the movement of the young working population from rural to urban industrialized communities, and across national boundaries, the care of the elderly is at a critical state. Because of the change of family structure, more Asian elderly may be living alone in the future and cannot expect much support from close relatives. Caring for an elderly with dementia at home will be problematic (Kua and Rawtaer 2019).

Culture and Health-Seeking Behavior

Cultural and religious beliefs of people in rural communities of the Asia-Pacific region determine their illness behavior and health-seeking tendency (Minas and Lewis 2017; Kua and Sunbaunat 2012). In these communities, the theories of health and illness are embedded in customs and religions. The role of the priest is linked to the art of healing. In many Asian countries, with the advances in modern medicine, the doctor has a lead role in health care, but in many rural communities especially with the elderly, the traditional healers are still consulted first.

The traditional healers in rural communities are popular because they share the same sociocultural beliefs about illness and health. An example is the "possession-trance" (Kua et al. 1986), a common culture-related phenomenon commonly observed in Asia, Africa, and Latin America. Because the possession-trance is not deemed an illness, the traditional healer is often consulted, and the person avoids the stigma of mental disorder.

The World Health Report (2001) recognizes traditional healers as important health agents for they contribute as active case finders, facilitators of referrals, and counselling providers. Their role is considered to be important for increasing access (especially reaching rural and remote communities), filling the gaps produced by the shortage of health professionals, and providing culturally competent care (using patients' language, respecting their values, notions of illness and treatment preferences) (Incayawar 2009). In many rural areas, psychiatric patients and relatives feel more at ease when visiting a traditional healer.

There is much to learn from the psychotherapeutic techniques of traditional healers in the management of mental disorders. Fundamental in the therapeutic relationship is trust, and an understanding of the cultural belief system is a sine qua non. In the provision of mental health care in rural communities, the pivotal role of the traditional healer is gradually being acknowledged by the local government in many Asia-Pacific countries.

In Indonesia, the psychiatrists work closely with the traditional healers and organize training programs to help them identify patients with psychosis to ensure early referral to the hospital for treatment. This collaborative effort is especially vital in countries where there is a perennial shortage of trained psychiatrists or mental health professionals, and this also reduces the stigma tainted by mental disorder. In these communities, it is sometimes the responsibility of the village chief or elder who advises on where to seek help. The family plays a crucial role in providing care, and the ethos of the community embraces the therapeutic role of the ubiquitous traditional healer.

Culture and Psychiatric Symptomatology

An intriguing question is whether the psychiatric presentations of patients with mental disorders in rural communities differ from those in the cities. In the past, transcultural disorders like "amok" and "koro" were more common in villages in Malaysia and Singapore (Kua 1991; Ng and Kua 1996) – both conditions are not as often encountered today. The more common phenomenon of "possession-trance" is reported in people living in both the city and villages.

In rural communities, sociocultural beliefs about illness and health are more prevalent and pervasive than in the cities. In the Chinese culture, the constellation of symptoms of a patient is explained as due to an excess of "ying" (coolness, negative, dark) or "yang" (warmness, positive, bright). The traditional healer is able to explain the symptoms using the belief system the patient is familiar with – a powerful therapeutic factor is the rapport between the patient and the healer. To the elderly Chinese, depressive symptoms like poor appetite, lethargy, or poor sleep are interpreted as due to a "weakness of mental energy." The traditional healer understands the ethos of the subculture, and consulting one also avoids the stigma of being labelled a "mental patient," as would happen when they see the doctor in the psychiatric clinic (Kua et al. 1993).

Presentations of mental disorders in rural communities are often colored by sociocultural beliefs, for example, a patient with schizophrenia may ascribe the auditory hallucination as the voices of a spirit from the forest or a bipolar patient may believe he is the village chief. In Central Asia, the countries of Kazakhstan, Uzbekistan, Tajikistan, and Kyrgyzstan are agrarian communities with nomadic people, and the presentations of mental disorders like schizophrenia, depression, and anxiety are influenced by the local culture and are not too different from the symptoms of patients in European countries (Cooper et al. 2008).

Culture and Mental Health-Care Services

A review paper from Latin America by Dr. Incayawar et al. (2010) provides us insight to rethink primary care psychiatry in places where there is no psychiatrist. In those rural communities scattered along the Andes mountains, mental health care is challenging not only because of the dearth of human resources but also because of the forbidding terrain. All the services are dependent on primary care doctors or nurses and rarely do they have a trained psychologist, occupational therapist, or social worker. It is incontrovertible that in the midst of scarcity, improvision and innovation are necessary.

The Quichua patients in Latin America depend on the families and traditional healers. In the highlands of Ecuador, where the majority of indigenous people speak Quichua, the patients have serious difficulty communicating with invariably mono-lingual Spanish-speaking doctors, nurses, and other health professionals. Many doctors lack adequate preparation to see at least 40% of the Ecuadorian population whose language and culture are not Hispanic – there is an important linguistic and cultural distance between the health practitioners and Quichua patients.

According to the World Health Organization, in some parts of the world, up to 80–85% of the population rely on traditional healers (WHO 2002). In the Andes, a widespread traditional health and mental health-care network exists within the indigenous peoples' communities. The traditional healers still enjoy a high prestige in these com-munities and are often considered to be their trusted doctors (Incayawar 2007, 2008). Some researchers in the Andean region have an interest in the use of medicinal plants as psychotropic drugs (Maldonado-Bouchard 2009). However, the therapeutic properties of the vast number of plants used for psychiatric purposes remain largely unknown.

Durie (2003), referring to the indigenous people in New Zealand, in which 15% is Maori, proposes that a "combination of conventional services and indigenous pro-grams is needed." In the face of the enormous mental health-care neglect, it is almost inevitable to foresee the need for a partnership between traditional healers and psychiatrists. The respectful healers-doctors collaboration is important in the devel-opment of effective, culturally sensitive services.

Matamua et al. (2013) from the South Pacific island of Samoa reported on the integration of the family or *aiga* – as an active partner in the provision of care using the Samoan cultural values to promote culturally appropriate family-focused com-munity mental health care. For example, when a nurse enters the home of a patient to discuss with the family, there is use of complimentary greetings and other cultural etiquette. The language spoken inspires compassion and respect allowing the family to express their feelings and honest thoughts about the patient. Good communication with the family leads to greater medication adherence and recovery and maintains the dignity of people with mental disorders.

The report on mental health care in the Pacific island of Fiji (Chang 2011) illustrates the issues faced by the country – challenges from a range of political, social, economic, and geographical issues. In addition, Fiji faces recurrent natural disasters (flooding, cyclones, and hurricanes as well as tsunami threats), and all these factors negatively impact on mental health, making mental health services and programs a much-needed priority. Primary care is an increasing important aspect of service delivery, as is the development of local education and training programs. The development of new mental health legislation and policy gives an indication that further progress will be made in providing clinical services. There is collaboration between the mental health service in Fiji and volunteers from Australia, New Zealand, Malaysia, and Taiwan.

In the East Malaysian state of Sabah on the island of Borneo, the advent of telemedicine service has become a valuable means to link remote communities which are not accessible by road. Telephone consultation between the district hospitals facilitates referrals to the specialists. The personnel behind the mental health services in Sabah are mainly the nurses and health-care personnel who review patients in the outpatient clinics and also do home visits if necessary. The psychiatrist is required to travel and provide consultation to some 18 district hospitals. These visits are to provide support for the medical staff in patient management and to teach the hospital staff as part of their continuing medical education program. The psychiatrist in Sabah has earned thousands of frequent flyer mileage per year providing consultation to the many mental health-care centers scattered around a landscape half the size of England (Chee 2010).

Providing psychogeriatric service in rural communities is challenging and exacting. In many villages in East Asia like China, Vietnam, Malaysia, Thailand, the Philippines, Cambodia, Laos, and Indonesia, there are an increasing number of elderly people. Caring for the frail elderly especially those with dementia can be demanding for families with few caregivers – many of whom are elderly themselves (Burns and Robert 2019). To assist the family doctors in East Asian countries, a 10-item questionnaire, Elderly Cognitive Assessment Questionnaire (ECAQ), has been constructed for the detection of cognitive impairment (Kua and Ko 1992). The ECAG can be administered by a nurse at a busy outpatient clinic and completed in 5 min. Currently, there are a few screening instruments, including the Mini-Mental State Examination (Folstein et al. 1975); however, the validity of these tests is doubtful in a different cultural setting where literacy is low.

With the advent of the Internet, providing mental health care in rural communities can be improved. In China a study has been conducted with e-learning using social media, for dementia-specific training of community nurses, who play a crucial role in early detection and diagnosis of dementia (Wang et al. 2017). This is a randomized controlled trial to improve nurses' knowledge, attitudes, and practice using an innovative mobile phone in primary care settings. Overall, the study showed acceptability and feasibility in improving nurses' attitude and knowledge on dementia with e-learning as a platform for effective educational experience.

Future of Mental Health Care in Rural Communities

In 2007, most of the world's countries signed the UN Declaration on the Rights of Indigenous Peoples, which states that "indigenous individuals have an equal right to the enjoyment of the highest attainable standard of physical and mental health. States shall take the necessary steps with a view to achieving progressively the full realization of this right" (UN 2009).

It is undeniable that the majority of people with mental disorders are not detected although they may have been to the health services. The underdiagnosis is partly because psychiatry in the undergraduate curriculum is often not emphasized and doctors especially in primary care have difficulties in recognizing the early signs and symptoms. There is a need to re-examine the curriculum in medical education. There is a gradual change in focus in some medical schools to post students to clinics in the rural district to see common psychiatric problems. Medical students are often given clinical teaching in the general hospital or mental hospital where cases are of moderate to severe degree. However, at primary care practice, the clinical presentations are usually mild and protean and may not fulfil the criteria of more advanced disorders as stipulated in the DSM (*Diagnostic and Statistical Manual of Mental Disorders*, American Psychiatric Association) or ICD (*International Classification of Diseases*, WHO).

Primary care psychiatry teaching in rural communities needs to go beyond diagnostic or therapeutic skills and to include time for leadership and administrative training (Kua 2010b). This has been emphasized in an editorial by Professor N Sartorius (2009) because at these mental health-care centers, the nurse or doctor has to provide leadership and network with the people in the community.

Primary prevention at the prenatal level, an adequate diet, and advice on drug or alcohol abuse which may affect the fetus are important. It may be necessary to help parents who have problems in childrearing to learn parenting skills and to anticipate the normal crises of childhood and adolescence. A powerful tool in primary prevention is health education, for example, to prevent alcohol abuse, it is vital to educate young people about the hazards of excessive drinking through the mass media or talks in schools.

Academic psychiatry in many Asia-Pacific universities is growing with quiet aplomb in its own talent. In the past 8 years, a few psychiatrists from Southeast Asia and Australia have banded together to form a Teachers of Psychiatry (TOP) club with a tagline, *psychiatry sans frontier*. The TOP club has organized many workshops on primary care psychiatry in the ASEAN countries (Kua 2004, 2011).

The rich experiences of traditional healers inform us that understanding the cultural mores is crucial in psychological therapy. Time is premium in the clinics in rural districts, and providing mental health care should not just be routine pharmacotherapy. There is still a place for brief psychological therapy that is more personalized and integrates culturally accepted techniques like mindfulness practice. There is now a growing interest in brief integrative personal therapy which is more appropriate for the community with an affirmation of cultural values (Feng et al. 2011; Kua and Mahendran 2014, 2019).

In the future there may be a need to think about best-practice mental health care in rural communities as suggested by Chee et al. (2009). The authors identified a number of key ingredients to facilitate constructive change such as understanding local factors and belief system to mobilize resources for culturally appropriate strategies and building on the strength of the local model.

Little is known about the efficacy, safety, or cost-effectiveness of traditional healers' involvement in the provision of mental health care. Research is needed to determine how mental health workers can better collaborate with traditional healers in order to improve access, diagnosis, and successful treatment for persons suffering from mental illness.

Galvanizing community support is critical to ensure the success of the mental health-care service. Because there is a woeful shortage of mental health professionals, it may be necessary to encourage voluntarism among students and working people who can be trained to provide emotional support and to assist some of the community mental health-care activities.

There will be more many challenges in rural communities in the Asia-Pacific region. Unique problems exist for governments attempting to provide services for populations dispersed across many small islands with limited connecting transportation. Political instability and reliance on tourism for economic growth can also limit the capacity to develop sustainable programs to meet societal needs.

The potential of technology in early detection, referral, and management should be emphasized. There can be e-counselling to provide support for many patients and their caregivers. There are now apps to monitor mood and teach relaxation techniques. Exploiting the benefits of artificial intelligence for mental health promotion will be crucial in the future. More importantly it is possible using e-learning to start training programs in preventive psychiatry for students, adults, and the elderly. The dementia and depression program in the community in Singapore is now using the e-learning platform (Rawtaer et al. 2015).

Because of a dearth of data from rural communities, there is a need to write and share experiences like what the doctors in remote Sabah, Andean mountains, and South Pacific islands have done. Information on mental health care in London and New York is different from the villages around the Asia-Pacific rim (Kua 1995). The invaluable clinical experiences of mental health professionals can add to our limited knowledge on caring for the mentally ill in the rural communities.

Conclusion

There is a perennial shortage of trained psychiatrists and other mental health professionals in rural communities in the Asia-Pacific countries. The family will need the necessary support, resources, and knowledge to care for the mentally ill at home. Besides the primary health-care team of doctors and nurses, the traditional healers have a pivotal role in these remote areas.

Future psychiatric education should focus on primary care doctors and health workers like nurses who can be trained to identify and treat early or mild mental disorders and to refer more complex problems to the specialists. There is also a need for a paradigm re-orientation in the teaching of psychiatry in medical schools with emphasis on the common psychiatric problems in primary care. Using the digital technology for information and communication on illness and services can certainly improve the care of the mentally ill in the many remote districts around the Asia-Pacific rim.

Cross-References

References

Burns A, Robert P (2019) Dementia care: international perspectives. Oxford University Press, Oxford
Chang O (2011) Mental health care in Fiji. Asia Pac Psychiatry 3:73–75
Chee KY (2010) Mental health service: where there is no psychiatrist – my experience in Sabah. Asia Pac Psychiatry 2:128
Chee N, Goding M, Fraser J (2009) Moving towards best-practice community mental health care in the Asia-Pacific. Asia Pac Psychiatry 1:38–42
Chiu HFK, Chan SSM, Caine ED (2012) Suicide prevention in the Asia Pacific region. Asia Pac Psychiatry 4:3–4
Cooper JE, Sartorius N, Nixon N, Solojenkina X (2008) Images of mental illness in Central Asia: a casebook with commentaries. Global Initiative for Psychiatry, Hilversum
Durie M (2003) Providing health services to indigenous peoples. BMJ 327:408–409
Feng L, Li J, Yap KB, Kua EH, Ng TP (2009) Vitamin B-12, apolipoprotein E genotype, and cognitive performance in community-living older adults: evidence of a gene-micronutrient interaction. Am J Clin Nutr 89:1263–1268
Feng L, Cao Y, Zhang Y, Wee ST, Kua EH (2011) Psychological therapy with Chinese patients. Asia Pac Psychiatry 3:167–172
Folstein MF, Folstein SE, McHugh PR (1975) Mini-mental state. A practical method of grading the cognitive state of patients for the clinician. J Psychiatr Res 12:189–198
Haro JM, Novicte D, Bertsch J, Karagianis J, Dossenbach M, Jones P (2011) Cross-sectional clinical and functional remission rates: worldwide schizophrenia outpatient health outcomes. Br J Psychiatry 199:194–201
Ho ECL, Chiu HFK, Chong MY, Yu X, Kundadak G, Kua EH (2012) Elderly suicide in Chinese populations. Asia Pac Psychiatry 4:5–9
Incayawar M (2007) Indigenous peoples of South America – inequalities in mental health care. In: Bhui K, Bhugra D (eds) Culture and mental health – a comprehensive textbook. Hodder Arnold, London, pp 185–190
Incayawar M (2008) Efficacy of Quichua healers as psychiatric diagnosticians. Br J Psychiatry 192:390–391
Incayawar M (2009) Future partnerships in global mental health – foreseeing the encounter of psychiatrists and traditional healers. In: Incayawar M, Wintrob R, Bouchard L (eds) Psychiatrist and traditional healers – unwitting partners in global mental health. Wiley-Blackwell, Chichester, pp 251–260
Incayawar M, Bouchard L, Maldonado-Bouchard S (2010) Living without psychiatrists in the Andes. Asia Pac Psychiatry 2:119–125

Kua EH (1991) Amok in nineteenth-century British Malaya history. Hist Psychiatry iii:429–436

Kua EH (1995) Ethics of refereeing re-examined. Addiction 90:1309–1322

Kua EH (2004) Psychiatry in Singapore. Br J Psychiatry 185:79–82

Kua EH (2010a) Psychiatric service with no psychiatrist. Asia Pac Psychiatry 2:117–118

Kua EH (2010b) Leadership in psychiatry: training future leaders. Asia Pac Psychiatry 2:173–174

Kua EH (2011) Academic psychiatry on the rise in Asia. Asia Pac Psychiatry 3:1–2

Kua EH (2017) Colours of ageing. Write Edition, Singapore

Kua EH, Ko SM (1992) A questionnaire to screening for cognitive impairment among elderly people in developing countries. Acta Psychiatr Scand 85:119–122

Kua EH, Mahendran R (2014) Mental health care of the elderly. Write Edition, Singapore

Kua EH, Mahendran R (2017) Epidemiology of mental disorders. In: Chiu H, Shulman K (eds) Mental health and illness of the elderly. Springer Nature, Singapore, pp 53–82

Kua EH, Mahendran R (2019) Mental health care in Singapore: current and future challenges. Taiwan J Psychiatry 33:6–12

Kua EH, Rawtaer I (2019) Singapore. In: Burns A, Robert P (eds) Dementia care – international perspectives. Oxford University Press, Oxford, pp 79–84

Kua EH, Sunbaunat K (2012) Mental health care. In: Detels R, Sullivan SG, Tan CC (eds) Public health in East and Southeast Asia. University of California Press, Berkeley

Kua EH, Sim P, Chee KT (1986) A cross-cultural study of the possession-trance in Singapore. Aust N Z J Psychiatry 20:361–364

Kua EH, Chew PH, Ko SM (1993) Spirit possession and healing among Chinese psychiatric patients. Acta Psychiatr Scand 88:447–450

Kua J, Tsoi WF, Wong KE, Yeo B, Kua EH (2003) A 20-year follow up study of schizophrenia in Singapore. Acta Psychiatr Scand 108:118–125

Leff J, Sartorius N, Jablensky A, Korten A, Ernberg G (1992) The international pilot study of schizophrenia: five-year follow-up findings. Psychol Med 22:131–145

Li QJ, Huang XY, Wen H, Liang XQ, Lei L, Wu JL (2012) Retrospective analysis of treatment effectiveness among patients in Mianyang municipality enrolled in the national community management program for schizophrenia. Shanghai Arch Psychiatry 24:131–139

Liu RT, Fuh JL, Wang SJ (1998) Prevalence and subtype of dementia in a rural Chinese population. Alzheimer Dis Assoc Disorder 12:127–134

Loh KS, Kua EH, Mahendran R (2017) Mental health and psychiatry in Singapore: from asylum to community care. In: Minas H, Lewis M (eds) Mental health in Asia and the Pacific – historical and cultural perspectives. Springer, New York, pp 193–204

Maldonado-Bouchard S (2009) South American indigenous knowledge of psychotropics – the need for culturally adapted intellectual property rights. In: Incayawar M, Wintrob R, Bouchard L (eds) Psychiatrists and traditional healers – unwitting partners in global mental health. Wiley-Blackwell, Chichester, pp 35–51

Maniam T (1988) Suicide and parasuicide in a hill resort in Malaysia. Br J Psychiatry 153:222–225

Matamua LSE, Aiilelei T, Tupou S, Latama P, Pisaina T, Ilse B (2013) Developing a culturally appropriate mental health care service for Samoa. Asia Pac Psychiatry 5:101–107

Minas H, Lewis M (2017) Mental health in Asia and the Pacific: historical and cultural perspectives. Springer Nature, New York

Ng BY, Kua EH (1996) Koro in ancient Chinese history. History of Psychiatry vii:563–570

Philips MR, Li X, Zhang Y (2002) Suicide rates in China, 1995–99. Lancet 359:835–840

Rawtaer I, Feng L, Fam J, Kua EH, Mahendran R (2015) Psychosocial interventions with art, music, tai-chi and mindfulness for subsyndromal depression and anxiety in older adults: a naturalistic study in Singapore. Asia Pac Psychiatry 7:240–250

Sartorius N (2009) Training psychiatrists for the future. Asia Pac Psychiatry 1:111–116

Tasman A, Sartorius N, Saraceno B (2009) Addressing mental health resource deficiencies in Pacific rim countries. Asia Pac Psychiatry 1:3–8

Tseng WS, Ebata K, Kim K, Krahl W, Kua EH, Lu Q, Shen Y, Tan ES, Yang MJ (2001) Mental health in Asia: social improvements and challenges. Int J Soc Psychiatry 47:8–23

Tsoi WF, Kua EH (1992) Predicting the outcome of schizophrenia ten years later. Aust N Z J Psychiatry 26:257–261

United Nations, Declaration on the Rights of Indigenous Peoples, IWGIA (2009). http://www. iwgia.org/sw248.asp. Accessed 15 Dec 2009. Archived by WebCite at http://www.webcitation. org/5m2sZaHow

Wang F, Xiao LD, Wang K, Li M, Yang Y (2017) Evaluation of a WeChat-based dementia-specific training program for nurses in primary care settings. Appl Nurs Res 38:51–59

WHO (2002) WHO traditional medicine strategy 2002–2005. World Health Organization, Geneva

World Health Organization (2001) The world health report 2001 – mental health: new understanding, New Hope. World Health Organization, Geneva

Yip SF, Lin KY, Hu JP, Song XM (2005) Suicide rates in China during a decade of rapid social changes. Soc Psychiatry Psychiatr Epidemiol 40:792–798

Yoga and Traditional Healing Methods in Mental Health

19

Shivarama Varambally and B. N. Gangadhar

Contents

S. Varambally · B. N. Gangadhar (✉)
NIMHANS Integrated Centre for Yoga, Department of Psychiatry, National Institute of Mental
Health and Neurosciences (NIMHANS), Bangalore, Karnataka, India
e-mail: ssv.nimhans@gmail.com; kalyanybg@yahoo.com

© Springer Nature Singapore Pte Ltd. 2020
S. Chaturvedi (ed.), *Mental Health and Illness in the Rural World*, Mental Health and
Illness Worldwide, https://doi.org/10.1007/978-981-10-2345-3_20

Abstract

Medications remain the primary approach for the treatment of most mental disorders in most centers around the world, with or without psychological interventions. Drug treatments have indeed helped many patients with psychiatric disorders, especially after the psychopharmacological revolution in the 1990s. However, there are significant limitations to current medications and psychological approaches that make persons with mental disorders seek non-pharmacological treatments as add-on or independent therapies, also known as traditional methods of healing or by the term "complementary and alternative medicine" (CAM). This is even more pronounced in the developing countries, especially rural populations. The wide use of CAM approaches by patients has led clinicians and researchers to also explore the efficacy and evidence base for these interventions. The CAM approaches include whole alternative medical systems, mind-body therapies, biologically based therapies, energy-based therapies, and body-based manipulative therapies.

Mind-body therapies have been particularly in focus over the last two decades, with increasing evidence suggesting that these interventions are efficacious in several mental disorders. Among mind-body therapies, yoga-based therapies have emerged as an important therapeutic option. Clinical trials with yoga have been conducted in depression, anxiety, and schizophrenia, with promising results. While the evidence is preliminary, the available data suggest that yoga has distinct benefits in patients with these disorders, and these effects are mediated by certain neurobiological processes that may have been deranged in the disease. Yoga may correct the balance and is therefore therapeutic.

There is tremendous scope for scientific study of yoga and other traditional healing methods in mental health conditions. This chapter attempts to synthesize the available knowledge in this important area of health, particularly focusing on yoga and meditative practices.

Keywords

Yoga · Mind-body therapies · Traditional healing methods · Mental health

Introduction

Modern medicine has made great strides in management of infectious diseases and acute conditions over the last two centuries. This has led to allopathy being seen as the "mainstream" method of medical treatment for all conditions, even in developing countries like India which had their own indigenous systems of medicine for thousands of years. While it is undoubtedly true that modern medicine has led to drastic reduction in mortality due to infectious diseases, it is equally true that it does not really have solutions for most chronic noncommunicable disorders ("lifestyle disorders") that are the bane of the current health systems worldwide. It is also true that a majority of the population in countries like India live in rural areas, which are grossly underserved in terms of health infrastructure. A combination of faith in traditional systems and lack of modern medical facilities has led to a situation where more than two-thirds of the population in India continue to use traditional healthcare methods as the primary approach to deal with their medical issues. This is even more marked in mental health, as modern psychiatric facilities in rural areas are almost nonexistent. Added to this, cultural beliefs and stigma associated with mental health issues are very prominent in rural India leading to poor awareness and long duration of untreated illnesses. Even when the need for treatment becomes acute, the first line of management chosen is most often faith healing or other traditional healing methods which are easily accessible to this population. It is therefore essential to examine the various traditional healing methods and the evidence for their efficacy in mental health conditions.

Current treatment approaches for mental disorders can be classified under two broad headings, namely, physical or somatic and psychosocial. Somatic treatments include psychotropic drugs (antipsychotics, antidepressants, mood stabilizers, anti-anxiety drugs). Electroconvulsive therapy (ECT); newer stimulation techniques such as transcranial magnetic stimulation (TMS), transcranial direct current stimulation (tDCS), vagal nerve stimulation (VNS), and psychosurgery; Cognitive behavioral therapy (CBT), psychodynamic psychotherapy, relaxation therapy, and rehabilitative therapies are some of the psychosocial approaches.

The somatic methods of treatment have become increasingly sophisticated, particularly after the "psychopharmacological revolution" in the 1990s which produced a significant number of new medications, and newer stimulation therapies developed in the last decade. However, though these approaches have increased the efficacy of treatments, this has not translated into better functioning or quality of life, especially in psychosis (Hegarty et al. 1994). Some of the newer antipsychotics also have problematic side effects, especially on the hormonal and metabolic systems, with the "metabolic syndrome" being a common problem in many chronic psychiatric patients. The "pipeline" of newer medications also seems to be drying up (Abbott 2010). Psychosocial approaches have certainly helped many patients with diagnoses such as depression, anxiety disorders, and somatoform disorders. However, these treatments are resource- and time-intensive, making them difficult to provide in large populations, especially in rural areas which are resource-poor.

In the light of the above, it is easy to see how complementary and alternative (CAM) approaches are popular among patients with psychiatric disorders, with some estimates showing that more than half of patients with depression have used CAM methods with or without modern medical treatment (Kessler et al. 2001). Most such traditional methods or practices are derived from local traditions and cultural contexts and can be considered under the rubric of traditional medicine (TM). The World Health Organization (WHO) estimates that between 65% and 80% of the world's healthcare services may be currently classified as traditional medicine and in many countries constitute the major form of treatment. According to the WHO, "traditional medicine is the sum total of knowledge, skills and practices based on the theories, beliefs and experiences indigenous to different cultures that are used to maintain health, as well as to prevent, diagnose, improve or treat physical and mental illnesses". It is a comprehensive term which includes systems such as traditional Chinese medicine, Indian Ayurveda, and Arabic Unani medicine and various forms of indigenous medicine. In many countries, these are referred to as "complementary," "alternative," or "non-conventional" medicine, and they are often used along with allopathic treatment. Other terms to describe these therapeutic approaches are integrative medicine and holistic medicine.

The wide use of TM in developing countries, particularly rural communities, is attributable to its accessibility and affordability and also because it is firmly embedded within wider belief systems. In Asia and Latin America, populations continue to use TM as a result of historical circumstances and cultural beliefs. In China, TM accounts for around 40% of all healthcare delivered. Even in many developed countries, CAM approaches are becoming more popular. The percentage of the population which has used CAM at least once is 48% in Australia, 70% in Canada, 42% in the United States, 38% in Belgium, and 75% in France (WHO 2002).

The US National Center for Complementary and Alternative Medicine (NCCAM) has organized CAM into five broad categories: mind-body therapies, whole medical systems, biologically based therapies, energy-based therapies, and body-based manipulative therapies (Wieland et al. 2011).

Mind-Body Therapies

The NIH has defined mind-body therapies as "interventions that use a variety of techniques designed to facilitate the mind's capacity to affect bodily functions and symptoms." Yoga, *Qigong*, and meditation are discussed in some detail below.

Yoga: Yoga is one of the most popular CAM practices worldwide. Yoga (derived from "yuj" which means "yoking" or "union" in Sanskrit) is an ancient lifestyle practice which was intended to help individuals consciously evolve spiritually. Early evidence of yoga practice dates back 5000 years ago in India, and it has been practiced as a spiritual and health system ever since. Although yoga has been defined variously by different texts/authors, one of the most succinct is the one attributed to Sage Patanjali (the pioneer who codified the practice and teaching of yoga): "Yogah chitta vritti nirodhah" (*yoga is the process of gaining mastery over the fluctuations of the mind*).The original aim of practicing yoga was union of the individual consciousness with the universal one, with better physical and mental health being a necessary step in the process. Sage Patanjali described the rationale and the methods to achieve the objective through the practice of yoga in his *Yoga Sutras* (Iyengar 1966). *Ashtanga yoga* as enunciated by Patanjali has eight limbs or components, namely, *Yama* (self-control), *Niyama* (observances), *asana* (assuming and maintaining certain postures), *pranayama* (regulation of breath), *pratyahara* (restraint of senses), *dharana* (single focused thought), *dhyana* (expansion of awareness or meditation), and samadhi (*meditative consciousness*). The first five have been called as *bahiranga* (external) yoga practices and the last three as *antharanga* (internal) practices.

According to Patanjali, these "limbs" are complementary to each other and may run sequentially or in parallel.

Even a cursory reading of the yoga texts make it clear that yoga was designed primarily as a way of life and not as a therapy for any particular health condition. However, over the years, distinct physical and mental effects have been documented with certain yoga practices. These benefits of yoga, both physical and mental, have led to widespread use of yoga-based practices for various disorders, although purists argue that use of selective elements of yoga without understanding of the philosophical underpinnings or change in lifestyle is not really "yoga." Asana, pranayama, and meditation are the most common components of yoga used for the purpose of treatment. Yoga has proven benefits in physical disorders such as diabetes and hypertension (Cramer et al. 2014; Tyagi and Cohen 2014), obesity, dyslipidemia, low back pain (Cramer et al. 2013), and many more. It has also been found to be effective in stress management, improving cognitive performance (including memory and executive functions), motor functioning, and sleep patterns in normal individuals.

Some of the proposed mechanisms by which yoga practices produce these effects include a generalized reduction in both cognitive and somatic arousal, changes in the hypothalamic pituitary axis and reduction in basal cortisol and catecholamine secretion, modulation of the autonomic nervous system (decreased sympathetic tone and increased parasympathetic tone), reductions in the metabolic rate, oxygen consumption, and salutary effects on cognitive activity and cerebral neurophysiology.

There is a significant amount of documented evidence in regard to the effects of yoga in healthy volunteers of all ages. Reduction of stress and thereby improving/promoting health and well-being are commonly reported after yoga practice. However, in this chapter we have limited ourselves to evaluating the applications of yoga-based interventions in clinical or quasi-clinical situations in mental health. We examine the evidence for yoga's therapeutic efficacy, its mechanism of action, limitations of research, and potential areas for future research.

Yoga-Based Interventions for Psychiatric Disorders

Different yoga-based practices have been used for prevention and management of stress-related mental disorders such as anxiety and depression worldwide for quite some time now. However, systematic research in this area is quite minimal, with most of the studies being in the last two decades. A review in 2005 found that there was some promise that yoga could be helpful in depression and anxiety, but could not find strong evidence to make any specific recommendations (Pilkington et al. 2005). Although most studies reviewed had positive results, several methodological concerns in many of the studies such as small sample sizes and lack of rigor in terms of randomization made it difficult to ascribe the benefits to the yoga practice.

Fortunately, the last two decades have seen a remarkable surge in the number of scientific studies of yoga as an intervention in mental disorders. Recent reviews focusing on yoga-based interventions in major depression (Ravindran and da Silva

2013) and psychosis (Bangalore and Varambally 2012) make a reasonable case for such strategies being an efficacious treatment approach, as an adjunct therapy or even as a sole treatment in some cases. Overall, the current evidence seems to indicate that a combination of conventional and complementary treatments may provide optimal results in many psychiatric conditions. The sections below examine the evidence for efficacy of yoga-based practices in common mental disorders.

Major Depression and Other Depressive Disorders

Depressive disorders are the most prevalent mental disorders worldwide, afflicting more than 400 million people. They cause high morbidity and burden, not only for the sufferers but also their families and society as a whole. It also leads to direct and indirect loss of life by suicide and alcohol/drug addiction, respectively. Patients who suffer from depression experience pervasive sad mood, loss of pleasure, easy fatigability, low self-esteem, cognitive slowing, reduced biological functions, and impaired occupational functioning.

Depressive disorders can be classified as follows:

1. Major depressive disorder (MDD) which can be further subtyped based on severity into mild, moderate, and severe type. Melancholic and psychotic symptoms may or may not be present.
2. Bipolar depression which resembles major depression but cycles with manic episodes.
3. Dysthymia, which is a chronic form of minor or neurotic depression and is very prevalent.
4. Miscellaneous conditions with comorbidity.

Current treatment approaches for depression include antidepressant medications, electroconvulsive therapy (ECT), cognitive behavioral therapy (CBT), and other psychotherapies and newer stimulation therapies such as transcranial magnetic stimulation (TMS), transcranial direct current stimulation (tDCS), and vagal nerve stimulation (VNS). However, these have several limitations, including suboptimal response and adverse effects.

Yoga-Based Therapies for Depression

Open trials: An open trial of Iyengar yoga, as adjunct to antidepressants, was found to significantly improve residual depressive symptoms and anxiety (Shapiro et al. 2007). An open trial of hatha yoga in psychiatric inpatients with varying mood disorders, psychotic disorders, and personality disorders found that one session of yoga, added on to existing medication, resulted in significant improvement in depressive and anxiety symptoms (Lavey et al. 2005).

Controlled trials: Among yoga-based practices, possibly the one with the largest evidence base as monotherapy in MDD is sudarshan kriya yoga (SKY), which is a part of the stress management package offered by the Art of Living Foundation headquartered in Bengaluru, India. A feeling of well-being and relief from dysphoria are well-documented effects in participants of SKY training workshops. SKY practice has also been shown to be associated with neurobiological effects like increased parasympathetic tone, calming of stress response systems, and neuroendocrine release of hormones (Brown and Gerbarg 2005b). There are five trials using SKY in depression, including two RCTs and three open trials. Janakiramaiah and colleagues randomized 45 consenting inpatients diagnosed with major depression into three groups in a 4-week clinical trial ($n = 15$ in each group). Patients were given daily SKY practice under supervision, three ECTs/week, or imipramine 150 mg/day, respectively. Weekly assessments on Hamilton Depression Rating Scale revealed that the SKY group experienced significant reductions in depression scores which was comparable to imipramine, although less than the ECT group (Janakiramaiah et al. 2000). Another study compared full SKY to partial SKY (which had normal breathing in place of cyclical breathing as in full SKY). This study showed that 75% of full SKY patients had responded as against 45% of partial SKY at the end of 4 weeks, although both groups had significant reduction in depression scores (Rohini et al. 2000). In a clinical trial of SKY for dysthymia by the same group, 46 outpatients were given SKY training for 1 week, after which they practiced at home for the next month. Twenty-five out of 37 patients who completed the trial showed remission by the end of first month of practice and maintained the same for next 3 months (Janakiramaiah et al. 1998).

Other yoga-based techniques have also been used in patients with depression. An early RCT used shavasana yoga as monotherapy and found significantly superior results compared to no treatment (Khumar et al. 1993). A controlled trial tested sahaja yoga as an adjunct to antidepressants, with the control group receiving only antidepressants. Assessments with HAM-D and Hamilton Rating Scale for Anxiety (HAM-A) before and after 8 weeks of intervention showed that both groups showed significant improvement, with the percentage of improvement being higher in the sahaja yoga group. More patients in the sahaja yoga group also achieved remission (Sharma et al. 2005). A multi-intervention RCT comparing yoga as a monotherapy with Ayurvedic medicine and wait-listed controls in elderly patients with MDD found yoga significantly superior in reducing depressive symptoms. These benefits were sustained up to 6 months of follow-up (Krishnamurthy and Telles 2007). Another RCT using Iyengar yoga as monotherapy found that it significantly improved mild depression and anxiety in adolescent patients as compared to wait list control (Woolery et al. 2004). A RCT included patients with varying depressive disorders who all received psychoeducation, with one group receiving yoga and meditation and another group receiving hypnosis, in addition to psychoeducation. Although there was no significant difference among groups in terms of symptom reduction, more patients in the yoga group achieved remission at 9-month follow-up (Butler et al. 2008).

A systematic review (Uebelacker et al. 2010) suggested that yoga may be a good way to augment current depression treatment strategies which are not efficacious or have significant adverse effect potential for many individuals. This paper also described plausible biological, psychological, and behavioral mechanisms by which yoga may have an impact on depression. However, the authors cautioned that results from the trials reviewed till that point of time should be viewed as very preliminary. While the results were encouraging, the trials as a group suffered from significant methodological limitations, and there was variation in the effect sizes of different comparative trials. Another detailed review (da Silva et al. 2009) enumerated nine RCTs, six open trials, and one case series in major depression.

The clinical guidelines released by the Canadian Network for Mood and Anxiety Treatments (CANMAT) for the management of major depressive disorder in adults in 2009 stated that there was reasonable evidence for the use of yoga in MDD (Ravindran et al. 2009). A meta-analysis in 2011 concluded that yoga was effective as an adjunct treatment for depression and anxiety (Cabral et al. 2012), and a systematic review in 2013 concluded that yoga has level 3 evidence of benefit as an adjunctive treatment for unipolar depression (Ravindran and da Silva 2013).

A recent trial allocated 137 outpatients with DSM-IV major depression to one of the three treatments as per their choice – yoga only, drugs only, or both. A validated generic yoga module (Naveen et al. 2013) was taught over a month (at least 12 sessions) for patients in the yoga groups. Patients were assessed at baseline, after 1 and 3 months on Depression and Clinical Global Impression (Severity) scales. Fifty-eight patients completed the study period with all assessments. The results showed that patients in all three arms obtained a reduction in depression scores as well as clinical severity. However, both yoga groups (with or without drugs) were significantly better than the drugs-only group. Higher proportion of patients remitted in the yoga groups compared with the drugs-only group. The authors concluded that the findings support a case for prescribing yoga for outpatients with non-suicidal depression as a sole therapy or as an adjunct with medications.

A summary of controlled yoga trials in depression is presented in Table 1.

Yoga-Based Therapies for Psychosis

Schizophrenia and other psychoses are considered the most severe illnesses treated by psychiatrists. They cause severe morbidity, disability, and disruption in personal, professional, and family life due to the severity of the illness itself, the associated lack of insight, as well as the fact that it strikes in the critical period of life for individuation and personality formation (adolescence and early adulthood). In spite of significant advances in understanding and treatment of psychosis, the prognosis remains difficult in nearly half of patients (Hegarty et al. 1994). Among the symptom dimensions, the positive symptoms (delusions, hallucinations, and formal thought disorder) are easily identifiable and respond well to available treatments. However, the negative symptoms (amotivation, anhedonia, emotional blunting, and poor insight) and cognitive deficits are also primary features of psychosis and in fact

Table 1 Summary of controlled trials of yoga in depression

Author/year	Sample	Yoga technique	Control	Duration of yoga	Results
Khumar et al. 1993	MDD (n = 50) RCT	Shavasana yoga (n = 25)	No treatment (n = 25)	4 weeks	Yoga significantly superior
Janakiramaiah et al. 2000	DSM IV MDD (n = 45) RCT; 3 groups	SKY (n = 15)	ECT (3/week) and adequate dose of imipramine (n = 15 each)	4 days/week for 4 weeks	Significant reductions in HDRS and BDI scores in all 3 groups; ECT superior, improvement similar in IMN and SKY groups
Rohini et al. 2000	DSM-IV MDD (n = 30)	SKY (n = 15)	Partial SKY (without cyclical breathing) (n = 15)	4 weeks	Reduced BDI scores in both groups; No difference between total and partial SKY at 4 weeks. *Remission*: 12 SKY subjects and 7 partial SKY subjects
Woolery et al. 2004	MDD (n = 28) RCT; 2 groups	Iyengar yoga monotherapy (n = 13)	Wait-list control (n = 15)	5 weeks	Yoga significantly superior in reducing depression
Sharma et al. 2005	DSM IV MDD (n = 30) RCT; 2 groups	Sahaj yoga meditation + antidepressants (n = 15)	Antidepressants (n = 15)	8 weeks	HDRS scores were significantly lower after 8 weeks in group I in comparison with group II
Krishnamurthy and Telles 2007	MDD (n = 69) RCT; 3 groups	Yoga monotherapy (n = 23)	Ayurvedic medicine and wait-list control (n = 23 each)	24 weeks	Yoga significantly superior
Butler et al. 2008	Total n = 46 DSM IV MDD (n = 23) Dysthymia (n = 23) RCT; 3 groups	Meditation + hatha yoga + psychoeducation (n = 15)	Group therapy with hypnosis + psychoeducation (n = 15) Psychoeducation alone (n = 16)	8 weekly sessions, 1 four-hour retreat and 1 booster session at 12 weeks	No difference between groups in HDRS scores at 6/9 months. Meditation + yoga group had higher remission rate (73%) than other 2 groups (62% and 36%) at 9 months
Gangadhar et al. 2013	DSM IV MDD (n = 137) Nonrandomized trial; 3 groups	Yoga module only (n = 23) Yoga + medication (n = 36)	Standard antidepressant medication (n = 78)	2 weeks daily + 2 weekly sessions; booster session in 2nd and 3rd month	58 Patients completed. All groups obtained reduction in depression scores. Both yoga groups significantly better than drugs-only group. Higher proportion of patients remitted in the yoga groups

may precede the onset of positive symptoms by months or years. These dimensions are more difficult to treat with available treatments (Buckley and Stahl 2007) and cause more disability. The negative symptoms have also been well correlated with real-life functioning and have a strong bearing on productivity of the individual. Along with neurocognitive deficits, specific deficits in social cognition have been demonstrated in these patients, which significantly influence real-world functioning and prognosis (Hofer et al. 2009; Kee et al. 2003). Facial emotion recognition deficits (FERD), which is one of the markers of dysfunction in social cognition, is increased in patients with schizophrenia.

Suboptimal response and significant adverse effects of antipsychotic medications have led to complementary treatment options emerging as a critical area of research, with various combinations of psychosocial therapies, including yoga. However, yoga as a therapy for psychosis has been evaluated scientifically only in the twenty-first century. This is possibly due to some reports that meditative practices may worsen or provoke psychotic symptoms (Walsh and Roche 1979), as also the perception that patients with psychosis may not be able to understand and follow yoga protocols. In view of these concerns, an important point to be noted in the studies listed below is that the components of all the yoga modules were mainly yogasana and pranayama. Meditative practices have been avoided in view of possible provocation of psychosis, as well as difficulty in verifying the actual practice of the procedure itself.

An early study included institutionalized patients with schizophrenia, who were able learn the yoga under supervision, and the yoga module produced some benefits in social and cognitive domains without causing disturbing side effects (Nagendra et al. 2000).A RCT using a specific yoga-based module for psychosis included 61 consenting outpatients with moderate symptoms (Duraiswamy et al. 2007). All patients were on stabilized antipsychotic therapy, which remained so over the duration of the study. Subjects were randomly allotted to either the yoga module or a standard set of physical exercises and were trained for 1 month (at least 12 hourly sessions of either yoga or physical exercises) by the same instructors. The patients were advised to continue the practices at home for next 3 months. Assessments of the patients' symptomatology were based on the Positive and Negative Syndrome Scale (PANSS) for schizophrenia (Kay et al. 1987), and social functioning was evaluated using the Social and Occupational Functioning Scale (SOFS) (Saraswat et al. 2006). The assessments were carried out at baseline and after 4 months by a psychiatrist uninvolved in treatment allocation. The results showed that the scores on negative syndrome and social dysfunction dropped in both groups over the 4 months, but the patients in the yoga group performed significantly better. However, this study had some limitations: a modest sample size, lack of a nonintervention control group, and lack of measures of cognition. The same group conducted an expanded RCT by the same group, which included 120 consenting patients stabilized on antipsychotic medications and used the same yoga and physical exercise modules (Varambally et al. 2012). This study which is the largest published RCT of yoga for psychosis till date had three arms – yoga, physical exercise, and wait list. The methodology was fairly similar to the earlier study, with patients receiving a minimum of 12 supervised sessions in the first month. The subjective judgment of learning yogasana well enough to be practiced at home (as assessed by the yoga instructors) was satisfactory in all patients. The patients were assessed on the PANSS, SOFS, and a standardized tool for social cognition, the Tool for Recognition of Emotions in Neuropsychiatric Disorders (TRENDS) (Behere et al. 2008) by an independent rater. Improvement was operationally defined as a drop of 15 in PANSS total score, a drop of 7 each in PANSS negative and positive scores, and a drop of 14 in SOFS total score. The results confirmed those of the earlier study, with more patients in the yoga group improving significantly than those in the other two groups, particularly in negative symptoms and socio-occupational functioning. The likelihood of improvement in yoga group in terms of negative symptoms was about five times greater than the wait list group and the exercise group on odds ratio analysis. With regard to the scores on TRENDS, the patients in the yoga group showed improved emotion recognition from baseline to the second month as well as to the end of the study. No significant change occurred in either exercise or wait-listed patients (Behere et al. 2011). On the basis of these studies, the recent NICE guidelines for management of schizophrenia recommended yoga as a complementary intervention but also emphasized the need for more systematic research in this direction (http://guidance.nice.org.uk/CG178).

A randomized controlled pilot study in the United States included 18 institutionalized patients with schizophrenia and allotted them to either 8 weeks of a yoga module which included postures, breathing exercises, and relaxation or a wait list group. The results showed significant improvements in both positive and negative symptom scores on the PANSS, as well as quality of life, in the yoga group as compared to the wait list group. Keeping in view the limitations of this study, the authors called for bigger and better controlled studies to confirm the findings (Visceglia and Lewis 2011). Another study from India evaluated a yoga module as a cognitive remediation technique in patients with schizophrenia and bipolar disorder (Bhatia et al. 2012). The results demonstrated the positive effects of the yoga module on several cognitive functions, especially in schizophrenia.

It is important to note that most of the above studies were done in chronic patients stabilized on antipsychotics. It is now well accepted that the duration of illness is an important predictor of prognosis, and interventions delivered earlier in the course of the illness are likely to be more effective in changing the outcome. A recent randomized controlled single blind trial of yoga in the acute phase of inpatient treatment in psychosis (Manjunath et al. 2013) used the same yoga modules as in the above study by Varambally (Varambally et al. 2012). Consenting inpatients with psychosis ($n = 88$) were included and were randomized into yoga therapy group ($n = 44$) and physical exercise group ($n = 44$). The interventions were provided as early as possible after admission. In most cases, the intervention was possible by the second week. Sixty patients completed the study period of 6 weeks. At the end of 6 weeks, patients in the yoga group had significantly lower mean scores on Clinical Global Impression Severity (CGIS), PANSS, and Hamilton Depression Rating Scale (HDRS). Repeated measure analysis of variance detected an advantage for yoga over exercise in reducing the clinical CGIS and HDRS scores. The study concluded that adding yoga intervention to standard pharmacological treatment is feasible and may be beneficial even in the early and acute stage of psychosis. However, studies are needed to test whether these effects translate into measurable improvements in prognosis. Such interventions may be a very valuable part of multimodal management programs for first-break psychosis and may help prevent relapses. A potential area of great promise is use of yoga-based interventions as a prophylactic measure in subjects with high risk for psychosis, for whom there are currently no proven preventive strategies.

A summary of controlled trials of yoga in psychosis is provided in Table 2.

Yoga-Based Therapies for Anxiety Disorders

Anxiety disorders are one of the commonest psychiatric disorders, and anxiety is the commonest symptom of presentation to mental health programs. Anxiety disorders can be classified under categories such as generalized anxiety disorder (GAD), panic disorder, phobic disorders, and others. Obsessive compulsive disorder (OCD), which was earlier grouped under anxiety disorders, is now a separate category in the DSM-V classification system (APA 2013). Medications help some symptoms in these patients, but most patients have incomplete response to medication alone. Psychological approaches, mainly focusing on cognitive-behavioral models and relaxation techniques, have demonstrated success in reducing anxiety symptoms. However, they require significant expertise and are resource and time-intensive and not really available in most rural settings.

Several neurophysiological markers of stress and anxiety like galvanic skin response and stress hormone levels have been shown to change with yoga practices. Also, most people who practice yoga and meditation experience a feeling of relaxation as well as significant relief from anxiety and stress. Therefore, it is not surprising that yoga has been tried as a treatment for anxiety disorders worldwide. However, research in this area is minimal, with available studies mainly being in adult populations with small sample sizes and effect sizes. A systematic review in 2005 (Kirkwood et al. 2005) concluded that there was not enough evidence to definitely conclude that yoga was an effective treatment for anxiety disorders. Another review published in the same year (Brown and Gerbarg 2005a) was more optimistic and suggested that there was sufficient evidence to consider SKY to be a beneficial, low-risk, and low-cost adjuvant to treatment of stress, anxiety, PTSD, and stress-related medical illnesses.

One of the yoga-based techniques with reasonable evidence in patients with anxiety disorders is mindfulness-based stress reduction (MBSR). An open trial of MBSR by Kabat-Zinn and others significantly improved anxiety and comorbid depressive symptoms in 20 of 22 patients with GAD or panic disorder, as monotherapy or as an add-on to medication (Kabat-Zinn et al. 1992). The treatment effects were evident at post-intervention and 3 months follow-up. Of note, a 3-year followup of this cohort (18 of the original 22 patients) reported that the gains were maintained after 3 years on the Hamilton and Beck anxiety scales as well as on the

Table 2 Summary of controlled trials of yoga in psychosis (all as adjunct to medications)

Author/year	Sample	Yoga technique	Control	Duration of yoga	Results
Nagendra 2000	Institutionalized patients with chronic schizophrenia ($n = 26$)	Integrated yoga therapy	Physical exercise	1 year	Both groups improved in anxiety and social avoidance; yoga group also improved in blunted affect, somatic concern, and attention
Xie et al. 2006	Schizophrenia ($n = 90$)	Yoga module with postures, breathing techniques, meditation, relaxation	Usual care	8 weeks	Physical and psychological functioning and quality of life improved in yoga group
Duraiswamy et al. 2007	Schizophrenia ($n = 61$)	Yoga module with asana and pranayama ($n = 31$)	Physical exercise ($n = 30$)	4 months	Patients in the yoga group significantly better in PANSS scores, functioning, and quality of life
Visceglia and Lewis 2011	Schizophrenia ($n = 18$)	Personalized yoga module ($n = 10$)	Waitlist ($n = 8$)	8 weeks; twice weekly sessions	Yoga group obtained significantly greater improvements in PANSS scores and perceived quality of life
Vancampfort et al. 2011	Schizophrenia or schizoaffective disorder ($n = 49$)	Single 30-minute yoga session (Hatha Yoga)	Single 20-min session of aerobic exercise or reading	Single session	Yoga and aerobic exercise groups had decreased state anxiety, decreased psychological stress, and increased subjective well-being compared to control condition
Bhatia et al. 2012	Schizophrenia ($n = 88$)	Yoga module with asana and pranayama ($n = 65$)	Treatment as usual ($n = 23$)	3 weeks	Yoga group showed greater improvement with regard to measures of attention; changes more prominent among men
Varambally et al. 2012	Schizophrenia ($n = 119$) RCT; 3 groups	Yoga module with asana and pranayama ($n = 46$)	Physical exercise ($n = 36$) Waitlist ($n = 37$)	4 months	More patients in yoga group improved in PANSS negative and total scores and functioning
Manjunath et al. 2013	Nonaffective psychosis ($n = 88$)	Yoga module with asana and pranayama ($n = 44$)	Physical exercise ($n = 44$)	6 weeks	Patients in the yoga group had (CGIS), PANSS total, and HDRS
Jayaram et al. 2013	Schizophrenia ($n = 43$)	Yoga module with asana and pranayama ($n = 15$)	Waitlist ($n = 28$)	4 weeks	Yoga group significantly improved in functioning and TRENDS scores; increase in plasma oxytocin levels as compared with the waitlist group

respective depression scales. The majority of subjects were reported to be compliant with the meditation practice. The authors concluded that such an intervention based on mindfulness meditation can have long-term beneficial effects for patients with anxiety disorders. A RCT of MBSR including 76 patients with panic disorder with or without agoraphobia (PD/AG), social anxiety disorder (SAD), and GAD reported that patients who completed treatment improved significantly on all outcome measures compared to wait-listed controls. The study reported medium to large effect sizes on measures of anxiety and a large effect size for symptoms of depression and also that the gains were maintained at 6 months follow-up (Vollestad et al. 2011). A systematic review and meta-analysis of mindfulness- and acceptance-based interventions (MABIs) for anxiety disorders published in 2012 evaluated 19 studies and reported an overall between-group effect size of 0.83 for anxiety symptoms and 0.72 for depression symptoms in controlled studies. The authors concluded that MABIs are associated with robust and substantial reductions in symptoms of anxiety and comorbid depressive symptoms but that more research is needed to determine their efficacy compared to current treatments and to clarify the contribution of processes of mindfulness and acceptance to the outcome (Vollestad et al. 2012).

Yoga-Based Therapies for Obsessive Compulsive Disorder (OCD)

This is also an area with minimal systematic research and very few studies. Two small trials of kundalini yoga (KY) have been published. In the first trial which was an open uncontrolled pilot study using KY as monotherapy or adjunct to fluoxetine, five of eight patients showed 55.6% improvement in the mean Yale-Brown Obsessive Compulsive Scale (Y-BOCS) score after a 12-month trial (Shannahoff-Khalsa and Beckett 1996). A subsequent RCT by the same group compared KY group and another group with relaxation response plus mindfulness meditation. The study reported that the KY group had greater and statistically significant improvement after intent-to-treat analysis. This RCT reported a large effect size ($d = 1.20$) (Shannahoff-Khalsa et al. 1999). The two treatment groups were then merged and received a further 12 months of KY, with positive effects persisting. Of importance is the report that most of the patients in both the above studies who were originally on medication were able to reduce or eliminate medication use by the end of the study. However, these findings have not been replicated. A recently published paper details the development, validation, and feasibility testing of a generic yoga module for OCD (Bhat et al. 2016).

Yoga-Based Therapies for Substance Use Disorders

Substance abuse is highly prevalent in both urban and rural settings by itself and comorbid with other psychiatric disorders. This is a truly a modern epidemic, as abuse and dependence of alcohol and other drugs contribute heavily to healthcare burden, road accidents, and increased prevalence of other mental illnesses and suicide. Current standard management approaches for substance use are multimodal programs including pharmacological, behavioral, and psychosocial components. Yoga practices also have been used as a part of such programs. An early study by Shaffer (Shaffer et al. 1997) found that a hatha-yoga-based module was comparable to conventional methadone treatment with traditional group psychotherapy in 61 randomly assigned clients on a variety of psychological, sociological, and biological measures. Sudarshan kriya (SK) and pranayam (P) as an intervention was used in a study on cancer patients with nicotine addiction after they had completed standard therapy. This study reported that SK and P helped to control the tobacco habit in 21% of individuals who were followed up to 6 months of practice (Kochupillai et al. 2005). Another trial from India used a 90-day residential group pilot treatment program for substance abuse that incorporated a comprehensive array of yoga, meditation, spiritual, and mind-body techniques (Khalsa et al. 2008). Results showed improvements on a number of psychological self-report questionnaires including the Behavior and Symptom Identification Scale and the Quality of Recovery Index. A recent study which included 18 patients with alcohol dependence evaluated the feasibility and efficacy of a 10-week yoga package as an adjunct therapy in a treatment program versus treatment as usual. The authors reported that the yoga-based intervention was a feasible and well-accepted adjunct treatment, but did not produce any significant effects in terms of reduction in alcohol consumption. They concluded that larger studies are needed to adequately assess the efficacy and long-term effectiveness of yoga as an adjunct treatment for alcohol dependence (Hallgren et al. 2014). More systematic research is needed to understand and match types of interventions to types of addiction, patients, and settings. A recent narrative review summarizes the philosophical origins, current evidence, and the future prospects of yoga and mindfulness in this important clinical area (Khanna and Greeson 2013).

Yoga-Based Therapies for Somatoform Disorders

For health systems worldwide, somatoform and pain disorders are a difficult and expensive group of common psychiatric disorders, as the response to medication as well as psychosocial interventions is unsatisfactory in most cases. Yoga has been used with quite good success as a therapy for several pain syndromes, showing particular efficacy in low back pain (Cramer et al. 2013). There is also promising preliminary data for migraine (John et al. 2007; Kisan et al. 2014) and fibromyalgia (Langhorst et al. 2013; Mist et al. 2013). A randomized study using yoga nidra as a treatment in patients of menstrual disorders with somatoform symptoms included 150 female patients with menstrual disorders. They were randomized to yoga nidra intervention with medication or only medication. Results showed significant improvement in pain symptoms, gastrointestinal symptoms, cardiovascular symptoms, and urogenital symptoms after 6 months of yoga nidra therapy in comparison to the control group (Rani et al. 2011). A recent open trial at a tertiary psychiatric center in India evaluated an integrated yoga therapy module in 64 patients with somatoform pain disorder. Yoga intervention led to a significant reduction in pain scores and improvement in anxiety, sleep, and quality of life in patients who completed the study (Sutar et al. 2016). The demonstration of abnormalities in inflammatory and neuroimmunological pathways in patients with somatoform disorders and evidence that yoga produces reduction in several inflammatory markers makes this a logical area for further research.

Yoga-Based Therapies for Psychiatric Disorders of Childhood

Studies have trialed yoga-based programs in children with anxiety spectrum disorders, depression, and autism, but most of these are case reports or case series (Birdee et al. 2009). A notable study included 24 children aged 3–16 years with a diagnosis of an autism spectrum disorder (ASD) who were taught an 8-week multimodal yoga, dance, and music therapy program based on the relaxation response (RR). The outcomes were measured by the Behavior Assessment System for Children, Second Edition (BASC-2), and the Aberrant Behavior Checklist (ABC). The study found robust changes on the BASC-2, primarily for 5- to 12-year-old children, and surprisingly even on the atypicality scale of the BASC-2, which measures some of the core features of autism.

The evidence for yoga-based programs is much stronger for attention-deficit hyperactivity disorder (ADHD), which is a major psychiatric disorder in this age group. Medications and behavioral interventions have been the standard of treatment, although the adverse effects of stimulant medications have made them unpopular among parents. Yoga and meditative practices have therefore been explored as sole or adjunct interventions in this disorder. A RCT (Jensen and Kenny 2004) included boys diagnosed with ADHD and stabilized on medication; they were randomized to either to a 20-session specific yoga program ($n = 11$) or a control group (cooperative activities; $n = 8$). The children in the yoga group obtained significant benefits on five subscales of the Conners' Parents Rating Scales (CPRS). The authors concluded that yoga may have merit as a complementary treatment for ADHD, particularly for its evening effect when medication effects are on the wane or absent. However, the limitations of this study including low statistical power and inconsistency of home practices make it necessary to caution that the results need to be replicated on larger groups with a more intensive supervised practice program. A study from Thailand used a meditation and imagery program in children with ADHD (Hassasiri et al. 2002) and found a significant impact with regard to lowering symptoms. Another study (Haffner et al. 2006) included 19 children (12 boys and 7 girls) diagnosed with ADHD and used a 2×2 crossover design comparing yoga training ($n = 8$) versus regular physical exercise ($n = 11$) over 8 weeks. Eight of the 19 children were also on medication, and seven received other complementary therapies as well. This study found that yoga training was superior to the physical exercise regimen on test scores on an attention task and parent ratings of ADHD symptoms, with effect sizes in the medium-to-high range (0.60–0.97). Furthermore, the training was particularly effective for children undergoing pharmacotherapy (MPH). Two dissertations from the United States (Kratter 1983; Moretti-Altuna 1987) used mantra meditation as an intervention in boys with ADHD. The first study by Kratter compared three intervention groups: meditation training versus relaxation training versus waiting list control ($n = 24$ boys). The medicated participants were distributed evenly between treatment groups using a modified randomization method. This study found that mantra meditation as well as relaxation training significantly decreased impulsivity and improved behavior at home as rated by the parents. The study by Moretti-Altuna ($n = 23$ boys) compared meditation training versus drug therapy versus standard therapy control, but did not find statistically significant difference between the meditation therapy group and the drug therapy group in either the teacher rating ADHD scale or the distraction test. A Cochrane review in 2010 (Krisanaprakornkit et al. 2010) concluded that there was insufficient evidence to support the effectiveness of any types of meditation for ADHD as the four randomized controlled trials that were published at that time were relatively small in size and limited in design, with inconsistent results.

Some recent studies have shown more promising results. An open-label exploratory study included nine patients with moderate to severe ADHD (eight were on medications) admitted in a child psychiatry ward (Hariprasad et al. 2013a). The participants were taught a specific yoga module by a trained therapist daily during their inpatient stay and assessed at the end of first, second, and third month by an independent rater. An average of eight yoga training sessions was given to subjects. They were able to learn yoga reasonably well. There was a significant improvement in the ADHD symptoms at the time of discharge. Two recent school-based studies with pragmatic methodology and good follow-up have provided better evidence for the efficacy of yoga-based multimodal programs in ADHD. The first study (Mehta et al. 2011) used a 6-week multimodal peer-mediated behavioral program that included yoga, for children between 6 and 11 years of age diagnosed with ADHD. The authors measured performance and behavioral scores using the Vanderbilt scale. The 1-hour program (twice weekly sessions) combining yoga, meditation, and play therapy was taught by trained high school volunteers. Seventy-six school children with ADHD were included. After 6 weeks of the program, 90.5% of included children showed reductions in performance-impairment score, a measure of academic performance. The researchers then followed up 69 of these children with weekly sessions for 1 year and showed that improvement on the performance impairment scores for ADHD was sustained through 12 months in 85% of the students. Almost all (92%) of the students also had improvements in their Vanderbilt scores as assessed by parents. The follow-up study results validated and confirmed the efficacy and cost-effectiveness of the program (Mehta et al. 2013). We may conclude that there is evidence to suggest that yoga therapy for ADHD is feasible in both hospital and school settings and may produce significant reduction in symptoms as a sole therapy or along with medication and as a part of multimodal intervention programs.

Yoga for Cognitive Disorders in the Elderly

This is a priority area for most health services around the world due to the rapidly increasing proportion of the population who are elderly and the lack of effective interventions for these disorders. There is preliminary evidence demonstrating improvements in cognitive parameters in healthy elderly individuals with yoga-based programs (Hariprasad et al. 2013b) and in subjects with primary psychiatric disorders such as depression (Sharma et al. 2006) and psychosis (Bhatia et al. 2012). Yoga practices may also have observable neurobiological effects on the structure of the brain, specifically areas relating to memory in the elderly brain (Hariprasad et al. 2013d). However, there are very few studies looking at the efficacy of yoga-based interventions for patients with specific cognitive disorders such as mild cognitive impairment or early dementia, and the available evidence is inconclusive (Balasubramaniam et al. 2013).

Yoga-Based Therapies for Insomnia

Sleep disorders are quite prevalent in the general population, particularly insomnia, and even more so in patients with psychiatric disorders. Insomnia and dependence on hypnotics are common problems in patients in primary care as well. Various non-pharmacological methods have been tried for insomnia, yoga-based programs being one of them.

In otherwise healthy subjects, there are several reports of improvement of sleep parameters with cyclic meditation (Patra and Telles 2009), silver yoga exercises in elderly subjects (Chen et al. 2009), and elderly subjects in old-age homes (Hariprasad et al. 2013c). Studies of Tibetan yoga in subjects with lymphoma (Cohen et al. 2004) and yoga in cancer survivors (Mustian et al. 2013) have also demonstrated benefits in sleep quality. A study in 20 subjects with chronic insomnia using a yoga-based module reported improvements in sleep efficiency, total sleep time, total wake time, sleep onset latency, and wake time after sleep onset at end treatment as compared with pretreatment values (Khalsa 2004). In a systematic review of CAM therapies for sleep, Sarris and Byrne found some evidence for acupressure, tai chi, and yoga in the treatment of chronic insomnia. This study was able to find only one RCT, which was a study comparing yoga and Ayurvedic medication in 120 elderly subjects with insomnia (Manjunath and Telles 2005). This study found medium-to-large effect sizes for benefits of yoga therapy on latency, duration, and quality of sleep. Thus, there have been some preliminary demonstrations but limited clinical evidence of the benefits of yoga and meditation-based interventions for insomnia. Further research support is needed to replicate and confirm the efficacy of yoga as a therapeutic agent in insomnia and other sleep disorders.

Biological correlates of the effect of yoga in mental disorders: Apart from effects on clinical symptoms, yoga has been shown to "correct" biological parameters which may be abnormal in several psychiatric disorders. Low pretreatment P300 event-related potential (ERP) amplitude was shown to "normalize" in patients with both MDD and dysthymia (Murthy et al. 1997). In patients with alcohol dependence and depressive symptoms, SKY reduced both depression scores and serum cortisol (Vedamurthachar et al. 2006). In a recent comparative trial (Gangadhar et al. 2013), serum BDNF rose in patients after 3 months of yoga therapy, and there was a significant positive correlation between fall in HDRS and rise in serum BDNF levels in the group receiving yoga alone (Naveen et al. 2013). There was also a significant correlation between rise in serum BDNF levels and reduction in serum cortisol levels in this group (Naveen et al. 2016). In patients with psychosis, along with improvement in social cognition, a significant elevation in serum levels of the "cuddling hormone" oxytocin has been demonstrated. Oxytocin levels rose significantly in the yoga group, nearly threefold, with no such change evident in the wait-listed group (Jayaram et al. 2013). This suggests the possibility that yoga causes oxytocin elevations, and this may mediate the benefits in social cognition as well as the well-being effect following yoga.

How do we explain the effects of yoga-based interventions in these diverse disorders? Current research seems to point to a unifying hypothesis of impaired neuroplasticity in these disorders. Impaired neuroplastic mechanisms have been postulated in depression (Duman and Monteggia 2006), schizophrenia (Voineskos et al. 2013), and other psychiatric disorders. Neuroplasticity has been in fact described as the mediator of treatment response in depression (Andrade and Rao 2011). In this light, there are several lines of evidence that yoga has salutary and remedial effects on neuroplastic mechanisms in the human brain. Studies in healthy populations have shown that yoga and meditative practices led to increase in cortical thickness (Lazar et al. 2005) and increased volume of gray matter in parts of the brain important for memory and cognition (Luders et al. 2009). Recent evidence indicates that yoga increases levels of neurotransmitters in certain brain regions (Streeter et al. 2007, 2010) and that yoga practices may facilitate neuroplasticity in disorders such as depression (Naveen et al. 2013). This may inform future research and may provide a viable explanation for the positive effects of yoga in psychiatric illnesses. A review of the research in yoga therapy for schizophrenia (Bangalore and Varambally 2012) summarizes some of these issues and provides directions for future research in this area.

Meditation and Mindfulness

Over the last few decades, there has been tremendous scientific interest in studying the effects of meditative practices on the human brain and physiology, especially with focus on Buddhist meditation. His Holiness the 14th Dalai Lama has initiated scientific exploration of these aspects through institutions such as the Mind and Life Institute and others. Meditation refers to a variety of practices aimed at focusing awareness and attention to voluntarily control mental processes (Marchand 2013). Meditation is an extremely complex process, and more than one function is responsible for its effects. Many different practices of meditation have existed across cultures, including Hinduism (Transcendental Meditation, Kundalini meditation, also see yoga), Islam (*tafakkur*), Sikhism (*simran*), Christianity (*contemplative prayer*), and the Tibetan meditative practice (*rig pa cog bzhag*).

Transcendental Meditation became very popular after it was introduced to the West by Maharishi Mahesh Yogi, a scholar of the ancient Indian Vedic tradition. The technique is simple and easily learned, requiring to be practiced for 20 min twice daily while sitting with eyes closed and repeating a "mantra," a sequence of specific sounds, to promote a natural shift of awareness to a wakeful but deeply restful state. A reduction in mental and physical activity occurs leading to a mental state called transcendental consciousness, which is different from usual waking, dreaming, or sleep states. This experience is deemed responsible for the restoration of normal function of various bodily systems, especially those involved in adapting to environmental stressors or challenges. Regular practice of Transcendental Meditation has been shown to improve cortical coherence, attention, and overall brain functioning. Clinical effects include reduced anxiety, pain, and depression, enhanced mood and self-esteem, and decreased stress (Varvogli and Darviri 2011).

Mindfulness or reality-directed observation: This originates from Buddhist philosophy and approaches such as *Vipassana* (often translated as "insight"). In modern sources, the approach is also referred to as "awareness meditation" or "mindfulness." Mindfulness is the process of attending to present moment sensations and experiences with a nonjudgmental stance. Its cultivation is described as conducive to full understanding of the three essential characteristics of all conditioned phenomena: impermanence, suffering, and no(n)-self. The latter implies that one's physical and mental constituents do not represent the eternal self or belong to oneself forever. Mindfulness has been shown to decrease ruminations, worry, and symptoms of depression and anxiety through both distinct and common emotion regulatory mechanisms (Desrosiers et al. 2013). Ong et al. (2008) used mindfulness meditation combined with cognitive behavioral therapy in 30 adults with insomnia and found improvement in several sleep-related measures. Mindfulness-based stress reduction and mindfulness-based cognitive therapy have been studied as clinical interventions and have strong evidence documenting their effectiveness in stress-related psychiatric disorders, some of which have been described in the section on yoga above.

Tai Chi and *Qi*gong: *Tai chi chuan* or *Tai Chi* involves a series of slow circular movements designed to increase the life force in the body. It is a moving form of meditation and is based on the search for perfect balance between *yin* and *yang* energies. Chinese *Qigong* has been practiced for more than 2000 years. *Qigong* means the skill or work (*gong*) of cultivating energy (*Qi*). "Still" *Qigong* is practiced as a motionless meditation with the emphasis on breath and intentional thoughts. "Moving" *Qi*gong involves external movements under the conscious direction of the mind. Both *Tai Chi* and *Qigong* involve sequences of flowing movements coupled with changes in mental focus, breathing, coordination, and relaxation. Both practices have a significant overlap of technique and are low-impact, moderate intensity aerobic exercises that are suitable across different age groups and health conditions. They are practiced worldwide in a variety of modern and traditional forms. However, many RCTs have used the Yang style or the Yang style short form to study the effectiveness of this practice.

Tai Chi and *Qigong* have been shown to promote relaxation, decrease sympathetic output, improve immune function and vaccine response, increase endorphin release, and reduce levels of inflammatory markers, adrenocorticotrophic hormone, and cortisol (Abbott and Lavretsky 2013). EEG studies have shown increased frontal α, β, and θ activity, suggesting increased relaxation and attentiveness. RCTs have shown that *Tai Chi* and *Qi*gong may improve bone density, cardiopulmonary health, arthritis, fibromyalgia, tension headaches, and other medical conditions. It is associated with improvements in psychosocial well-being including reduced stress, anxiety, depression, increased self-esteem, and improvement in health-related quality of life (HRQOL). It improves sleep quality, improves balance, and reduces falls, owing to which it has been recommended as treatment of mild depression in geriatric population. Tai Chi also seems to reduce balance impairments in patients with mild-to-moderate Parkinson disease. In patients with substance use disorders, *Qigong* participants experienced a higher treatment completion rate, improved anxiety scores, greater reductions in cravings, and lower relapse rates. Patients with traumatic brain injury also had improved mood and self-esteem after *Qigong*, with improved HRQOL.

Challenges and Methodological Issues Relating to Mind-Body Therapies in Psychiatric Disorders

There are several methodological challenges in the study of mind-body therapies in mental illness; some of them are quite unique due to their nature being inherently different from other interventions in medicine. The main issues are finding appropriate placebo control procedures and difficulties with blinding in such research. For example, many different schools of yoga follow different systems of nosology, and some insist that yoga should be looked at as a complete lifestyle and not used piecemeal as therapy. This mandates a need for broad agreement among yoga schools and cooperative research with medical professionals for building evidence. These concerns have been discussed in detail and future directions for research have been provided in a recent review (Gangadhar and Varambally 2011). Another recent article analyzed the links between yoga and spirituality, which is an important component that needs to be factored into research in yoga for psychiatric disorders (Varambally and Gangadhar 2012). Such issues are relevant for other mind-body therapies as well. In the research on meditative practices, verifiability of the actual practice is problematic, although researchers have used EEG and other methods to assess brain functioning during the practice. Functional magnetic resonance imaging (fMRI) and functional near-infrared spectroscopy (fNIRS) are now being used to study the brain during activity, although the understanding of the brain patterns during such practices is still preliminary.

Whole Medical Systems

Ayurveda is a medical system that originated in India about 4000 BC and is believed to be one of the most comprehensive medical systems in the world. It is similar to many of the traditional health systems in its beliefs about the energy points in the body and a vital force (*prana*) that must be in balance to maintain health. It bases diagnosis on examination of pulse, urine, warmth, or coldness of body and for treatment uses diet, medicines, purification, enemas, and, in some cases, bloodletting. Current forms of this medical system show Buddhist and Hindu contributions and are highly respected, involving considerable medical training from recognized Ayurvedic medical schools all over India and some other countries. Ayurvedic concepts may be applicable to conditions such as anxiety, neurosis, and depressive disorders.

Modern Chinese medicine relies on herbs in addition to other methods such as acupuncture, massage, diet, and exercise, to correct so-called imbalances in the body.

Tibetan medicine or *Sowa-Rigpa*: Traditional Tibetan medicine, also known as Sowa-Rigpa medicine, is a century-old traditional medical system that employs a complex approach to diagnosis, incorporating techniques such as pulse analysis and urinalysis, and utilizes behavior and dietary modification, medicines composed of natural materials (e.g., herbs and minerals) and physical therapies (e.g., Tibetan acupuncture, moxibustion, etc.) to treat illness. The Tibetan health system is as old as the seventh century AD and has elements of Arabic, Indian, and Chinese health systems, with its practice closely related to religion and magic.

Homeopathy was discovered in the early nineteenth century by Samuel Hahnemann, a German physician. Its pharmacopoeia has more than 2000 medicines, and medications are derived from plants, minerals, and animals, while they are dispensed as tinctures or pills with lactose fillers in infinitesimally dilute solutions. The selection of medicine is based on the law of similars – *Similia similibus curantur* (let like be cured by like). For instance, a medicine which produces nausea would be used in minute concentrations to treat nausea.

Naturopathy: According to the manifesto of British Naturopathic Association, "Naturopathy is a system of treatment which recognizes the existence of the vital curative force within the body." It therefore advocates aiding the human system to remove the cause of disease, i.e., toxins by expelling the unwanted and unused matters from human body for curing diseases. It includes hydrotherapy ("water cure") which was made famous by Vincent Priessnitz in the nineteenth century. Louis Kuhne provided a theoretical base for his method through his principle of Unity of Disease and Treatment, illustrated in his book *New Science of Healing*. However, many consider naturopathy as a philosophy rather than a system of medicine.

Other health systems such as Siddha, Unani, Balinese medicine, and various forms of tribal medicine are practiced in circumscribed parts of the world.

However, many of these systems do not have specific diagnostic approaches for mental health disorders and evidence for efficacy in such disorders is minimal.

Biologically Based Therapies

Herbs are probably the oldest known system of healthcare, its historical origins attributed to China in approximately 4000 BC, but, certainly, plants have been used medicinally worldwide throughout prehistory. Herbal medicine uses plant products to maintain health and treat illness. A Greco-Roman medical text by Dioscorides, *De Materia Medica*, in the first century, describes the use of more than 500 plants to treat disease, beginning a tradition followed in European medicine to this day.

Many such natural substances are claimed to increase the "non-specific" resistance of an organism to adverse influences and stress, and the term "adaptogen" has been used to describe them (Panossian et al. 1999). The concept of adaptogens as "medicine for the healthy" or in helping the body cope with stress is a great deal similar to many remedies common in Chinese herbology. Ayurveda, the Indian system of medicine, has a similar concept of *rasayana*. Some of the herbal medicines and adaptogens which have been found to have usefulness in mental health are St. John's wort, *Ginkgo biloba*, saffron, passionflower, valerian, ginseng, huperzine, snowdrop, kava, green tea, chamomile, arctic root (*Rhodiola rosea*), *and Schisandra* (*Schisandra chinensis*). Some of these have reasonable evidence for efficacy in clinical depression and anxiety as adjuvants and sometimes as a sole therapy in mild cases, particularly St. John's wort (Modabbernia and Akhondzadeh 2013; Gerbarg and Brown 2013).

Energy-Based Therapies

The concept of subtle energy and methods of its use for healing has been described by numerous cultures for thousands of years. These vital energy concepts (which include the Indian term *prana*, the Chinese term *chi*, and the Japanese term *qi*) all refer to so-called subtle or nonphysical energies that permeate existence and have specific effects on the body-mind of all conscious beings. Similar concepts in the West are reflected in the concepts of Holy Spirit or spirit and can be dated back to writings in the Old Testament as well as the practice of laying on of hands. According to the National Institutes of Health (NIH), energy-based therapies "are intended to affect energy fields that purportedly surround and penetrate the human body."

Despite differences in ontologies of these proposed forces, a common thread is the techniques that attempt to use subtle energy to stimulate one's own healing process. These are clearly reflected in internal (intrapersonal), movement-oriented practices described above such as *yoga*, *Tai Chi*, and internal *Qigong* and are often noted as part of the experience of meditation and prayer. In addition, different cultures have developed external (interpersonal) practices that purport to specifically use subtle energies for the process of healing another. These include local or proximal practices such as external *Qigong*, *pranic* healing, and laying on of hands, where a healer transmits or guides energy to a recipient who is physically present, as well as distance practices where a healer sends energy to a recipient in a different physical location, such as intercessory prayer or distance healing. Thus, the energy-based therapies share in common (1) the idea that the therapeutic effects rest on manipulation of some energy fields associated with the human body and (2) the manipulation of that energy is done by a therapist, with the patient as passive participant.

Acupressure and Acupuncture: Traditional Chinese medicine has been used to treat mental illness since 1100 BC. A basic tenet of Chinese medicine is that energy (*qi* or *chi*) flows along specific pathways (meridians) that have about 350 major points (acupoints) whose manipulation corrects imbalances by stimulating or removing blockages to energy flow. Another fundamental concept is the idea of two opposing energy fields (yin and yang) that must be in balance for health to be sustained. In acupressure, the pressure points are manipulated by the fingers; in acupuncture, sterilized silver or gold needles are inserted into the skin to varying depths and are rotated or left in place for varying periods to correct any imbalance of *qi*. The needles may be manipulated manually or stimulated by electrical impulses. The generally sympatho-inhibitory effects of acupuncture in animals and humans – dependent on needle location and acupuncture type – may be mediated largely through neurotransmitter systems in an opioid-dependent manner. The opioid system is hypothesized to be aberrant in anxiety and stress-related disorders. Other techniques such as moxibustion (burning small cones of dried, powdered *Artemisia vulgaris* (moxa) leaves held above the point to be warmed or placed on the skin and removed before overheating occurs), pressure, heat, and laser are also used. A variety of mental health conditions may benefit from the use of acupuncture or moxibustion, such as pain and addictions to alcohol and illicit drugs; symptoms of PTSD, depression, and anxiety.

Body-Based/Manipulative Therapies

Methods of spinal manipulation have dated back to times of Hippocrates and Galen. Acupressure, as described above, would also fit in this subclass of CAM treatments. Massage, as a form of therapy, is described below.

Massage: Massage is a treatment that involves manipulation of the soft tissues and the surfaces of the body. It was prescribed for the treatment of disease more than 5,000 years ago by Chinese physicians, and Hippocrates believed it to be an important method of healing. Archaeological evidence of massage has been found in many ancient civilizations including China, India, Japan, Korea, Egypt, Rome, Greece, and Mesopotamia. Sanskrit records indicate that massage had been practiced in India long before the beginning of recorded history. It is said to increase blood circulation, to improve the flow of lymph, to soothe sore muscles, and to have a tranquilizing effect on the mind.

Other Traditional Healing Practices

A multitude of natural and supernatural therapies have been tried and experienced among different cultures to improve health and reduce disease. In most forms of supernatural therapies, the therapeutic goal is achieved through the ritual of prayer, testimony, sacrifice, reliving experience, or even spirit possession. Healing mechanisms used in these practices are assurance, suggestions, and generation of conviction. Some examples would be Shamanism, Zar ceremonies, spirit dancing ceremony, Christian religious healing, etc. The pervasive interest in faith healing, the curative anecdotes of television evangelists and spiritual gurus, and the millions of hopeful individuals visiting religious shrines in search of relief give witness to the continuing interest in and prevalence of prayer and spirituality in the process of healing.

There are several clinical reports and reviews advocating the use of shared prayer, silent prayer, and distant prayer in nursing care. There is a large body of epidemiological research that indicates that religious beliefs and practices are negatively correlated with substance abuse and positively correlated with health status (Chatters 2000). Also, programs like the 12-step program of Alcoholics Anonymous have successfully incorporated prayer and spirituality in the treatment of addictive behavior. Subjects receiving religion-based cognitive behavioral therapy do better, regardless of the religiosity of the therapist conducting the therapy. The core and effectiveness of different methods of religious and magical healing seem to lie in their ability to arouse hope and the innate healing ability in the individual. The folk healing practices and modern psychotherapy share a number of non-specific therapeutic mechanisms.

Conclusion

Although modern medicine has been in practice for more than two centuries and has spread through the world, a review of preventive and therapeutic approaches for mental health issues shows that a high proportion of people, especially in rural communities, continue to use various traditional healing practices based on local culture and beliefs. Of these, the mind-body therapies have gained significant currency in the last few decades, with yoga and meditative practice being intensively researched for different mental health conditions. Current evidence indicates that yoga-based interventions have shown efficacy in depression, schizophrenia, anxiety disorders, somatoform disorders, attention deficit hyperactivity in children, and mild cognitive impairment in the elderly, and there is preliminary evidence of usefulness in other mental health conditions such as substance abuse and insomnia. Mind-body therapies have most often been used as adjuncts to modern medical treatments and occasionally as sole therapy in some cases. Recent research has also uncovered several biological correlates of these therapies, suggesting that these have neuroplastic effects in the brain as well as several cascading hormonal and biochemical changes in the body. Other traditional healing approaches such as Ayurveda and other whole medical systems, biologically based therapies, energy-based therapies, and body-based manipulative therapies are in use in many parts of the world, and more careful research is needed to document the effects of these therapies in specific mental health conditions. The need of the hour is developing integrated mental health services which include modern medical methods and traditional healing practices and cooperative research by professionals of all these disciplines. This can help provide the best care for the increasing number of people suffering from mental health disorders.

Acknowledgments The authors wish to acknowledge Dr. Sneha Karmani for her inputs to the chapter

References

Abbott A (2010) Schizophrenia: the drug deadlock. Nature 468(7321):158–159

Abbott R, Lavretsky H (2013) Tai Chi and Qigong for the treatment and prevention of mental disorders. Psychiatr Clin North Am 36(1):109–119

Andrade C, Rao NS (2011) How antidepressant drugs act: a primer on neuroplasticity as the eventual mediator of antidepressant efficacy. Indian J Psychiatry 52(4):378–386

APA (2013) Diagnostic and statistical manual of mental disorders, 5th edn. American Psychiatric Association, Washington, DC

Balasubramaniam M, Telles S, Doraiswamy PM (2013) Yoga on our minds: a systematic review of yoga for neuropsychiatric disorders. Front Psychiatry 3:117

Bangalore NG, Varambally S (2012) Yoga therapy for schizophrenia. Int J Yoga 5(2):85–91

Behere RV, Raghunandan VN, Venkatasubramanian G (2008) TRENDS: a tool for recognition of emotions in neuro-psychiatric disorders. Indian J Psychol Med 30:32–38

Behere RV, Arasappa R, Jagannathan A, Varambally S, Venkatasubramanian G, Thirthalli J et al (2011) Effect of yoga therapy on facial emotion recognition deficits, symptoms and functioning in patients with schizophrenia. Acta Psychiatr Scand 123(2):147–153

Bhat S, Varambally S, Karmani S, Govindaraj R, Gangadhar BN (2016) Designing and validation of a yoga-based intervention for obsessive compulsive disorder. Int Rev Psychiatry 28(3):327–333

Bhatia T, Agarwal A, Shah G, Wood J, Richard J, Gur RE et al (2012) Adjunctive cognitive remediation for schizophrenia using yoga: an open, non-randomized trial. Acta Neuropsychiatr 24(2):91–100

Birdee GS, Yeh GY, Wayne PM, Phillips RS, Davis RB, Gardiner P (2009) Clinical applications of yoga for the pediatric population: a systematic review. Acad Pediatr 9(4):212–220 e211–219

Brown RP, Gerbarg PL (2005a) Sudarshan Kriya Yogic breathing in the treatment of stress, anxiety, and depression. Part II – clinical applications and guidelines. J Altern Complement Med 11 (4):711–717

Brown RP, Gerbarg PL (2005b) Sudarshan Kriya yogic breathing in the treatment of stress, anxiety, and depression: part I-neurophysiologic model. J Altern Complement Med 11(1):189–201

Buckley PF, Stahl SM (2007) Pharmacological treatment of negative symptoms of schizophrenia: therapeutic opportunity or cul-de-sac? Acta Psychiatr Scand 115(2):93–100

Butler LD, Waelde LC, Hastings TA, Chen XH, Symons B, Marshall J et al (2008) Meditation with yoga, group therapy with hypnosis, and psychoeducation for long-term depressed mood: a randomized pilot trial. J Clin Psychol 64(7):806–820

Cabral P, Meyer HB, Ames D (2012). Effectiveness of yoga therapy as a complementary treatment for major psychiatric disorders: a meta-analysis. Prim Care Companion CNS Disord 13(4)

Chatters LM (2000) Religion and health: public health research and practice. Annu Rev Public Health 21:335–367

Chen KM, Chen MH, Chao HC, Hung HM, Lin HS, Li CH (2009) Sleep quality, depression state, and health status of older adults after silver yoga exercises: cluster randomized trial. Int J Nurs Stud 46(2):154–163

Cohen L, Warneke C, Fouladi RT, Rodriguez MA, Chaoul-Reich A (2004) Psychological adjustment and sleep quality in a randomized trial of the effects of a Tibetan yoga intervention in patients with lymphoma. Cancer 100(10):2253–2260

Cramer H, Lauche R, Haller H, Dobos G (2013) A systematic review and meta-analysis of yoga for low back pain. Clin J Pain 29(5):450–460

Cramer H, Lauche R, Haller H, Steckhan N, Michalsen A, Dobos G (2014) Effects of yoga on cardiovascular disease risk factors: a systematic review and meta-analysis. Int J Cardiol 173(2): 170–183

da Silva TL, Ravindran LN, Ravindran AV (2009) Yoga in the treatment of mood and anxiety disorders: a review. Asian J Psychiatr 2(1):6–16

Desrosiers A, Vine V, Klemanski DH, Nolen-Hoeksema S (2013) Mindfulness and emotion regulation in depression and anxiety: common and distinct mechanisms of action. Depress Anxiety 30(7):654–661

Duman RS, Monteggia LM (2006) A neurotrophic model for stress-related mood disorders. Biol Psychiatry 59(12):1116–1127

Duraiswamy G, Thirthalli J, Nagendra HR, Gangadhar BN (2007) Yoga therapy as an add-on treatment in the management of patients with schizophrenia – a randomized controlled trial. Acta Psychiatr Scand 116(3):226–232

Gangadhar BN, Varambally S (2011) Yoga as therapy in psychiatric disorders: past, present, and future. Biofeedback 39(2):60–63

Gangadhar BN, Naveen GH, Rao MG, Thirthalli J, Varambally S (2013) Positive antidepressant effects of generic yoga in depressive out-patients: a comparative study. Indian J Psychiatry 55(Suppl 3):S369–S373

Gerbarg PL, Brown RP (2013) Phytomedicines for prevention and treatment of mental health disorders. Psychiatr Clin North Am 36(1):37–47

Haffner J, Roos J, Goldstein N, Parzer P, Resch F (2006) The effectiveness of body-oriented methods of therapy in the treatment of attention-deficit hyperactivity disorder (ADHD): results of a controlled pilot study. Z Kinder Jugendpsychiatr Psychother 34(1):37–47

Hallgren M, Romberg K, Bakshi AS, Andreasson S (2014) Yoga as an adjunct treatment for alcohol dependence: a pilot study. Complement Ther Med 22(3):441–445

Hariprasad VR, Arasappa R, Varambally S, Srinath S, Gangadhar BN (2013a) Feasibility and efficacy of yoga as an add-on intervention in attention deficit-hyperactivity disorder: an exploratory study. Indian J Psychiatry 55(Suppl 3):S379–S384

Hariprasad VR, Koparde V, Sivakumar PT, Varambally S, Thirthalli J, Varghese M et al (2013b) Randomized clinical trial of yoga-based intervention in residents from elderly homes: effects on cognitive function. Indian J Psychiatry 55(Suppl 3):S357–S363

Hariprasad VR, Sivakumar PT, Koparde V, Varambally S, Thirthalli J, Varghese M et al (2013c) Effects of yoga intervention on sleep and quality-of-life in elderly: a randomized controlled trial. Indian J Psychiatry 55(Suppl 3):S364–S368

Hariprasad VR, Varambally S, Shivakumar V, Kalmady SV, Venkatasubramanian G, Gangadhar BN (2013d) Yoga increases the volume of the hippocampus in elderly subjects. Indian J Psychiatry 55(Suppl 3):S394–S396

Hassasiri A, Dhammakhanto K, Wongpunya S (2002) Manual of meditation and imagery training for attention deficit children: age 5–11. Paper presented at the 8th International Congress of Department of Mental Health, Thailand

Hegarty JD, Baldessarini RJ, Tohen M, Waternaux C, Oepen G (1994) One hundred years of schizophrenia: a meta-analysis of the outcome literature. Am J Psychiatry 151(10):1409–1416

Hofer A, Benecke C, Edlinger M, Huber R, Kemmler G, Rettenbacher MA et al (2009) Facial emotion recognition and its relationship to symptomatic, subjective, and functional outcomes in outpatients with chronic schizophrenia. Eur Psychiatry 24(1):27–32

Iyengar BK (1966) Light on yoga. Harper Collins India, New Delhi

Janakiramaiah N, Gangadhar BN, Nagavenkatesha Murthy P, Shetty TK, Subbakrishna DK, Meti BL et al (1998) Therapeutic efficacy of Sudarshan Kriya Yoga (SKY) in dysthymic disorder. NIMHANS J 17:21–28

Janakiramaiah N, Gangadhar BN, Naga Venkatesha Murthy PJ, Harish MG, Subbakrishna DK, Vedamurthachar A (2000) Antidepressant efficacy of Sudarshan Kriya Yoga (SKY) in melancholia: a randomized comparison with electroconvulsive therapy (ECT) and imipramine. J Affect Disord 57(1–3):255–259

Jayaram N, Varambally S, Behere RV, Venkatasubramanian G, Arasappa R, Christopher R et al (2013) Effect of yoga therapy on plasma oxytocin and facial emotion recognition deficits in patients of schizophrenia. Indian J Psychiatry 55(Suppl 3):S409–S413

Jensen PS, Kenny DT (2004) The effects of yoga on the attention and behavior of boys with attention-deficit/hyperactivity disorder (ADHD). J Atten Disord 7(4):205–216

John PJ, Sharma N, Sharma CM, Kankane A (2007) Effectiveness of yoga therapy in the treatment of migraine without aura: a randomized controlled trial. Headache 47(5):654–661

Kabat-Zinn J, Massion AO, Kristeller J, Peterson LG, Fletcher KE, Pbert L et al (1992) Effectiveness of a meditation-based stress reduction program in the treatment of anxiety disorders. Am J Psychiatry 149(7):936–943

Kay S, Fiszbein A, Opler R (1987) The positive and negative syndrome scale for schizophrenia (PANSS). Schizophr Bull 13:261–276

Kee KS, Green MF, Mintz J, Brekke JS (2003) Is emotion processing a predictor of functional outcome in schizophrenia? Schizophr Bull 29(3):487–497

Kessler RC, Soukup J, Davis RB, Foster DF, Wilkey SA, Van Rompay MI et al (2001) The use of complementary and alternative therapies to treat anxiety and depression in the United States. Am J Psychiatry 158(2):289–294

Khalsa SB (2004) Treatment of chronic insomnia with yoga: a preliminary study with sleep-wake diaries. Appl Psychophysiol Biofeedback 29(4):269–278

Khalsa SB, Khalsa GS, Khalsa HK, Khalsa MK (2008) Evaluation of a residential Kundalini yoga lifestyle pilot program for addiction in India. J Ethn Subst Abus 7(1):67–79

Khanna S, Greeson JM (2013) A narrative review of yoga and mindfulness as complementary therapies for addiction. Complement Ther Med 21(3):244–252

Khumar SS, Kaur P, Kaur S (1993) Effectiveness of Shavasana on depression among university students. Indian J Clin Psychol 20(2):82–87

Kirkwood G, Rampes H, Tuffrey V, Richardson J, Pilkington K (2005) Yoga for anxiety: a systematic review of the research evidence. Br J Sports Med 39(12):884–891; discussion 891

Kisan R, Sujan M, Adoor M, Rao R, Nalini A, Kutty BM et al (2014) Effect of yoga on migraine: a comprehensive study using clinical profile and cardiac autonomic functions. Int J Yoga 7(2): 126–132

Kochupillai V, Kumar P, Singh D, Aggarwal D, Bhardwaj N, Bhutani M et al (2005) Effect of rhythmic breathing (Sudarshan Kriya and Pranayam) on immune functions and tobacco addiction. Ann N Y Acad Sci 1056:242–252

Kratter J (1983) The use of meditation in the treatment of attention deficit disorder with hyperactivity. St. John's University, New York

Krisanaprakornkit T, Ngamjarus C, Witoonchart C, Piyavhatkul N (2010) Meditation therapies for attention-deficit/hyperactivity disorder (ADHD). Cochrane Database Syst Rev (6):CD006507

Krishnamurthy MN, Telles S (2007) Assessing depression following two ancient Indian interventions: effects of yoga and ayurveda on older adults in a residential home. J Gerontol Nurs 33(2):17–23

Langhorst J, Klose P, Dobos GJ, Bernardy K, Hauser W (2013) Efficacy and safety of meditative movement therapies in fibromyalgia syndrome: a systematic review and meta-analysis of randomized controlled trials. Rheumatol Int 33(1):193–207

Lavey R, Sherman T, Mueser KT, Osborne DD, Currier M, Wolfe R (2005) The effects of yoga on mood in psychiatric inpatients. Psychiatr Rehabil J 28(4):399–402

Lazar SW, Kerr CE, Wasserman RH, Gray JR, Greve DN, Treadway MT et al (2005) Meditation experience is associated with increased cortical thickness. Neuroreport 16(17):1893–1897

Luders E, Toga AW, Lepore N, Gaser C (2009) The underlying anatomical correlates of long-term meditation: larger hippocampal and frontal volumes of gray matter. Neuroimage 45(3):672–678

Manjunath NK, Telles S (2005) Influence of Yoga and Ayurveda on self-rated sleep in a geriatric population. Indian J Med Res 121(5):683–690

Manjunath RB, Varambally S, Thirthalli J, Basavaraddi IV, Gangadhar BN (2013) Efficacy of yoga as an add-on treatment for in-patients with functional psychotic disorder. Indian J Psychiatry 55(Suppl 3):S374–S378

Marchand WR (2013) Mindfulness meditation practices as adjunctive treatments for psychiatric disorders. Psychiatr Clin North Am 36(1):141–152

Mehta S, Mehta V, Shah D, Motiwala A, Vardhan J, Mehta N et al (2011) Multimodal behavior program for ADHD incorporating yoga and implemented by high school volunteers: a pilot study. ISRN Pediatr 2011:780745

Mehta S, Shah D, Shah K, Mehta N, Mehta V, Motiwala S et al (2013) Peer-mediated multimodal intervention program for the treatment of children with ADHD in India: one-year followup. ISRN Pediatr 2012:419168

Mist SD, Firestone KA, Jones KD (2013) Complementary and alternative exercise for fibromyalgia: a meta-analysis. J Pain Res 6:247–260

Modabbernia A, Akhondzadeh S (2013) Saffron, passionflower, valerian and sage for mental health. Psychiatr Clin North Am 36(1):85–91

Moretti-Altuna GE (1987) The effects of meditation versus medication in the treatment of attention deficit disorder with hyperactivity. St. John's University, New York

Murthy PJ, Gangadhar BN, Janakiramaiah N, Subbakrishna DK (1997) Normalization of P300 amplitude following treatment in dysthymia. Biol Psychiatry 42(8):740–743

Mustian KM, Sprod LK, Janelsins M, Peppone LJ, Palesh OG, Chandwani K et al (2013) Multicenter, randomized controlled trial of yoga for sleep quality among cancer survivors. J Clin Oncol 31(26):3233–3241

Nagendra HR, Telles S, Naveen KV (2000) An integrated approach of Yoga therapy for the management of schizophrenia. Final Report submitted to Dept. of ISM & H, Ministry of Health and Family Welfare, Government of India

Naveen GH, Thirthalli J, Rao MG, Varambally S, Christopher R, Gangadhar BN (2013) Positive therapeutic and neurotropic effects of yoga in depression: a comparative study. Indian J Psychiatry 55(Suppl 3):S400–S404

Naveen GH, Varambally S, Thirthalli J, Rao M, Christopher R, Gangadhar BN (2016) Serum cortisol and BDNF in patients with major depression-effect of yoga. Int Rev Psychiatry 28(3):273–278

Ong JC, Shapiro SL, Manber R (2008) Combining mindfulness meditation with cognitive-behavior therapy for insomnia: a treatment-development study. Behav Ther 39(2):171–182

Panossian A, Wikman G, Wagner H (1999) Plant adaptogens. III. Earlier and more recent aspects and concepts on their mode of action. Phytomedicine 6(4):287–300

Patra S, Telles S (2009) Positive impact of cyclic meditation on subsequent sleep. Med Sci Monit 15(7):CR375–CR381

Pilkington K, Kirkwood G, Rampes H, Richardson J (2005) Yoga for depression: the research evidence. J Affect Disord 89(1–3):13–24

Rani K, Tiwari SC, Singh U, Agrawal GG, Srivastava N (2011) Six-month trial of Yoga Nidra in menstrual disorder patients: effects on somatoform symptoms. Ind Psychiatry J 20(2):97–102

Ravindran AV, da Silva TL (2013) Complementary and alternative therapies as add-on to pharmacotherapy for mood and anxiety disorders: a systematic review. J Affect Disord 150(3):707–719

Ravindran AV, Lam RW, Filteau MJ, Lesperance F, Kennedy SH, Parikh SV et al (2009) Canadian Network for Mood and Anxiety Treatments (CANMAT) clinical guidelines for the management of major depressive disorder in adults. V. Complementary and alternative medicine treatments. J Affect Disord 117(Suppl 1):S54–S64

Rohini V, Pandey RS, Janakiramaiah N, Gangadhar BN, Vedamurthachar A (2000) A comparative study of full and partial sudarshan kriya yoga in major depressive disorder. NIMHANS J 18:53–57

Saraswat N, Rao K, Subbakrishna DK, Gangadhar BN (2006) The Social Occupational Functioning Scale (SOFS): a brief measure of functional status in persons with schizophrenia. Schizophr Res 81(2–3):301–309

Shaffer HJ, LaSalvia TA, Stein JP (1997) Comparing Hatha yoga with dynamic group psychotherapy for enhancing methadone maintenance treatment: a randomized clinical trial. Altern Ther Health Med 3(4):57–66

Shannahoff-Khalsa DS, Beckett LR (1996) Clinical case report: efficacy of yogic techniques in the treatment of obsessive compulsive disorders. Int J Neurosci 85(1–2):1–17

Shannahoff-Khalsa DS, Ray LE, Levine S, Gallen CC, Schwartz BJ, Sidorowich JJ (1999) Randomized controlled trial of yogic meditation techniques for patients with obsessive-compulsive disorder. CNS Spectr 4(12):34–47

Shapiro D, Cook IA, Davydov DM, Ottaviani C, Leuchter AF, Abrams M (2007) Yoga as a complementary treatment of depression: effects of traits and moods on treatment outcome. Evid Based Complement Alternat Med 4(4):493–502

Sharma VK, Das S, Mondal S, Goswampi U, Gandhi A (2005) Effect of Sahaj Yoga on depressive disorders. Indian J Physiol Pharmacol 49(4):462–468

Sharma VK, Das S, Mondal S, Goswami U, Gandhi A (2006) Effect of Sahaj Yoga on neuro-cognitive functions in patients suffering from major depression. Indian J Physiol Pharmacol 50(4):375–383

Streeter CC, Jensen JE, Perlmutter RM, Cabral HJ, Tian H, Terhune DB et al (2007) Yoga Asana sessions increase brain GABA levels: a pilot study. J Altern Complement Med 13(4):419–426

Streeter CC, Whitfield TH, Owen L, Rein T, Karri SK, Yakhkind A et al (2010) Effects of yoga versus walking on mood, anxiety, and brain GABA levels: a randomized controlled MRS study. J Altern Complement Med 16(11):1145–1152

Sutar R, Desai G, Varambally S, Gangadhar BN (2016) Yoga-based intervention in patients with somatoform disorders: an open label trial. Int Rev Psychiatry 28(3):309–315

Tyagi A, Cohen M (2014) Yoga and hypertension: a systematic review. Altern Ther Health Med 20(2):32–59

Uebelacker LA, Epstein-Lubow G, Gaudiano BA, Tremont G, Battle CL, Miller IW (2010) Hatha yoga for depression: critical review of the evidence for efficacy, plausible mechanisms of action, and directions for future research. J Psychiatr Pract 16(1):22–33

Varambally S, Gangadhar BN (2012) Yoga: a spiritual practice with therapeutic value in psychiatry. Asian J Psychiatr 5(2):186–189

Varambally S, Gangadhar BN, Thirthalli J, Jagannathan A, Kumar S, Venkatasubramanian G et al (2012) Therapeutic efficacy of add-on yogasana intervention in stabilized outpatient schizophrenia: randomized controlled comparison with exercise and waitlist. Indian J Psychiatry 54(3):227–232

Varvogli L, Darviri C (2011) Stress management techniques: evidence-based procedures that reduce stress and promote health. Health Sci J 5(2):74–89

Vedamurthachar A, Janakiramaiah N, Hegde JM, Shetty TK, Subbakrishna DK, Sureshbabu SV et al (2006) Antidepressant efficacy and hormonal effects of Sudarshana Kriya Yoga (SKY) in alcohol dependent individuals. J Affect Disord 94(1–3):249–253

Visceglia E, Lewis S (2011) Yoga therapy as an adjunctive treatment for schizophrenia: a randomized, controlled pilot study. J Altern Complement Med 17(7):601–607

Voineskos D, Rogasch NC, Rajji TK, Fitzgerald PB, Daskalakis ZJ (2013) A review of evidence linking disrupted neural plasticity to schizophrenia. Can J Psychiatry 58(2):86–92

Vollestad J, Sivertsen B, Nielsen GH (2011) Mindfulness-based stress reduction for patients with anxiety disorders: evaluation in a randomized controlled trial. Behav Res Ther 49(4):281–288

Vollestad J, Nielsen MB, Nielsen GH (2012) Mindfulness- and acceptance-based interventions for anxiety disorders: a systematic review and meta-analysis. Br J Clin Psychol 51(3):239–260

Walsh R, Roche L (1979) Precipitation of acute psychotic episodes by intensive meditation in individuals with a history of schizophrenia. Am J Psychiatry 136(8):1085–1086

Wieland LS, Manheimer E, Berman BM (2011) Development and classification of an operational definition of complementary and alternative medicine for the Cochrane collaboration. Altern Ther Health Med 17(2):50–9. PMID: 21717826

Woolery A, Myers H, Sternlieb B, Zeltzer L (2004) A yoga intervention for young adults with elevated symptoms of depression. Altern Ther Health Med 10(2):60–63

Psychopharmacology in Rural Settings

20

Samir Kumar Praharaj and Chittaranjan Andrade

Contents

S. K. Praharaj
Department of Psychiatry, Kasturba Medical College, Manipal, Manipal Academy of Higher
Education, Manipal, Karnataka, India
e-mail: samir.kp@manipal.edu; samirpsyche@yahoo.co.in

C. Andrade (✉)
Department of Psychopharmacology, National Institute of Mental Health and Neurosciences,
Bangalore, Karnataka, India
e-mail: candrade@psychiatrist.com

© Springer Nature Singapore Pte Ltd. 2020
S. Chaturvedi (ed.), *Mental Health and Illness in the Rural World*, Mental Health and
Illness Worldwide, https://doi.org/10.1007/978-981-10-2345-3_17

Abstract

A large proportion of psychiatric patients in developing countries live in rural areas. The practice of psychopharmacology in rural settings is limited by availability, accessibility, and affordability of the medications. Sometimes, second-line medications may need to be prescribed when the first-line medications are not available. Also, cultural factors as well as other practical realities may constrain the use of medications. Therefore, evidence-based medicine may need to be tailored to suit to these ground realities.

Keywords

Psychopharmacology · Psychotropic drugs · Rural

Introduction

There are special limitations to the practice of psychopharmacology in rural settings. In most parts of the world, and more especially in developing countries, people who live in rural areas are poor and cannot afford expensive drugs; for the same reason, they cannot afford expensive medical investigations related to their treatments; and because they may reside far from medical facilities, they may not be able to make frequent medical visits.

Thus, prescribing psychotropic medications in rural settings is limited by the familiar three As: availability, accessibility, and affordability. As examples, newer psychotropics may be recommended as first-line agents but may not be available locally or, if available, may be unaffordable. Therapeutic drug monitoring facilities may not be available. Other monitoring, requiring frequent clinic or hospital visits, may not be feasible. Certain interventions may be limited by a fourth A: acceptability, when cultural taboos, more prevalent among the uneducated, hold sway.

The limitations of psychopharmacological practice in rural settings are therefore many. So, for very practical reasons, older rather than newer medications, generic rather than branded medications, or medications that are considered second-line may sometimes need to be prescribed. In other words, prescribing in rural settings requires a judicious compromise between an evidence-based approach and practical realities.

In this context, a special note is necessary. There is no convincing evidence that generic medications are in any way inferior to innovator brands; therefore, clinicians can prescribe generics, wherever available, to reduce treatment costs.

Another special note is that, because frequent follow-up visits may not be feasible, clinicians may need to provide detailed instructions for actions such as how and when to uptitrate medications, what to do in case adverse events occur, and so on. Anticipatory prescriptions may need to be issued to deal with such contingencies.

In other regards, psychopharmacological practice in rural settings is little different from those in urban settings. Whereas this chapter provides a summary of treatment-related issues, practitioners who need more information will need to refer to larger texts.

Pharmacokinetic and Pharmacodynamic Principles in Psychopharmacology

Drug pharmacokinetics includes issues related to the absorption, transport, distribution, metabolism, and elimination of drugs. The absorption of drugs is affected by factors such as route of administration, drug formulation (e.g., sustained-release formulation), timing of administration in relation to food, food-drug and drug-drug interactions, local factors such as gastric acidity, and others. Transport of drugs may depend on plasma protein binding. Distribution of the drug into the central nervous system and other tissues depends on the lipophilicity of the drug. Steady-state levels are achieved after about five half-lives. Metabolism usually takes place in two phases. Phase I metabolism includes oxidative, reductive, and hydrolytic processes, and phase II metabolism is most commonly by conjugation through glucuronidation. Most psychotropic drugs undergo phase I metabolism by cytochrome P450 enzymes, and the bulk of such metabolism occurs in hepatic cells. A few psychotropics are excreted unchanged in urine.

Drug pharmacodynamics refers to the mechanism of action of medications that result in the observed clinical effects or adverse effects. Psychotropic drugs act on neurotransmitter or neuromodulator synthesis, release, transport, breakdown, and/or receptor binding. They may act on ion channels. They may have a single mechanism or multiple mechanisms; the same mechanism may be responsible for both efficacy and adverse effects, or different mechanisms may explain different actions of the drug.

There is perhaps little reason to consider that pharmacokinetic or pharmacodynamic aspects of prescribing in rural settings are any different from prescribing in urban settings; so this section is not considered further, and readers are referred to standard textbooks of pharmacology and psychopharmacology.

Schizophrenia and Other Psychotic Disorders

First-Line Antipsychotics

The mainstay of treatment of psychotic disorders is with antipsychotic drugs, either typical/first-generation antipsychotics or atypical/second-generation drugs. These groups of medications are principally dopamine D2 receptor antagonists; the atypical antipsychotics additionally have 5HT2 serotonin receptor antagonist activity (see Table 1 for the differences). It may be noted that there are exceptions to the group characteristics in each group of drugs.

The atypical antipsychotics are usually preferred as first-line medications. This is because these drugs are as effective as the typical antipsychotics for positive symptoms and are less likely to induce extrapyramidal symptoms (EPS), more especially tardive dyskinesia. Of note, because the atypical drugs produce less EPS, they also produce less secondary negative symptoms. Furthermore, because they have less antidopaminergic effects in the prefrontal cortex, they are less likely to induce secondary cognitive impairment.

It is possible that different atypical antipsychotics differ in efficacy. The superiority of clozapine in treatment-refractory schizophrenia is well known. However, a

Table 1 Differences between the typical and atypical antipsychotics

Characteristics	Typical antipsychotics	Atypical antipsychotics
Receptor profile	Dopamine antagonist	Serotonin-dopamine antagonist
Propensity for EPS	High	Low
Hyperprolactinemia	High	Low
Metabolic adverse effects	Low	High

meta-analysis (Leucht et al. 2013) has shown that in schizophrenia patients, in general, clozapine and amisulpride top the list for efficacy, with olanzapine and risperidone following. Newer drugs such as asenapine, lurasidone, and iloperidone were low down on the list in comparison with other drugs. Amisulpride, olanzapine, and clozapine also fared well with regard to low risk of all-cause discontinuation, and haloperidol fared the worst.

Clozapine may also have anti-suicidal properties in schizophrenia; some evidence suggests that certain other atypical antipsychotics, such as olanzapine, quetiapine, ziprasidone, aripiprazole, and asenapine, may also have anti-suicidal effects (Pompili et al. 2016).

For rural practice, therefore, it seems reasonable to keep risperidone, olanzapine, and aripiprazole as first-line drugs, depending on availability, cost, and patient matching for adverse effect profile. If prescription of a typical antipsychotic is considered, perphenazine or haloperidol may be choices, in that order.

Many patients with acute psychosis require sedation to control for agitation. In such situation parenteral formulations (e.g., haloperidol or olanzapine), liquid preparations (e.g., risperidone or haloperidol), or mouth dissolving tablets (e.g., olanzapine or aripiprazole) may be useful. Sometimes, combining parenteral benzodiazepines (e.g., lorazepam) with haloperidol is a useful strategy to reduce the dose requirement of antipsychotics for controlling agitation. Similarly, combining promethazine injection with haloperidol reduces the dose requirement as well as prevents acute dystonia. Wherever available and affordable, intramuscular preparations of aripiprazole, ziprasidone, and olanzapine may be used as alternative strategy for the treatment of acute agitation because of a lower risk of dystonia. However, the benefits are offset by higher cost and the risk for akathisia. Occasionally, sedating antipsychotics such as chlorpromazine or quetiapine may be used for control of acute symptoms, specifically in young patients with psychosis.

Equivalent Doses of Antipsychotics

The chlorpromazine equivalent doses for antipsychotics are available (see Table 2) and can be used in switching between antipsychotics. A trial of antipsychotic drug at dose of 600 mg of chlorpromazine equivalent for 4–6 weeks is generally considered adequate. However, practically, if a patient with acute psychosis does not show improvement after 2 weeks at an adequate dose, chances are that there won't be a treatment response later, and a change of antipsychotics may be considered. In those with less severe symptoms, waiting for a longer period may be feasible.

Usually, medication is started at a low dose and uptitrated into the target range with the appropriate dose for each patient being discovered through trial and error, based on efficacy and tolerability. The speed of uptitration will depend on a balance between the need to minimize adverse effects through slow uptitration and the need to rapidly control symptoms by getting the dose into the therapeutic range as quickly as possible. First-episode patients tend to require lower doses than patients with established schizophrenia.

Table 2 Chlorpromazine equivalent doses (100 mg) for first-generation antipsychotics (FGAs) and second-generation antipsychotics (SGAs)

FGAs	Equivalent dose	SGAs[a]	Equivalent dose
Haloperidol	2 mg	Risperidone	1.5 mg
Trifluoperazine	5 mg	Olanzapine	7.5 mg
Pimozide	2 mg	Quetiapine	100 mg
Fluphenazine	2 mg	Asenapine	5 mg
Flupenthixol	3 mg	Ziprasidone	20 mg
Sulpiride	200 mg	Iloperidone	4 mg
Perphenazine	10 mg	Lurasidone	20 mg
Zuclopenthixol	25 mg	Aripiprazole	5 mg
Loxapine	10 mg	Clozapine	50 mg
Molindone	10 mg		
Thioridazine	100 mg		

[a]Strictly exact conversion between SGAs and FGAs is not possible, and it is a rough approximation

Table 3 Minimum effective dose (MED) and maximum recommended doses (MRD) for first-generation antipsychotics (FGAs) and second-generation antipsychotics (SGAs)

FGAs	MED	MRD	SGAs	MED	MRD
Haloperidol	2	20	Risperidone	2	16
Chlorpromazine	200	1000	Amisulpride	400	1200
Trifluoperazine	10–15	30	Olanzapine	5	20
Pimozide	2–4	10	Quetiapine	150	800
Fluphenazine	5–10	20	Asenapine	10	20
Flupenthixol	3	18	Ziprasidone	40	160
Sulpiride	400	2400	Iloperidone	4	24
Perphenazine	8–12	32	Lurasidone	40	160
Zuclopenthixol	50	150	Aripiprazole	10	30

All values in mg/day

It is generally prudent to prescribe antipsychotics in monotherapy, though some patients may require a sedating drug such as quetiapine, to augment a nonsedating one such as aripiprazole, until the illness stabilizes. Switching antipsychotics should be considered if the patient fails to respond to an adequate dose or the highest tolerated dose after 4–6 weeks. The minimum effective doses and maximum recommended doses are summarized in Table 3.

Course of Treatment

The treatment of schizophrenia should be initiated as early as possible because longer durations of untreated psychosis are associated with poorer treatment outcomes and worse long-term prognosis.

After initiation of antipsychotic treatment, about 70–80% of patients can expect to achieve complete or near complete remission of positive symptoms. Patients with first-episode psychosis who recover to premorbid levels with antipsychotic treatment are usually advised to continue treatment for at least 1–2 years. Patients who remain symptomatic at threshold or subthreshold levels will need to continue on antipsychotic medication for longer. Vulnerable patients and those at high risk of relapse will need to be continued on treatment for longer. If the diagnosis of schizophrenia is certain, the antipsychotics should be continued indefinitely.

During maintenance therapy, medications are best continued at the same dose that elicited treatment response. Lowering the dose may be reasonable only if the dose was admittedly high for better symptom control during the acute phase; in such cases, doses should be lowered only if required to reduce adverse effect. It would be injudicious to lower the dose to levels below those that are considered minimally effective.

As discussed earlier, discontinuation of antipsychotics may be considered in some stable, recovered first-episode psychosis patients after 1–2 years, following a thorough risk-benefit analysis and discussion with the patient. If it is decided to withdraw antipsychotics, withdrawal should be very gradual (over a period of at least 3 months), and with close monitoring, as it is known that abrupt or rapid withdrawal doubles the risk of relapse. If any stressful situation is anticipated, it would be prudent to continue antipsychotics till that period passes.

About 50% of schizophrenia patients relapse within 3 months and 90% within a year of nonadherence to their medication regime. The risk of relapse is 20–30% in those who adhered to their prescription. That is, antipsychotics are effective for preventing relapse, but a proportion of patients will relapse while on medications.

Monitoring of Adverse Effects

Before starting antipsychotics, a full physical examination should be conducted, and weight and body mass index should be recorded. Baseline investigations that are necessary, especially with regard to a future risk of metabolic syndrome, include fasting glucose and serum lipid profile. In middle-aged adults, and when prescribing drugs that can prolong the QTc interval, an electrocardiogram is also desirable. These assessments and investigations need to be repeated periodically, usually at 3–6-month intervals, after starting antipsychotics.

Acute dystonia, parkinsonian symptoms, and akathisia are adverse effects that may arise during the initial days to weeks of initiation of typical antipsychotic drugs. The risk of these adverse effects can be reduced by the use of oral anticholinergic drugs in prophylaxis; as an example, trihexyphenidyl (4 mg/day) or procyclidine (5 mg/day) can be administered in the initial month, after which the need for continued anticholinergic medication can be reassessed.

Should acute dystonia occur, treatment with 25–50 mg of intramuscular or intravenous promethazine will elicit dramatic response; oral anticholinergics are also effective, but the onset of action is slower. As the chances of recurrence are very high after an initial episode of acute dystonia, especially when the antipsychotic is continued, all patients who develop dystonia should continue on oral anticholinergic drugs for at least the next few days if not longer.

Acute akathisia may develop within days of starting antipsychotics and is characterized by subjective as well as objective restlessness. Akathisia is deeply distressing to the patient. Of note, akathisia may occur with many atypical antipsychotics (such as aripiprazole and risperidone) besides the typical drugs; the risk is higher when dose uptitration is more rapid. Akathisia responds to benzodiazepines and to beta-blockers. For example, clonazepam (0.25–1.0 mg/day) or propranolol (20 mg/day, uptitrated to about 40–80 mg/day) is usually effective treatment for akathisia.

Drug-induced parkinsonism occurs with typical antipsychotics but also with many atypicals, including risperidone, ziprasidone, and others; the risk is dose-dependent. The syndrome responds to anticholinergic medications. The neuroleptic malignant syndrome is a rare antipsychotic drug adverse effect.

Long-term treatment with typical antipsychotics may lead to tardive syndromes such as tardive dyskinesia (common) and tardive dystonia (rare). Tardive dystonia and tardive dyskinesia are rare with atypical antipsychotics.

Metabolic adverse effects are common with atypical agents, especially with clozapine, olanzapine, and risperidone. These adverse effects include weight gain, dyslipidemia, and type 2 diabetes mellitus. Drugs such as aripiprazole, ziprasidone, and lurasidone are less likely to cause weight gain and metabolic syndrome. Adjuvant metformin (1000–1500 mg/day) can attenuate weight gain and reduce the risk of metabolic syndrome.

Hyperprolactinemia is seen mostly with typical antipsychotics but also with some atypical drugs such as amisulpride, risperidone, and paliperidone. Persistent elevation of prolactin may lead to sexual dysfunction, menstrual irregularities, and breast engorgement with galactorrhea; reduced bone mineral density is a long-term possibility. A switch to non-prolactin-elevating drug (e.g., olanzapine or quetiapine) or addition of aripiprazole may be considered.

Treatment-Resistant Schizophrenia

Nonresponse to at least two antipsychotics administered in adequate doses for adequate durations is a generally accepted definition of treatment-resistant schizophrenia. Approximately a third of patients have treatment resistance. Clozapine is the drug of choice for treatment-refractory schizophrenia. However, some patients with treatment resistance may respond to other second-generation antipsychotics as well (Andrade 2016), though the rate of response may be lower.

In refractory patients, clozapine is initiated at a dose of 25 mg/day and is increased gradually by 25 mg/day until the target dose is reached (usually, about 300 mg/day). A more rapid dose uptitration risks the development of clozapine-related cardiomyopathy. If the response at 300 mg/day is inadequate, the dosage may be increased gradually up to 600 mg/day, as tolerated. The maximum dose of clozapine used could be up to 900 mg/day, if tolerated. A trial of clozapine may need to be continued for up to 6 months before it is considered ineffective.

Monitoring During Clozapine Treatment

Clozapine treatment is associated with a large number of adverse effects. Clozapine is very sedating. It can cause postural hypotension. It can cause constipation. It increases appetite, weight, and the risk of the metabolic syndrome. Nighttime hypersalivation is common.

The use of clozapine is also associated with a 1 in 200–500 risk of a life-threatening adverse effect, agranulocytosis. Patients receiving clozapine therefore require close hematological monitoring (see guideline in Table 4).

At doses that are higher than 500 mg/day, the risk of seizures increases sharply and may necessitate the use of prophylactic anticonvulsants, such as sodium valproate or phenytoin. At lower doses, if the patient develops seizure, the dose may be reduced if clinically feasible or can be continued at the same level under cover of anticonvulsant treatment. Cardiomyopathy and myocarditis are other serious adverse effects with clozapine; the risk is highest with rapid dose uptitration and in the initial 6–8 weeks.

Anticholinergic adverse effects such as constipation are seen at higher doses of clozapine and may require laxative treatment. Sialorrhea usually attenuates with anticholinergic drugs (e.g., glycopyrrolate, 2–4 mg at night).

Augmentation in Clozapine-Resistant Patients

About 40–70% patients do not respond sufficient to clozapine. These patients receive clozapine augmentation with aripiprazole (15–30 mg/day) or amisulpride (400–800 mg/day). The augmentation may reduce clozapine metabolic and

Table 4 Clozapine hematological monitoring

ANC range	Treatment recommendation	ANC monitoring[a]
Normal range (ANC >1500/μL)	Initiate treatment	Weekly for 6 months, once every 2 weeks for 6 months, monthly thereafter
Mild neutropenia (ANC 1000–1500/μL)	Continue treatment	Thrice weekly till ANC >1500/μL, then last normal range monitoring
Moderate neutropenia (ANC 500–1000/μL)	Interrupt treatment	Daily until ANC >1000/μL, thrice weekly till ANC >1500/μL, weekly for 4 weeks, then last normal range monitoring
	Resume when ANC >1000/μL	
Severe neutropenia (ANC <500/μL)	Discontinue treatment	Daily until ANC >1000/μL, thrice weekly till ANC >1500/μL
	Do not rechallenge unless benefit outweighs risk	If rechallenged, resume as normal range monitoring

ANC absolute neutrophil count
[a]For benign ethnic neutropenia (BEN), clozapine can be initiated for ANC >1000/μL, continue same level of monitoring for mild neutropenia, and continue treatment with moderate neutropenia with frequent monitoring

Table 5 Comparison of typical and atypical long-acting injectable antipsychotics

	Dosage	Comment
Atypical antipsychotics		
Risperidone microspheres	25–50 mg/2 weeks	Drug release delayed for 2–3 weeks
Paliperidone palmitate	39–234 mg/month	Loading dose required (234 mg)
Olanzapine pamoate	210–300 mg/2 weeks	3-h postinjection monitoring needed
Aripiprazole lauroxil	300–400 mg/month	Test dose not required
Typical antipsychotics		
Haloperidol decanoate	50–200 mg/month	High EPS
Flupenthixol decanoate	20–400 mg/month	High EPS
Zuclopenthixol decanoate	200–600 mg/month	High EPS
Fluphenazine decanoate	12.5–100 mg/month	Dose-dumping effect, high EPS

hypersalivation adverse effects besides improving efficacy. Augmenting with a large list of other drugs has been studied but has not shown promise.

Long-Acting Antipsychotics

Long-acting injections (LAIs) of antipsychotics are indicated in schizophrenia when poor medication adherence is observed or anticipated. Various LAI formulations of antipsychotics are summarized in Table 5. Olanzapine LAI requires monitoring for the initial few hours because of concerns about a postinjection sedation-delirium syndrome. Depot preparations of typical antipsychotics are widely available and affordable to most and may be used as alternatives. Haloperidol, flupenthixol, and zuclopenthixol depot injections are preferred over fluphenazine depot, because of the "dose-dumping" effect of the latter leading to prominent extrapyramidal symptoms. Most typical antipsychotic depots necessitate the use of prophylactic anticholinergic medications. Oral antipsychotics need to be continued for a few weeks to a few months when a depot antipsychotic is initiated until LAI steady state is achieved, after which the oral medications can be discontinued.

Antipsychotics for Delirium

Atypical antipsychotics in low doses (e.g., risperidone 1–2 mg/day) are generally preferred to treat delirium because of the higher risk of extrapyramidal and other adverse effects with neuroleptics.

Antipsychotics for Behavioral and Psychological Symptoms of Dementia (BPSD)

Low doses of atypical antipsychotics are effective against delusions, hallucinations, and other behavioral and psychological symptoms of dementia. It may be noted here that antipsychotics, though effective against BPSD, are associated with an increased risk of adverse cardiovascular and cerebrovascular events, including mortality, especially during the initial month after treatment initiation, in patients with dementia. The risk is greater with typical than with atypical antipsychotics. Therefore, psychosocial interventions should be preferred for the treatment of BPSD, and antipsychotics should be initiated only if psychosocial interventions fail.

Antipsychotics in Pregnancy and Lactation

Antipsychotics may need to be initiated or continued during pregnancy and lactation in patients with schizophrenia because untreated psychosis poses greater risks to the mother and unborn child than antipsychotic use. Whereas weight gain and the risk of gestational diabetes may be increased with the use of certain atypical antipsychotics during pregnancy, there is no convincing evidence for that these drugs are associated with teratogenicity. All the antipsychotics, except for clozapine, can be used safely during breastfeeding.

Antipsychotics in Children and the Elderly

Atypical antipsychotics such as olanzapine and risperidone have been found to be superior to haloperidol for the treatment of psychosis in children and adolescents. Clozapine is the treatment of choice for resistant symptoms in children and adolescent patients. Children and adolescents are more prone than adults to develop metabolic adverse effects with atypical antipsychotics. Elderly patients are more prone to develop adverse effects with typical antipsychotics, particularly anticholinergic effects, cardiovascular effects, extrapyramidal symptoms, and tardive dyskinesia. Olanzapine and risperidone have been found to effective and relatively safe in elderly patients with psychosis (Colijn et al. 2015).

Mood Disorders (Depressive Disorders)

Unipolar Antidepressants: First-Line Drugs

Antidepressant drugs are the treatment of choice for unipolar depression of moderate or severe intensity. There are several classes of antidepressants (see Table 6) that act on monoamine neurotransmitters in the brain. The targets may be a single monoamine, such as the selective serotonin reuptake inhibitors (SSRIs) and norepinephrine reuptake inhibitors (NRIs) or two or more monoamines, such as the serotonin-norepinephrine reuptake inhibitors (SNRIs) and the norepinephrine-dopamine reuptake inhibitors (NDRIs).

Antidepressant drugs appear to differ in efficacy. The clearest theme is that drugs with dual (serotonin-norepinephrine) mechanisms are slightly superior to SSRIs.

Table 6 Classes of antidepressants

Antidepressant class	Drugs (with dose in mg/d)	Comment
Selective serotonin reuptake inhibitor (SSRI)	Sertraline (50–200), citalopram (20–40), escitalopram (10–20), fluoxetine (20–80), fluvoxamine (50–300), paroxetine (10–40)	Gastrointestinal and sexual adverse effects, pharmacokinetic drug interactions
Serotonin-norepinephrine reuptake inhibitor (SNRI)	Venlafaxine (75–300), desvenlafaxine (50–100), duloxetine (60–120), milnacipran (50–100), levomilnacipran (40–120)	May have an activating effect; useful against pain syndromes
Norepinephrine-dopamine reuptake inhibitor (NDRI)	Bupropion (150–450)	Seizure risk at higher doses
Norepinephrine reuptake inhibitor (NRI)	Reboxetine (8–12), atomoxetine[a] (18–40)	Less effective as monotherapy
Serotonin antagonist and reuptake inhibitor (SARI)	Trazodone (200–600), nefazodone (300–600)	Priapism is rarely seen with trazodone
Noradrenergic and specific serotonergic antidepressants (NaSSAs)	Mirtazapine (15–45)	Sedation, weight gain
Tricyclic antidepressant (TCA)	Amitriptyline (150–300), nortriptyline (50–150), imipramine (150–300), desipramine (150–300), clomipramine (150–250), dothiepin (150–300), doxepin (150–300)	Anticholinergic effects, postural hypotension, sedation
Monoamine oxidase inhibitor (MAOI)	Phenelzine (30–60), tranylcypromine (20–60), isocarboxazid (20–40), moclobemide (reversible inhibitor) (300–600), selegiline (5–10)	Precautions with tyramine-containing food, pharmacodynamic drug interactions
Melatonin receptor agonist (MRA)	Agomelatine (25–50)	Useful if insomnia is present
Dopamine agonist	Pramipexole[a] (1.5–4.5), ropinirole[a] (9–24)	Depression in Parkinson's disease

[a]Not a primary antidepressant

A network meta-analysis (Cipriani et al. 2009) found that venlafaxine, mirtazapine, escitalopram, and sertraline were the most effective antidepressants, and reboxetine was the least effective. Among the tricyclic antidepressants (TCA), amitriptyline has a small efficacy advantage.

The above notwithstanding, different patients respond better to different antidepressants and/or may tolerate different antidepressants differently, so finding the best antidepressant for a patient may be a matter of trial and error or may follow past experiences of the patient with antidepressant drugs. In psychopharmacological lore, serotonergic drugs may treat core depression better, and noradrenergic drugs may improve cognitive symptoms and have an activating effect in patients with retardation symptoms.

In general, SSRIs are preferred as first-line antidepressants because they are effective and well tolerated, and escitalopram and sertraline are good choices as first-line treatments; the other SSRIs are associated with an increased risk of pharmacokinetic drug interactions and are so best reserved for later use in rural settings.

An antidepressant is initiated at minimum tolerated dose and then titrated upward to a minimum therapeutic dose and continued for 3–4 weeks for response (improvement by 50%). The response rate for the first trial of an antidepressant is usually in the 55–75% range. It takes about 6–8 weeks of treatment with adequate doses of antidepressants to achieve remission. During early treatment, severe insomnia or anxiety may require additional use of benzodiazepines for short periods. In those with prominent guilt, agitation, or frank psychotic symptoms, addition of antipsychotics may be required along with antidepressants.

If the patient shows response with lower therapeutic doses of medications, the same dose may be continued. Some patients may need higher doses of medications to achieve remission, and the decision to raise the dose can be taken after 2–4 weeks, depending on the initial response.

If there is remission of symptoms, antidepressants are continued at the same dose for 9–12 months to reduce the risk of relapse and to reduce the risk of emergence of a new episode. If the patient has had more than one episode in the past, and especially when the episodes have been closer together rather than separated by long years in time, it may be prudent to continue maintenance antidepressant pharmacotherapy indefinitely, as prophylaxis.

If an antidepressant is withdrawn with the expectation that the patient may remain well even without antidepressant maintenance therapy, treatment discontinuation is best done gradually, over a period of several months, with monitoring for illness recurrence.

Second- and Third-Line Drugs

Studies suggest that patients who do not respond to one SSRI may respond to another SSRI. However, it is generally suggested that patients who do not respond to drugs from one class (see Table 6) are best prescribed a drug from another class. Venlafaxine and mirtazapine are good choices as second-line drugs for patients who do not tolerate or respond to the first drug. Dothiepin is perhaps the best tolerated among the TCA. However, these are not rules, and clinicians may follow their personal preferences based on experience.

In general, the (other) tertiary amine TCA are poorly tolerated at therapeutic doses and are presently infrequently prescribed for depression. A recent study in India showed that amitriptyline and clomipramine are associated with very poor 6-month response and retention in depressed patients, relative to drugs such as escitalopram, desvenlafaxine, and sertraline. Secondary amines such as nortriptyline and desipramine are more noradrenergic and less serotonergic, and data from meta-analysis suggests that they are less effective than the tertiary amines.

The nonselective, irreversible monoamine oxidase inhibitors (MAOI) require dietary restrictions and are associated with potentially serious drug interactions; their use is best limited to experienced practices, and they are therefore not recommended for use in rural settings. Selective MAOI have limited efficacy in depression relative to the monoamine reuptake inhibitors. Newer drugs such as agomelatine and vilazodone have good tolerability but do not seem to share the same clinical efficacy in clinical experience, the good evidence base notwithstanding.

Adverse Effects

Different classes of antidepressant drugs have different kinds of adverse effects. The SSRIs are commonly associated with nausea and sometimes with vomiting, gastric acidity, or diarrhea, especially in the initial 1–2 weeks of treatment; the symptoms are usually mild and pass away on their own, even without treatment. If nausea is problematic, the temporary use of an antiemetic tides over the crisis.

SSRIs and other drugs that potently inhibit the reuptake of serotonin can cause decreased libido and difficulty in attaining orgasm or even anorgasmia.

TCAs and other drugs that have anticholinergic action can cause dry mouth, constipation, blurring of vision, impaired accommodation, tachycardia, sweating, difficulty in passing urine (in men), and erectile dysfunction. The problems are greater in the elderly. A high anticholinergic burden in the elderly can also risk cognitive impairment and even frank delirium.

Drugs with strong noradrenergic reuptake inhibition can increase heart rate and blood pressure; this is greatest with venlafaxine. These drugs can also produce effects that are similar to anticholinergic effect, mimicking sympathetic opposition to parasympathetic activity.

TCAs and mirtazapine are sedating; and they can increase appetite and weight. TCAs may produce postural hypotension and syncope. Antidepressants that produce sedation or syncope increase the risk of falls and fractures in the elderly. TCAs and citalopram can dose-dependently prolong the QTc interval.

Most antidepressants, and more specifically the monoamine reuptake inhibitors, may produce increased restlessness and anxiety in the initial weeks of treatment; in some patients, this may even manifest as increased suicidality. Patients should therefore be warned about the risk and monitored during the initial weeks; benzodiazepines may have a calming effect in such patients.

Different drugs may have different specific effects. For example, clomipramine and bupropion dose-dependently lower the seizure threshold; bupropion, mirtazapine, and vilazodone are relatively free of sexual adverse effects; and so on.

TCAs have a narrow therapeutic index and carry a high risk of fatality at overdoses of 1 g and above. SSRIs may be associated with the serotonin syndrome in overdose. Citalopram and escitalopram carry a high risk of seizures in overdose. The newer antidepressants, however, are generally far safer in overdose than the TCAs, and mortality with overdose of newer antidepressants is uncommon, unless the overdose is very high or the patient overdoses with more than one drug.

Treatment-Resistant Depression (TRD)

Therapeutic options in patients with failed antidepressant trials are summarized in Table 7. There are broadly two strategies: switching and augmentation. Switching to other antidepressants is not an evidence-based strategy in TRD.

Perhaps the best evidence for antidepressant augmentation is with atypical antipsychotics, with aripiprazole (5–20 mg/day) and quetiapine (100–300 mg/day) being the drugs of choices; other superior augmentation agents appear to be lithium (0.6–0.8 mEq/L), triiodothyronine (50 ug/day), or thyroxine (up to 50% hyperthyroxinemic) (Zhou et al. 2015).

Still other agents have been recommended in literature for antidepressant augmentation in partially or completely refractory patients, and these include buspirone (30–60 mg/day), L-methylfolate (15 mg/day), psychostimulants such as methylphenidate (20–50 mg/day) and modafinil (100–200 mg/day), and dopaminergic drugs such as pramipexole (up to 2 mg/day).

Augmentation of SSRIs or SNRIs with other antidepressants such mirtazapine (15–45 mg/day) or bupropion (300–450 mg/day) has been found to be effective for TRD in some individuals. Also, modafinil (100–200 mg/day) or armodafinil (50–100 mg/day) may be used to augment SSRIs, specifically if fatigue is a prominent symptom.

Subanesthetic ketamine (0.5 mg/kg), administered by different routes, can result in dramatic antidepressant benefits even in patients who are antidepressant refractory (Andrade 2017a, b); however, it may be challenging to administer ketamine in rural settings.

Antidepressants in Pregnancy and Lactation

Antidepressant use during pregnancy has been associated with an increased risk of many adverse outcomes. These include spontaneous abortion, major and minor congenital malformations, preterm delivery, low birth weight, poor neonatal adaptation syndrome, persistent pulmonary hypertension of the newborn, and various

Table 7 Therapeutic options in antidepressant failure

	Strategy	Description
1	Maximizing dose (optimizing strategy)	Nonresponse as a result of inadequate treatment trials has been termed "pseudoresistance." Dose adjustments up to maximum tolerated doses may be helpful
2	Switch to another antidepressant	Switching to another SSRI or SNRI, TCA, or newer antidepressant
3	Combine two antidepressants	Usually from different antidepressant classes and hence with a different mechanism of action and different neurotransmitter targets (e.g., combining with mirtazapine or bupropion)
4	Adjunctive medications	Augmentation with lithium, thyroid hormones, buspirone, psychostimulants, atypical antipsychotics

neurodevelopmental disorders, including learning delays, attention deficit hyperactivity disorder, and autism. However, it is not clear to what extent the risk is due to the antidepressant versus the genetics and the behavioral changes associated with the condition (usually, depression) for which the antidepressants are prescribed.

Whatever the source of the risk, it is generally agreed that the risk is small and that there are definite risks associated with untreated depression, as well; in consequence, there is an increasing trend among reproductive psychiatrists across the world to continue antidepressant monotherapy in effective doses through all trimesters of pregnancy. Furthermore, given that most antidepressants cross into breast milk in small to negligible concentrations, and given the absence of clear indication of risk associated with such antidepressant exposure in infants, there is an increasing trend to encourage breastfeeding regardless of maternal antidepressant use.

Antidepressants in Children and the Elderly

SSRIs require to be used with caution in children and adolescents because there are reports of emergence of suicidal ideation during treatment. Fluoxetine and escitalopram are the only approved antidepressants for depressed children and adolescents.

Elderly patients are vulnerable to adverse effects with medications in general, and target doses should therefore be lowered. Elderly patients are especially vulnerable to anticholinergic adverse effects and to antidepressant-associated hyponatremia.

Antidepressants in the Medically Ill

While treating depression in patients with medical illness, it is useful to consider the use of drugs that are effective against both depression and the medical illness (Rackley and Bostwick 2012). For example, SNRIs are effective against both depression and painful diabetic peripheral neuropathy. Or, SSRIs are effective against depression and may reduce the risk of ischemic heart disease events through their antiplatelet action (Andrade et al. 2013).

It is also important to select drugs that do not worsen the medical condition. For example, drugs with strong adrenergic or anticholinergic action could worsen symptoms in men with benign prostatic hypertrophy. Finally, it is important to select drugs that do not interact pharmacodynamically or pharmacokinetically with the medical prescription. In this regard fluoxetine, paroxetine, fluvoxamine, and duloxetine are associated with an increased risk of pharmacokinetic interactions with many medical drugs; escitalopram is associated with a low risk.

A special consideration is the selection of drugs that have negligible hepatic metabolism for depressed patients with moderate to severe liver disease; candidate drugs include milnacipran and levomilnacipran. These drugs, however, would need to be used with caution in patients with moderate to severe renal disease.

Mood Disorders (Bipolar Disorder)

The treatment of bipolar disorder consists of antimanic and antidepressant treatments for manic/mixed and depressive episodes, respectively; medications that are effective in the acute phase are often continued during the continuation phase (after symptomatic remission), in the same doses, for a period of 6–9 months. To prevent manic or depressive relapse, mood stabilizers are used during the maintenance phase of treatment.

Antimanic Drugs

Drugs which are effective in mania include all antipsychotics (typical as well as atypical), lithium, sodium valproate, carbamazepine, and oxcarbazepine. Antipsychotics are associated with a faster antimanic response. In practice, an atypical antipsychotic in combination with a classical mood stabilizer is commonly used as first-line treatment for acute mania, especially when agitation, severe excitement, or psychotic symptoms are prominent. Benzodiazepines, though not specifically antimanic, can be added to reduce the antipsychotic dose requirement. Baseline investigations are recommended prior to starting lithium, sodium valproate, carbamazepine, or oxcarbazepine (see Table 8); however, in practice, investigations are infrequently carried out before initiating valproate or carbamazepine and may be impractical in rural settings.

Lithium has a narrow therapeutic index and requires monitoring of serum levels. Typically, a 12 h ($\pm$30 min) level of 0.8–1.2 mEq/L is therapeutic for acute mania. The starting dose of immediate-release lithium in adults is usually 600–900 mg (lower in the elderly) in a single nighttime dose. Serum levels are obtained after 5–7 days, when steady state is achieved. During the initial weeks, monitoring for adverse effects and signs of toxicity is recommended. In the absence of facilities for serum lithium estimation, some primary care practitioners empirically use a dose of

Table 8 Baseline investigations prior to starting antimanic drugs

Lithium	Sodium valproate	Carbamazepine (CBZ)
1. RFT-BUN, creatinine	1. LFT	1. CBC – every 2 weeks for 2 months, then every 4–6 months. Discontinue CBZ if WBC <3000/mm^3, RBC <4 × 10^6/mm^3, Hb <11 mg/dL, ANC <1500/mm^3, or platelets <100,000/mm^3
2. TSH	2. CBC	2. LFT
3. Serum electrolytes	3. Pregnancy test[a]	3. ECG-QTc >440 s increased risk for arrhythmia
4. WBC count		4. Serum electrolytes – Na$^+$
5. ECG		5. Pregnancy test[a]
6. Pregnancy test[a]		

RFT renal function test, *CBC* complete blood count, *TSH* thyroid-stimulating hormone, *ECG* electrocardiogram, *WBC* white blood cell, *RBC* red blood cell, *ANC* absolute neutrophil count
[a]Women who may possibly conceive

Table 9 Management of common, early adverse effects of lithium

Adverse effect	Comments	Management
Nausea	10–20%, initial weeks	Take after meals, divided dosing, sustained-release preparations
Diarrhea	10%, first 6 months	Change to immediate-release preparations
Polyuria	70%	Lower dose, avoid toxicity, once-daily dosing
		Amiloride 5 mg twice daily
Tremor	25%, postural, early in treatment	Reduce caffeine, lower dose
		Propranolol 20–320 mg/day (usually <120 mg/day)
		Primidone, clonazepam, vitamin B6
Weight gain	70%, initial 1–2 years	Diet control, exercise
		Metformin 500 mg twice daily

900 mg at night and titrate the dose based on adverse effects such as lithium tremor. All patients must be educated regarding possible toxicity during dehydration. They must consume more water while on lithium and discontinue lithium for few days if they develop severe vomiting or diarrhea. The most common adverse effects of lithium and their management are summarized in Table 9.

For valproate, a loading-dose strategy is well tolerated and is associated with early response. The usual starting dosage is 20 mg/kg/day. The dosage can be increased to 30 mg/kg/day, depending on clinical response. Target serum levels are 75–125 ug/mL.

Carbamazepine is usually started at 200–400 mg/day and increased to a target of 600–800 mg/day. A typical dosing schedule is 200 mg twice daily, with increase by 100 mg/day every 3 days. Dose increase may sometimes be needed after 2–3 weeks when carbamazepine levels decrease because of autoinduction. Oxcarbazepine may be used for the treatment of mania, albeit with a higher risk of hyponatremia. Oxcarbazepine 300 mg is equivalent to carbamazepine 200 mg.

All typical and atypical antipsychotics, including the newer agents, have anti-manic effect in monotherapy. Among all antipsychotics, risperidone, olanzapine, and haloperidol should be considered as best of the available options for the treatment of manic episodes (Cipriani et al. 2011). In the treatment of acute mania, specifically in psychotic mania, the combination of an antipsychotic with lithium or valproate has the best response.

Antidepressant Treatment of Bipolar Depression

Conventional antidepressants, used to treat major depressive disorder, are also effective in bipolar depression; however, these antidepressants increase the risk of manic switch, cycle acceleration, and roughening of the course of illness. Furthermore, some studies suggest that the efficacy is poorer and that mood stabilizers such as lithium may be just as good but without the stated adverse outcomes. Medications used as antidepressants in bipolar disorder are therefore different from the traditional unipolar antidepressants (see Table 10).

Table 10 Bipolar antidepressants

Drug	Dose	Comment
Quetiapine	300–600 mg/day	Higher dose is more effective; however, tolerability is a problem
Olanzapine	10–20 mg/day	Administered along with fluoxetine
Lurasidone	20–120 mg/day	Some patients may respond to lower doses (20–60 mg/day)
Valproate	20–30 mg/kg/day	Not very effective
Lithium	0.8–1 meq/L	Delayed effect (6–8 weeks), may require higher doses
Lamotrigine	200–400 mg/day	Slow upward titration needed to reduce the risk of serious, life-threatening skin rash

Table 11 Adverse effects associated with long-term use of lithium

Adverse effect
Impaired renal function
Decreased urinary concentrating ability by 15% of normal maximum
Clinical hypothyroidism
Increased blood calcium
Increased parathyroid hormone
Weight gain

Mood Stabilizers for Prophylaxis

Lithium is effective in the prevention of manic recurrence at serum levels of 0.6–1 mEq/L; it is less effective in the prevention of depressive episodes (Severus et al. 2014). It is the only mood stabilizer with evidence of anti-suicidal effect in bipolar disorder. Lithium may be combined with another mood stabilizer for greater benefit. Adverse effects with long-term use of lithium are summarized in Table 11.

It is quite common for lithium or other monotherapy in bipolar patients to fail and for such patients to require combination mood stabilizer treatment for the optimal prophylaxis of episodes; the combination should address both manic and depressive poles of illness. In this context, lithium, valproate, and carbamazepine are more effective against manic relapse, and lamotrigine is more effective against depressive relapse. The atypical antipsychotics are effective against relapse of both poles of illness.

Carbamazepine and oxcarbazepine are third-line antimanic and mood stabilizer drugs because of their adverse effect profile and the risk of drug interactions. Lamotrigine dosing requires special mention (see Table 12).

Clozapine may be considered if combination therapies are ineffective and patients have treatment-resistant bipolar disorder. It has to be started at low dose with monitoring of adverse effects similar to that in schizophrenia, up to target doses of 300–600 mg/day. Sometimes, a combination of lithium and clozapine may be considered, and lithium may mitigate the clozapine-related leukopenia.

Table 12 Lamotrigine dosing guideline

	Dose titration	Final dose
Lamotrigine monotherapy	25 mg in weeks 1 and 2 → 50 mg in weeks 3 and 4 → 100 mg in week 5 → 200 mg in week 6 → target dose	200–400 mg
Lamotrigine with UGT enzyme inhibitors (e.g., valproate)	12.5 mg in weeks 1 and 2 → 25 mg in weeks 3 and 4 → 50 mg in week 5 → 100 mg in week 6 → target dose	100–200 mg
Lamotrigine with UGT enzyme inducers (e.g., carbamazepine, phenytoin)	50 mg in weeks 1 and 2 → 100 mg in weeks 3 and 4 → 200 mg in week 5 → 400 mg in week 6 → target dose	400–800 mg

Mood Stabilizers in Pregnancy and Lactation

Valproate is associated with a clear and unacceptably high, dose-dependent risk of teratogenicity associated with maternal use during pregnancy; the risk may be as high as 10% in women dosed at >1 g/day. Lithium is associated with an approximately 1% extra risk of cardiac defects. With the other mood stabilizers, the risks are clinically small.

In general, the use of valproate is strongly discouraged during pregnancy; other mood stabilizers, including lithium, may be continued all though pregnancy if the patient is symptomatic or if there is a meaningful risk of relapse or recurrence of illness; however, decision-making should be individualized in consultation with the patient and family. When treatment is continued, high-dose folic acid (4–5 mg/day) should be advised to reduce the risk of neural tube defects associated with older anticonvulsant mood stabilizers.

Dosing with lithium and lamotrigine needs to be optimized during pregnancy, with dose uptitration required as pregnancy progresses beyond the fifth month. With lithium, uptitration must be based on serum monitoring. With lamotrigine, a gradual uptitration by 50–100% may be required when serum level guidance is unavailable. After delivery, dosing must be brought back to baseline to avoid toxicity.

During breastfeeding, lithium is the only mood stabilizer the use of which is problematic; this is because serum levels in breastfed infants can be up to 50% that of maternal levels and lithium adverse effects and even toxicity may occur.

Common Drug Interactions with Mood Stabilizers

Thiazides and loop diuretics are known to reduce renal clearance of lithium and are best avoided (Finley 2016). If thiazides need to be prescribed, it is prudent to reduce lithium daily dosage by 25–50% and adjust the dose based on serum levels. Other medications that raise serum lithium levels include antihypertensives (ACE inhibitors and angiotensin receptor antagonists) and nonsteroidal anti-inflammatory drugs. Rare neurotoxicity is seen with lithium in combination with typical antipsychotics and carbamazepine. Rarely, lithium can precipitate serotonin syndrome when used along with serotonergic antidepressants.

Valproate is a UGT enzyme inhibitor and potentially increases serum concentrations of drugs metabolized by glucuronidation. Valproate can increase blood levels of phenobarbital, phenytoin, lamotrigine, carbamazepine, and TCAs (Fleming and Chetty 2005). Carbamazepine, being an enzyme inducer, decreases the blood levels of lamotrigine, valproate, warfarin, haloperidol, and estrogen (in oral contraceptive pills, leading to contraceptive failure).

Anxiety Disorders

Among anxiety disorders, except for specific phobias which is managed best with non-pharmacological approaches, all benefit from pharmacotherapy. Although benzodiazepines are still commonly used as anxiolytics by nonpsychiatrists, antidepressants are currently the preferred drugs of choice for most anxiety disorders because they are less sedating (newer antidepressants) and nonaddictive.

Antidepressants as Antianxiety Medications

The first-line treatment for generalized anxiety disorder, panic disorder, and social anxiety disorder includes SSRIs or SNRIs. Such patients may experience jitteriness or worsening of anxiety during the initial days of treatment. Therefore, it is prudent to start antidepressants at low doses and gradually increase to therapeutic doses across about 2 weeks. Sometimes, adding a benzodiazepine may be helpful, during the weeks of uptitration. Clonazepam, for example, at doses of 0.5–1.0 mg/day in a single nighttime dose or in divided doses may be used for 3–4 weeks while initiating SSRIs or SNRIs. Response to SSRIs or SNRIs usually develops across 4–6 weeks. Some patients with panic disorder may require doses higher than conventional antidepressant doses (e.g., 40–60 mg of paroxetine or 200 mg of desvenlafaxine) to show full benefit.

If treatment is effective, it is best continued even after remission. Although the optimal duration of maintenance treatment is not known, the medications are best continued for at least 1 year. If drug discontinuation is planned, it should be done gradually to minimize discontinuation symptoms as well as the increased risk of relapse. Benzodiazepine taper, in particular, may need to be effected across up to 3 months.

Benzodiazepines as Anxiolytics

Benzodiazepines are very effective in the treatment of common anxiety disorders and were formerly used in long-term treatment. However, these drugs may cause an early morning hangover, daytime sedation, impairment in cognition (including anterograde amnesia), impairment in psychomotor reflexes, falls and fractures (especially in the elderly), and dependence. The use of these drugs is therefore increasingly being discouraged, except for crisis or short-term use.

Clonazepam is generally started at 0.5–1 mg/day and increased up to 2 mg/day; some patients may require doses up to 4 mg/day to control panic attacks. Alprazolam can be started at 0.75 mg/day and gradually increased to 3–6 mg/day in panic disorder; this drug, however, is strongly addictive and should be considered only if other options do not work well. Diazepam (5–10 mg/day) can be very useful in patients with musculoskeletal symptoms associated with anxiety, especially severe or refractory tension headache.

Azapirones as Antianxiety Agents

Buspirone, a $5HT_{1A}$ partial agonist, has been used as antianxiety medication specifically in generalized anxiety disorder and social anxiety disorder (Loane and Politis 2012). It may take 2–4 weeks for antianxiety effects to develop. Buspirone is initiated at 5 mg two to three times a day, and the dose is uptitrated to 30 mg/day; higher doses (up to 60 mg/day) are less frequently required. Buspirone is not helpful for substitution in patients who are discontinuing benzodiazepines. Buspirone is well tolerated by medically ill patients and elderly patients. It does not affect psychomotor performance or cognition and is safe in persons with respiratory illnesses. Buspirone can also be used to treat anxiety associated with alcohol use disorder.

Second-Line Medications in Treatment-Resistant Patients

TCAs can be used as second-line anxiolytic agents. They are usually started at low doses (e.g., 25 mg of imipramine, amitriptyline, or dothiepin) and gradually increased to therapeutic doses that are similar to or lower than conventional antidepressant doses. Low-dose atypical antipsychotics may be used as augmenting agents in the treatment of generalized anxiety and other anxiety disorders.

Sometimes, anticonvulsants such as pregabalin (300–600 mg/day) and gabapentin (600–3600 mg/day) may be used as primary or adjunctive medications for resistant anxiety symptoms. For the treatment of anxiety, pregabalin is initiated at 150 mg/day, and after a few days, the dose is increased to 300 mg/day. If there is little response, the dose can be gradually increased by 150 mg every few days to the maximum of 600 mg/day. Gabapentin may be effective at lower doses (600 mg) for social anxiety, but many require between 900 and 2700 mg/day for controlling panic disorder.

Beta-blockers such as propranolol attenuate autonomic symptoms of anxiety and are useful to address anticipatory anxiety and social phobia. For example, single dose of propranolol 10–40 mg may be taken half hour before an examination to reduce the peripheral symptoms of anxiety.

Obsessive-Compulsive Disorder (OCD)

SSRIs are first-line medications for OCD. No SSRI is superior to any other SSRI; however, clomipramine, a nonselective serotonin reuptake inhibitor (SRI), may be associated with larger effect sizes, especially in children. Anti-OCD SSRI doses are typically higher than usual antidepressant doses and are close to or even higher than the maximum recommended antidepressant dose (Bloch et al. 2010). Suggested daily doses are 40–60 mg for paroxetine, 60–80 mg for fluoxetine, 20–30 mg for escitalopram, 100–300 mg for sertraline, and 150–300 mg for fluvoxamine. Venlafaxine may also be used to treat OCD.

Choosing a drug is based on the same considerations as in depression. If there is no response to the first SSRI, a switch to another SSRI is recommended. If there is no response, again, clomipramine may be the next best option. The maximum dose of clomipramine is 250 mg/day; at higher doses, the risk of seizures increases.

An ideal anti-OCD trial should last for 2–3 months because response is often slow. If there is full response with the SSRI/SRI, the same dose is continued as maintenance treatment for at least 1–2 years and perhaps indefinitely. However, if there is only partial response, SSRI/SRI augmentation is suggested.

Augmentation is best effected with the use of an atypical antipsychotic drug such as risperidone or aripiprazole. Antipsychotic augmentation should be started at very low doses, and doses should only gradually be increased. Target doses are lower than those used in the treatment of psychosis. The duration of an augmentation trial is 4–8 weeks. Antipsychotic augmentation is particularly helpful in patients with comorbid tic disorders.

Other augmentation approaches include lithium (300–600 mg/day), buspirone (up to 60 mg/day), clonazepam (up to 5 mg/day), clonidine (0.1–0.6 mg/day), and trazodone (100–200 mg/day), all of which increase serotonergic activity. There is some evidence to suggest that glutamatergic medications such as memantine and riluzole may also be helpful as augmenting agents.

Pharmacotherapy for OCD is most effective when combined with cognitive behavior therapy.

Substance Use Disorders

Alcohol and nicotine are the most commonly abused substances. Treatment involves both pharmacological and non-pharmacological approaches. Initial medications are used to manage withdrawal symptoms. Following detoxification, pharmacotherapy is aimed at relapse prevention.

Managing Alcohol Withdrawal Syndrome (AWS)

The first-line treatment for alcohol withdrawal involves benzodiazepines; these drugs reduce the risk of withdrawal seizures and delirium. In most patients, long-acting benzodiazepines such as chlordiazepoxide or diazepam are preferred to minimize breakthrough symptoms and control agitation. In the elderly and in those with hepatic disease, short-acting benzodiazepines such as lorazepam or oxazepam are preferred. Possible dosing schedules include a loading-dose regimen, fixed-dose regimen, and symptom-triggered treatment (see Table 13). A symptom-triggered regimen is not preferred in patients with history of complicated withdrawal. In some

Table 13 Dosing schedules for alcohol withdrawal syndrome (AWS)

Loading dose/front loading	Fixed-dose	Symptom-triggered
High doses of long-acting benzodiazepines are used to quickly achieve initial sedation with a self-tapering effect over time due to their pharmacokinetic properties	This involves administration of a specific dose of medication at regular intervals	Regular assessment of withdrawal symptoms using the CIWA-Ar, with dose adjusted accordingly
Diazepam 10–20 mg or chlordiazepoxide 100 mg can be repeated every 1–2 h till adequate sedation is achieved (usually three doses are required)	Chlordiazepoxide up to 125 mg or diazepam up to 60 mg per day may be initiated and continued for 2–3 days for stabilization of the withdrawal syndrome and then gradually tapered over a period of 7–10 days (by 10–20% per day)	Diazepam 5–10 mg or chlordiazepoxide 25–100 mg if CIWA-Ar >8, assess after 1 h, and if symptoms persist, doses are repeated hourly until the score is below 8. Once stable, patients can be assessed every 4–8 h for additional therapy

CIWA-Ar Clinical Institute Withdrawal Assessment for Alcohol, revised

patients with benzodiazepine-resistant withdrawal symptoms, barbiturates such as phenobarbitone may be useful.

All patients should also receive high-dose thiamine (100 mg), preferably parenterally, regardless of their nutritional status. In those with clinical features of Wernicke's encephalopathy, much higher daily doses of thiamine, up to 1500 mg, are recommended. In practice, multivitamin injections with high-dose thiamine are administered for 3–5 days, followed by oral dosing. Many patients will additionally require intravenous fluids and correction of electrolyte imbalances that are common during alcohol withdrawal. Hypokalemia is corrected with oral potassium supplementation. Although hypomagnesemia and hypophosphatemia are common, they do not usually require supplementation.

Other agents that have been used for alcohol withdrawal include clonidine (0.4–0.6 mg/day in two to four divided doses) and baclofen (10–30 mg/day). However, these agents do not prevent withdrawal seizures and should not be used if there is a past history of complicated withdrawal. Carbamazepine has also been used for the treatment of alcohol withdrawal. However, benzodiazepines are superior to all other agents and should be used as first-line treatment of alcohol withdrawal.

Medications for Alcohol Relapse Prevention

Several medications have been found to be useful for relapse prevention in patients with alcohol dependence. They can be grouped as deterrent (e.g., disulfiram) and anticraving agents (e.g., naltrexone, acamprosate, topiramate, baclofen).

Supervised disulfiram therapy is one of the low-cost options for relapse prevention in motivated patients. Disulfiram is an oral aldehyde dehydrogenase inhibitor. Consumption of alcohol while on disulfiram results in accumulation of acetaldehyde, a metabolic breakdown product of alcohol, resulting in symptoms such as flushing, nausea, vomiting, tachycardia, dizziness, hypotension, and even syncope, convulsions, coma, and death. This is known as the disulfiram-ethanol reaction (DER). The symptoms appear approximately 5–15 min after alcohol ingestion and last for 30 min to several hours; the intensity varies with the amount of alcohol. Fear of the DER motivates an abstinence from alcohol. Disulfiram is commonly dosed at 250 mg/day; higher doses may be required in some patients who do not experience DER with lower doses. Contraindications for the use of disulfiram include hepatitis, peripheral neuropathy, uncontrolled hypertension or diabetes, and psychosis or depression. If DER occurs, antihistaminergic agents such as parenteral diphenhydramine or promethazine help control the symptoms, in addition to other supportive measures.

Among anticraving agents, naltrexone and acamprosate have the highest level of evidence for relapse prevention in alcohol dependence (Donoghue et al. 2015). Acamprosate may be slightly more efficacious in promoting abstinence, and naltrexone slightly more efficacious in reducing heavy drinking and craving (Maisel et al. 2013). Naltrexone may be prescribed at 50 mg/day or 100 mg every 2 days, or 150 mg every 3 days, because of its long half-life. In those with moderate to severe hepatic impairment, naltrexone is contraindicated. Acamprosate is dosed at 666 mg thrice daily (lower doses in the elderly and in those with renal impairment). The advantage of acamprosate is that it can be administered safely to patients with moderate to severe hepatic impairment. It is generally well tolerated, but some patients develop gastrointestinal adverse effects such as anorexia, diarrhea, and flatulence.

Topiramate (100–300 mg/day) can be started without requiring alcohol detoxification. This drug may reduce problem drinking and can be helpful for patients with drinking obsessions and automaticity of drinking (Guglielmo et al. 2015). Topiramate should be started at 25 mg/day and gradually increased (25 mg every week) to the target dose. For patients with comorbid epilepsy, bipolar disorder, or binge pattern of drinking, antiepileptic drugs such as carbamazepine or sodium valproate may help. In those with comorbid mood or anxiety disorder who have frequent relapses related to dysphoric craving, antidepressants such as SSRIs or SNRIs are helpful.

Baclofen (60–180 mg/day) has been used safely in those with hepatic impairment. Ondansetron (16 mg/day) has shown some efficacy in treatment of heavy drinking in those with early-onset alcohol dependence.

Nicotine Replacement Strategies and Anticraving Medications

For acute treatment of nicotine withdrawal, nicotine replacement therapy (NRT) is recommended. Several formulations of nicotine are available (see Table 14). Nicotine gums are widely available and can be initiated in most patients. Patients need to be educated about the "park and chew" method. Patients who do not accept gums, or those with temporomandibular joint problems, can use nicotine lozenges or other

Table 14 Nicotine replacement therapy

	Preparation	Description
1	Gum	2–4 mg, 9–16 per/day, "park and chew" method. Usually one gum per hour is advised, may be taken for 3 months, tapered off
2	Lozenge	2–4 mg, start with 9 per/day (maximum 20 per/day), usually one every 1–2 h for 6 weeks, then tapered off over 6 weeks
3	Patch	Starting dose is 21 or 14 mg patch per day, later reduced to 7 mg patch, continued for 6–14 weeks
4	Nasal spray	Single dose (0.5 mg) to each nostril, 1–3 times per hour. Effective daily dose is 15–20 sprays (8–10 mg) per day
5	Inhaler	Dosage is 6–16 cartridges per day, each cartridge has 80 inhalations (total 4 mg of nicotine), up to 6 months, taper over last 3 months

preparations. A scheduled dosing is preferred rather than as required treatment. If available, nicotine patches deliver a continuous release of medication. Sometimes, a combination of nicotine gum and patch is more effective in reducing withdrawal than either treatment alone.

Bupropion and nortriptyline have been found to be effective for long-term smoking cessation, independent of their antidepressant effect (Hughes et al. 2014). Bupropion is prescribed at 150–300 mg daily for treatment of nicotine dependence. It has been safely used for treatment of nicotine dependence in patients with schizophrenia without risk of exacerbation of psychosis (Tsoi et al. 2010). Nortriptyline increases the quit rate at doses of 75–100 mg/day, but its use is limited by sedation and anticholinergic adverse effects. Varenicline, a nicotinic acetylcholine receptor partial agonist, has been found to be more effective than bupropion in smoking cessation (Cahill et al. 2016). Varenicline should be started 1 week before the quit date. The dose recommended is 0.5 mg for 3 days, then 0.5 mg twice daily for the next 4 days, followed by 1 mg twice daily for 12 weeks. However, cost and availability are limiting factors for its usage in rural settings.

Medically Unexplained Symptoms and Somatoform Disorders

Medically unexplained symptoms (MUS) are persistent bodily symptoms that are not explained by any structural pathology and are very common, sometimes accounting for up to 45% of general practice consultations (Chew-Graham et al. 2017). These patients are diagnosed as somatoform disorders by psychiatrists. Many of them have comorbid or overlap conditions such as fibromyalgia or myofascial pain syndrome, chronic fatigue syndrome, tension-type headache, migraine, irritable bowel syndrome, temporomandibular syndrome, etc. Treatment of MUS/somatoform disorders is primarily non-pharmacological using explanatory models and reattribution techniques. However, several medications have been trialed for these conditions, specifically for somatoform pain disorders, including fibromyalgia. Many such patients have comorbid depression and/or anxiety symptoms and benefit from antidepressant therapy.

Newer antidepressants have been found to be effective for the treatment of MUS/somatoform disorders. SNRIs such as duloxetine 30–120 mg/day and milnacipran 100–200 mg/day are effective and well tolerated, but the effect sizes for pain symptoms and subjective improvement are low (Calandre et al. 2015). Venlafaxine has also been reported to reduce pain symptoms in some studies.

If newer antidepressants are not available, low-dose TCAs are good alternative and may be equally effective. Sometimes, they are the preferred agents when there is associated insomnia. Typically, amitriptyline or nortriptyline is initiated at 25 mg/day and gradually increased up to 75 mg/day. Other TCAs such as imipramine, dothiepin, and clomipramine have also been found to be helpful. Most patients show partial response with medications, specifically reduction in persistent pain symptoms.

SSRIs such as citalopram, escitalopram, fluoxetine, paroxetine, and sertraline have been used at antidepressant doses, but the effect sizes for clinical improvement are low (Calandre et al. 2015). Hypochondriasis sometimes responds to high doses of SSRIs (80 mg/day of fluoxetine or 300 mg/day of fluvoxamine) for 10–12 weeks. Patients not responding to one SSRI may respond to another SSRI.

Similar responses are seen with SSRIs for the treatment of body dysmorphic disorder. High-dose pregabalin in doses from 300 to 450 mg/day, and possibly gabapentin, reduces persistent pain symptoms in fibromyalgia. Tramadol, a weak agonist of μ-opioid receptors and a serotonin and noradrenaline reuptake inhibitor, also improves fibromyalgia (Calandre et al. 2015).

The role of antipsychotics in the treatment of somatoform disorder is less clear. Levosulpiride in divided doses of 50–100 mg/day has been reported to reduce symptoms in these conditions. Anticonvulsants such as topiramate have been found to reduce pain symptoms in those with multisomatoform disorder.

Most of patients with somatoform disorders are undertreated. Those with excessive worry and preoccupation may benefit from anxiolytics and antidepressants. Antipsychotics may be used in those with overvalued ideas or delusions as in some cases of body dysmorphic disorder and hypochondriacal disorder.

Chronic Insomnia

Insomnia is a common condition characterized by difficulty in initiating or maintaining sleep or poor quality of sleep. A short course of hypnotic medication is recommended for acute insomnia. However, for chronic insomnia, medications may need to be prescribed for a longer time and have been safely used up to 1 year. The first-line hypnotics include non-benzodiazepine GABA receptor agonists (also called Z-drugs); if unavailable, short-acting benzodiazepines or other medications can be used.

First-Line Medications

The first-line medications for insomnia include the Z-drugs, benzodiazepines, and ramelteon (see Table 15). The four Z-drugs include zolpidem, zaleplon, zopiclone, and eszopiclone. These medications target the GABA-A receptor complex and have preferential affinity for the α-1 subunit. For sleep-onset insomnia, medications with shorter half-life (zaleplon or zolpidem) are preferred, whereas for sleep-maintenance insomnia, those with longer half-life (zolpidem sustained release or eszopiclone) are needed. Zaleplon can be used specifically for the treatment of middle-of-night awakenings because of its short half-life. Owing to the low risk of adverse effects, including daytime spillover effects, Z-drugs are considered as the first-line agents for chronic insomnia.

Several benzodiazepines, such as estazolam, temazepam, triazolam, and flurazepam, have also been approved for the treatment of chronic insomnia. In the absence of benzodiazepines approved as hypnotics, other agents such as lorazepam, oxazepam, clonazepam, or diazepam can be used. The strategy is to use long-acting benzodiazepines if comorbid anxiety is present. In other situations, either short-acting or intermediate-acting benzodiazepines are preferred for sleep-onset and sleep-maintenance insomnia, respectively. A limitation of the benzodiazepines is that tolerance may develop to their hypnotic action and discontinuation of the benzodiazepine can be hard.

Ramelteon, a melatonin MT_1 and MT_2 receptor agonist, is approved for chronic insomnia at doses of 4–8 mg/day. Ramelteon is well tolerated, even in the elderly. However, benefits with this melatonergic drug are modest; the drug is not a classical hypnotic in the true meaning of the word.

Table 15 First-line medications for chronic insomnia

		Dose	Half-life	Comment
Non-benzodiazepines (Z-drugs)				
1	Zaleplon	5–10 mg	1–1.5 h	Sleep-onset insomnia
2	Zolpidem	5–10 mg	1.5–2.6 h	Sleep-onset insomnia
3	Zolpidem-controlled release	6.25–12.5 mg	2.8 h	Sleep-maintenance insomnia
4	Eszopiclone	1–3 mg	6 h	Sleep-maintenance insomnia
Short-acting benzodiazepines				
1	Estazolam	1–2 mg	12–20 h	Sleep-maintenance insomnia
2	Temazepam	15–30 mg	8–12 h	Sleep-maintenance insomnia with anxiety symptoms
3	Triazolam	0.25–0.5 mg	2–5 h	Sleep-onset insomnia
4	Flurazepam	15–30 mg	50–200 h	Sleep-maintenance insomnia with anxiety symptoms
Melatonin agonists				
1	Ramelteon	8 mg	1–2.6 h	Sleep-onset insomnia

Second-Line Medications

Sedating tricyclic antidepressants may be used as hypnotics in low doses, e.g., dothiepin 25–50 mg, amitriptyline 10–50 mg, and imipramine 25–50 mg. Recently, very low-dose doxepin (3–6 mg) was approved for insomnia (Yeung et al. 2015); however, long-term studies are lacking.

Commonly used antihistaminergic agents such as diphenhydramine (50–100 mg) and promethazine (25–100 mg) may also be prescribed for chronic insomnia. Trazodone 25–200 mg normalizes sleep and has been used as a hypnotic medication, especially for insomnia in alcohol-dependent persons (Kolla et al. 2011). Trazodone is started at 25–50 mg and increased gradually by 25–50 mg up to a maximum dose of 200 mg/day for the treatment of insomnia. There are rare reports of priapism with trazodone, and patients should be advised to consult physicians if an erection persists for more than 4 h. Low-dose mirtazapine (7.5 mg) and quetiapine (25–200 mg) have been reported to be effective for the treatment of insomnia. Gabapentin (100–600 mg) and pregabalin (150–300 mg) may be useful in insomnia, specifically for those with comorbid neuropathic pain symptoms.

For those on regular doses of benzodiazepines for insomnia and developing loss of efficacy, taper is planned over several months. It is advisable to add non-benzo-diazepine medication (e.g., trazodone) to the existing treatment and start taper only after there is improvement in insomnia for at least 2 weeks.

Delirium

Delirium is a common condition in general hospital settings, specifically in those with comorbid medical or surgical conditions. More often than not, delirium is multifactorial in etiology. Among the most common causes are electrolyte imbalance, renal and hepatic derangements, infections, head injury, epilepsy, alcohol withdrawal, and medication adverse effects. In persons with delirium, the underlying cause has to be identified and treated. Anticholinergic medications worsen delirium. Therefore, anticholinergic burden should be reduced by discontinuing unnecessary medications having anticholinergic properties. In addition, most patients with delirium require additional pharmacotherapy, specifically those with hyperactive delirium. Pharmacotherapy primarily rests on restoring the imbalance between dopamine and acetylcholine.

First-Line Treatment

For the short-term, symptomatic treatment of delirium, the drug of choice is low-dose antipsychotic; and this is more efficacious than benzodiazepines (Meagher et al. 2013). Treatment of alcohol withdrawal is an exception though, where benzodiazepines are the drug of choice. It is prudent to start with a low dose of antipsychotic drug and uptitrate according to clinical response. Typically, oral haloperidol is initiated at 1–2 mg (even lower in the elderly), risperidone 0.25–1 mg, or quetiapine 25–50 mg, with repeat doses until the patient is calm or until adverse effects appear. Most patients will respond to 2 mg haloperidol or the equivalent thereof. If the sleep-wake cycle is reversed and sedation is required, nighttime dosing with quetiapine or olanzapine is helpful. Higher doses of quetiapine (>200 mg) or olanzapine (>10 mg) are not recommended as they contribute to anticholinergic burden.

For the acute control of symptoms, parenteral haloperidol may be used at half the usual oral doses, i.e., 2.5 mg intramuscularly. Intravenous haloperidol may precipitate arrhythmias and is not recommended for routine use. Although parenteral lorazepam is also widely used, it may exacerbate delirium and is better avoided unless delirium is associated with alcohol withdrawal (Meagher et al. 2013). Prophylaxis with antipsychotic drugs can reduce the risk of perioperative delirium in elderly patients by almost 50% (Teslyar et al. 2013).

Dementia

Treatment of dementia depends on the subtype and the clinical stage. The most common dementias include Alzheimer's disease, vascular dementia, and mixed dementia. Non-pharmacological approaches are the mainstay in management of dementia, even when behavioral and psychological symptoms are present. However, pharmacotherapy is indicated for the management of cognitive and severe, troublesome, and refractory behavioral symptoms. Medications are not indicated in those with mild cognitive impairment as there is no demonstrated improvement in cognitive symptoms or delay in progression of illness.

First-Line Medications

Cholinesterase inhibitors and memantine are treatment options for Alzheimer's dementia and are also useful for vascular dementias. The three cholinesterase inhibitors available are donepezil, rivastigmine, and galantamine (see Table 16). All the three medications increase cholinergic transmission through inhibition of acetylcholinesterase and/or butyrylcholinesterase. Cholinesterase inhibitors are the first-line medications for mild to moderate dementia. The clinical efficacy of these

Table 16 Cholinesterase inhibitors

	Mechanism	Dosage
Donepezil	Reversible AChE inhibition	Start 5 mg, increase to 10 mg, then 23 mg
Rivastigmine	Pseudo-irreversible AChE and BuChE inhibition	Start 1.5 mg bid, increase gradually up to 6 mg bid
Galantamine	Reversible AChE inhibition with modulation of nicotinic acetylcholine receptors	Start 4 mg bid, increase gradually up to 12 mg bid

AChE acetylcholinesterase, *BuChE* butyrylcholinesterase

three cholinesterase inhibitors is similar, which is small to modest at best and which includes a delay in the progression of dementia by about 6 months (Wong 2016).

Memantine is an NMDA receptor antagonist that protects against excitotoxic effects. It is indicated in those with moderate to severe dementia; however, the effects are small to modest. A combination of memantine with cholinesterase inhibitors may be used in moderate to severe dementia.

These medications improve cognitive as well as behavioral symptoms to some extent. It is recommended to start these medications early when the diagnosis is made as delayed treatment may not yield same cognitive benefit (Wong 2016). The choice of cholinesterase inhibitors is based on tolerability and cost. These medications have some efficacy in patients with Parkinson's disease dementia and Lewy body dementia as well, but not in frontotemporal dementia.

Common adverse effects of cholinesterase inhibitors include gastrointestinal symptoms (such as nausea, vomiting, diarrhea, and abdominal pain), anorexia, and muscle cramps. The gastrointestinal symptoms can be minimized by giving medications with food. These medications need to be used cautiously in those with cardiac conduction defects as they can precipitate bradycardia and heart block, leading to syncopal attacks. Memantine is usually better tolerated than cholinesterase inhibitors. Common adverse effects include headache, confusion, dizziness, and constipation.

A fair trial with cholinesterase inhibitors, memantine, or a combination should extend for at least 6 months. If there is positive response, which includes stabilization or slowing of the progression of cognitive symptoms, treatment may be continued for a longer period until dementia enters its late stages. If it is planned to discontinue medications, a slow taper is advised as behavioral symptoms may worsen with abrupt discontinuation. If behavioral symptoms reappear with discontinuation, the medications can be restarted at lower dose.

Treatment of Risk Factors

For vascular dementia and also Alzheimer's disease, treatment of risk factors such as hypertension, diabetes, obesity, and smoking are pertinent; these are also prevention strategies for dementia. In vascular dementia, calcium channel blockers and drugs affecting the renin-angiotensin system may have additional effect which is independent of the effects on blood pressure (Moretti et al. 2011).

Medications for Behavioral Symptoms

Many patients will require additional pharmacotherapy for behavioral symptoms as the benefits of cholinesterase inhibitors and memantine are modest. Appropriate pharmacotherapy may be considered for major depression, psychosis, and aggression (Kales et al. 2014). For agitation and psychotic symptoms, antipsychotics such as risperidone, olanzapine, or quetiapine have been useful. Anticonvulsant medications such as valproate or carbamazepine are useful for the treatment of agitation, aggression, disinhibition, and manic-like symptoms in dementia patients. Antidepressants such as SSRIs and SNRIs are indicated in dementia for depressed mood, anxiety, agitation, and apathy. Mirtazapine and trazodone may reduce insomnia and agitation in those with dementia. In general, pharmacotherapy does not work as well in dementia as it does for the same indications in patients without dementia.

Attention Deficit Hyperactivity Disorder (ADHD)

Medications for ADHD can be classified as stimulants and non-stimulants.

Stimulant Medications

The drug of choice in ADHD is a stimulant such as amphetamine or methylphenidate. About 70–80% of patients respond to psychostimulants, and the effect sizes are large. The doses need to be uptitrated based on efficacy and adverse effects. Benefits may be seen within a week of starting treatment. At adequate doses, core ADHD symptoms attenuate within 30 min of dosing. Dosing is timed to reduce symptoms during school or work hours.

Various formulations of stimulants are available for use (see Table 17). Extended-release preparations improve adherence and ease of administration but do not have dose peaks that are associated with peak benefits (and adverse effects).

Common acute adverse effects of psychostimulants include tachycardia, tremor, and headache. Common adverse effects during maintenance treatment include insomnia, irritability, reduced appetite, and slowed growth in children. Tics may exacerbate, cardiac arrhythmias may occur, psychosis may develop, and there may be seizures; however, these are all rare (Groenman et al. 2017). The use of sustained-release formulations helps reduce adverse effects associated with drug peaks. Drug holidays may be considered during weekends and during school holidays.

Baseline assessment and regular monitoring of height and weight are desirable when stimulants are prescribed to children and adolescents. A baseline ECG may also be performed.

Table 17 Examples of stimulant medications

Drug	Preparations	Duration of action	Dosage
Methylphenidate	Tablets, sustained-release tablets, OROS capsule, beaded capsule, transdermal patch	4–12 h	10–60 mg
D-Methylphenidate	Tablet, beaded capsule	4–12 h	5–40 mg
Amphetamine	Tablet, beaded capsule	6–10 h	5–40 mg
D-Amphetamine	Tablet, spansule capsule	4–10 h	5–40 mg
Lisdexamfetamine	Capsule	10 h	30–70 mg

OROS osmotic release oral system

Stimulant treatment in early years does not appear to increase the risk for substance-related disorders in adulthood. The safety of stimulants for period of up to 2 years appears satisfactory, but little data is available for safety in the very long term (Groenman et al. 2017).

Non-stimulant Medications

Alternatives to stimulant medications include atomoxetine and α_2-agonists. These medications can be used when stimulants are unavailable, contraindicated, poorly tolerated, or ineffective. Atomoxetine is a norepinephrine reuptake inhibitor that has a medium effect size in ADHD. The target therapeutic dose for atomoxetine is 1.2 mg/kg/day. It is rapidly absorbed and reaches peak plasma concentration within 1 h (food delays absorption to 3 h). Unlike stimulants, atomoxetine can be given in the evening, as well. However, once-daily dosing suffices for most patients.

Clonidine, a α_2-agonist, is useful to control hyperactivity symptoms; sedation is a troublesome adverse effect. It can be combined with stimulants to counteract stimulant-related insomnia. Guanfacine is a more selective α_2-agonist with less sedation and a longer duration of action.

Other medications used to treat ADHD, especially in adults, include bupropion and modafinil. TCAs have also been used, but have more adverse effects and may not be superior to atomoxetine.

If there is poor response to monotherapy, a combination of medications may help. A medication with strong dopaminergic effect (e.g., a stimulant) may be combined with a medication with predominant noradrenergic effect (e.g., clonidine or atomoxetine).

Medications for ADHD need to be continued into the long term with periodic attempts to reduce dosage to detect spontaneous improvement. Some patients may require continued treatment into adulthood for persistent attention deficit symptoms.

Mental Retardation (or Intellectual Disability)

Treatment of Comorbid Psychiatric Disorders

Mental retardation (MR) increases the risk of other neuropsychiatric disorders such as epilepsy, psychosis, or merely nonspecific disturbances of mood and behavior. Treatment of such comorbid conditions is similar to that of the independent disorders; however, the dosage of medications should be as low as possible to reduce the risk of adverse effects.

Lithium, valproate, and carbamazepine have been used to treat disturbances of mood and behavior in patients with MR (Aman et al. 2000). For psychosis, including schizophrenia, typical and atypical antipsychotics, including clozapine, have shown good response. Patients with ADHD symptoms require use of stimulants such as methylphenidate; however, the response is poorer as compared to those without MR

(Aman et al. 2000). For conduct disorder in children with MR, haloperidol, risperidone, and lithium are the ones that are best studied. For the treatment of depression, SSRIs, SNRIs, and TCAs have been used successfully.

Treatment of Challenging Behaviors

Challenging behaviors in patients with mental retardation include aggression, self-injury, stereotypy, and inappropriate social behavior. Treatment for most of these includes the use of behavioral therapy principles. There is very little systematic study of pharmacotherapy for these conditions. Antipsychotics are the most widely prescribed medications for challenging behaviors. The treatment pattern has changed from typical to atypical antipsychotics, with risperidone being the most frequently used medication. However, one review questioned the superiority of risperidone for the treatment of challenging behaviors (Singh et al. 2005).

Sexual Disorders

Among sexual disorders, erectile dysfunction and premature ejaculation are the two most common conditions that respond to pharmacological interventions. Several medications can adversely affect sexual function, and discontinuation of these may sometimes lead to improvement.

Drugs for Erectile Dysfunction (ED)

Erectile dysfunction needs to be distinguished from reduced desire. In addition to a thorough medical and sexual history, investigations such as fasting blood glucose and lipids, serum free testosterone, serum prolactin, and thyroid function tests are indicated. Some patients may require specialized procedures including intracavernosal injection of vasoactive drugs or penile Doppler ultrasonography. These patients may require referral to higher centers having such facilities.

For erectile dysfunction, the gold standard, first-line treatment of choice is a PDE_5 inhibitor, unless contraindicated (Hawksworth and Burnett 2015). The comparative pharmacology of PDE_5 inhibitors is summarized in Table 18. These medications restore the natural response of the body to sexual stimulation through smooth muscle relaxation in the corpus cavernosum, thus helping in obtaining and maintaining an

Table 18 PDE_5 inhibitors

Drug	Onset of action	Duration of action	Dosage	Interaction with food
Sildenafil	15–60 min	4 h	50–100 mg	Delays onset
Tadalafil	15–120 min	36 h	10–20 mg	No effect
Vardenafil	15–60 min	4 h	10–20 mg	Delays onset

erection; however, sexual desire and stimulation are necessary for erection to occur. For a therapeutic trial, a 50 mg dose of sildenafil is used. If there is little or no response, then 100 mg sildenafil may be tried. Common adverse effects include facial flushing, headache, rhinitis, and visual disturbances. Major contraindications include treatment with nitrates, unstable angina, or recent myocardial infarction.

Tadalafil has the longest duration of action (up to 36 h) and may provide more opportunities for successful intercourse and repeated attempts after a single dose. Food rich in fat tends to delay the effects of sildenafil and vardenafil, whereas there is no such effect with tadalafil. Some patients may require higher doses of PDE_5 inhibitors for the desired effect.

Nonresponse to PDE_5 Inhibitors

Although PDE_5 inhibitors are highly efficacious, some patients, including those with neurological conditions, severe arterial disease, penile trauma with degeneration of the erectile tissue, or unstable diabetes, may not respond. Patients with testosterone deficiency may also fail to respond to PDE_5 inhibitors; these patients may be converted to PDE_5 inhibitor responders by testosterone replacement therapy. Trazodone, which occasionally induces priapism, has also been used in doses of 100–200 mg at bedtime for ED. Other options include topical or transurethral application of alprostadil or intracavernous self-injection with papaverine, phentolamine, and alprostadil (Hawksworth and Burnett 2015); however, cost and availability of these in rural settings may limit their usage.

Drugs for Premature Ejaculation (PME)

An intravaginal ejaculatory latency time (IELT) of less than 1 min is used as a cutoff to diagnose PME. The treatment options available are "daily" and "on demand" (intake of drugs a few hours before intercourse) drug treatment.

Daily drug treatment: Delayed ejaculation is an adverse effect of SSRIs and clomipramine, and this has been used therapeutically for PME. The doses of SSRIs that are found to be useful in PME include paroxetine 20 mg, sertraline 50 mg, fluoxetine 20 mg, citalopram 20 mg, and fluvoxamine 100 mg (Castiglione et al. 2016). Among the long-acting SSRIs, paroxetine 20 mg produces longest ejaculatory delay. Clomipramine can also be used at doses of 10–20 mg/day for PME. The daily intake of serotonergic antidepressants has advantages over on-demand intake. While using a daily drug strategy, sexual contact may take place at any time of the day with higher chance of ejaculation delay and does not interfere with the desirable spontaneity of having sexual contact.

On-demand drug treatment: As compared with daily treatment with SSRIs, on-demand use of serotonergic antidepressants a few hours before intercourse leads to less ejaculation delay, specifically in those with lifelong PME. The short-acting SSRI, dapoxetine 30–60 mg, is indicated as "on-demand" basis for PME

(Russo et al. 2016). It is usually taken 1–3 h prior to intercourse and is tolerated well. If dapoxetine is not available, any traditional long-acting SSRI can be used effectively. For "on-demand" use, SSRIs need to be taken 3–6 h prior to sexual intercourse to be effective. Clomipramine 20–40 mg can also be used as "on-demand" basis, 4–6 h prior to coitus.

PDE$_5$ inhibitors can also be used for treatment of PME, specifically if it coexists with ED. Sometimes, a combination of PDE$_5$ inhibitors with an SSRI is required for the desired effect. Another medication that has some efficacy in PME is tramadol 50–100 mg, used "on demand." Other strategies for the treatment of PME include use of anesthetizing creams and sprays to delay ejaculation.

Conclusion

The first-line treatment options for various disorders should be the preferred agents. However, alternate medications are sometimes equally effective and more affordable. Hence, treatment decisions need to be individualized and involve risk-benefit analyses. Sometimes, doses less or more than what is recommended in guidelines are warranted depending on the clinical situation, to account for inter-individual variations. Even evidence-based psychopharmacology can be an art.

References

Aman MG, Collier-Crespin A, Lindsay RL (2000) Pharmacotherapy of disorders in mental retardation. Eur Child Adolesc Psychiatry 9(Suppl 1):S98–S107

Andrade C (2016) Antipsychotic drugs in schizophrenia: relative effects in patients with and without treatment resistance. J Clin Psychiatry 77:e1656–e1660. https://doi.org/10.4088/JCP.16f11328

Andrade C (2017a) Ketamine for depression, 1: clinical summary of issues related to efficacy, adverse effects, and mechanism of action. J Clin Psychiatry 78:e415–e419. https://doi.org/10.4088/JCP.17f11567

Andrade C (2017b) Ketamine for depression, 4: in what dose, at what rate, by what route, for how long, and at what frequency? J Clin Psychiatry 78:e852–e857. https://doi.org/10.4088/JCP.17f11738

Andrade C, Kumar CB, Surya S (2013) Cardiovascular mechanisms of SSRI drugs and their benefits and risks in ischemic heart disease and heart failure. Int Clin Psychopharmacol 28:145–155. https://doi.org/10.1097/YIC.0b013e32835d735d

Bloch MH, McGuire J, Landeros-Weisenberger A, Leckman JF, Pittenger C (2010) Meta-analysis of the dose-response relationship of SSRI in obsessive-compulsive disorder. Mol Psychiatry 15:850–855. https://doi.org/10.1038/mp.2009.50

Cahill K, Lindson-Hawley N, Thomas KH, Fanshawe TR, Lancaster T (2016) Nicotine receptor partial agonists for smoking cessation. Cochrane Database Syst Rev 5:CD006103. https://doi.org/10.1002/14651858.CD006103.pub7

Calandre EP, Rico-Villademoros F, Slim M (2015) An update on pharmacotherapy for the treatment of fibromyalgia. Expert Opin Pharmacother 16:1347–1368. https://doi.org/10.1517/14656566.2015.1047343

Castiglione F, Albersen M, Hedlund P, Gratzke C, Salonia A, Giuliano F (2016) Current pharmacological management of premature ejaculation: a systematic review and meta-analysis. Eur Urol 69:904–916. https://doi.org/10.1016/j.eururo.2015.12.028

Chew-Graham CA, Heyland S, Kingstone T, Shepherd T, Buszewicz M, Burroughs H et al (2017) Medically unexplained symptoms: continuing challenges for primary care. Br J Gen Pract 67:106–107. https://doi.org/10.3399/bjgp17X690785

Cipriani A, Furukawa TA, Salanti G, Geddes JR, Higgins JP, Churchill R et al (2009) Comparative efficacy and acceptability of 12 new-generation antidepressants: a multiple-treatments meta-analysis. Lancet 373:746–758. https://doi.org/10.1016/S0140-6736(09)60046-5

Cipriani A, Barbui C, Salanti G, Rendell J, Brown R, Stockton S et al (2011) Comparative efficacy and acceptability of antimanic drugs in acute mania: a multiple-treatments meta-analysis. Lancet 378:1306–1315. https://doi.org/10.1016/S0140-6736(11)60873-8

Colijn MA, Nitta BH, Grossberg GT (2015) Psychosis in later life: a review and update. Harv Rev Psychiatry 23:354–367. https://doi.org/10.1097/HRP.0000000000000068

Donoghue K, Elzerbi C, Saunders R, Whittington C, Pilling S, Drummond C (2015) The efficacy of acamprosate and naltrexone in the treatment of alcohol dependence, Europe versus the rest of the world: a meta-analysis. Addiction 110:920–930. https://doi.org/10.1111/add.12875

Finley PR (2016) Drug interactions with lithium: an update. Clin Pharmacokinet 55:925–941. https://doi.org/10.1007/s40262-016-0370-y

Fleming J, Chetty M (2005) Psychotropic drug interactions with valproate. Clin Neuropharmacol 28:96–101

Groenman AP, Schweren LJ, Dietrich A, Hoekstra PJ (2017) An update on the safety of psychostimulants for the treatment of attention-deficit/hyperactivity disorder. Expert Opin Drug Saf 16:455–464. https://doi.org/10.1080/14740338.2017.1301928

Guglielmo R, Martinotti G, Quatrale M, Ioime L, Kadilli I, Di Nicola M et al (2015) Topiramate in alcohol use disorders: review and update. CNS Drugs 29:383–395. https://doi.org/10.1007/s40263-015-0244-0

Hawksworth DJ, Burnett AL (2015) Pharmacotherapeutic management of erectile dysfunction. Clin Pharmacol Ther 98:602–610. https://doi.org/10.1002/cpt.261

Hughes JR, Stead LF, Hartmann-Boyce J, Cahill K, Lancaster T (2014) Antidepressants for smoking cessation. Cochrane Database Syst Rev 1:CD000031. https://doi.org/10.1002/14651858.CD000031.pub4

Kales HC, Gitlin LN, Lyketsos CG, Detroit Expert Panel on Assessment and Management of Neuropsychiatric Symptoms of Dementia (2014) Management of neuropsychiatric symptoms of dementia in clinical settings: recommendations from a multidisciplinary expert panel. J Am Geriatr Soc 62:762–769. https://doi.org/10.1111/jgs.12730

Kolla BP, Mansukhani MP, Schneekloth T (2011) Pharmacological treatment of insomnia in alcohol recovery: a systematic review. Alcohol Alcohol 46:578–585. https://doi.org/10.1093/alcalc/agr073

Leucht S, Cipriani A, Spineli L, Mavridis D, Orey D, Richter F et al (2013) Comparative efficacy and tolerability of 15 antipsychotic drugs in schizophrenia: a multiple-treatments meta-analysis. Lancet 382:951–962. https://doi.org/10.1016/S0140-6736(13)60733-3

Loane C, Politis M (2012) Buspirone: what is it all about? Brain Res 1461:111–118. https://doi.org/10.1016/j.brainres.2012.04.032

Maisel NC, Blodgett JC, Wilbourne PL, Humphreys K, Finney JW (2013) Meta-analysis of naltrexone and acamprosate for treating alcohol use disorders: when are these medications most helpful? Addiction 108:275–293. https://doi.org/10.1111/j.1360-0443.2012.04054.x

Meagher DJ, McLoughlin L, Leonard M, Hannon N, Dunne C, O'Regan N (2013) What do we really know about the treatment of delirium with antipsychotics? Ten key issues for delirium pharmacotherapy. Am J Geriatr Psychiatry 21:1223–1238. https://doi.org/10.1016/j.jagp.2012.09.008

Moretti A, Gorini A, Villa RF (2011) Pharmacotherapy and prevention of vascular dementia. CNS Neurol Disord Drug Targets 10:370–390

Pompili M, Baldessarini RJ, Forte A, Erbuto D, Serafini G, Fiorillo A et al (2016) Do atypical antipsychotics have antisuicidal effects? A hypothesis-generating overview. Int J Mol Sci 17:E1700. https://doi.org/10.3390/ijms17101700

Rackley S, Bostwick JM (2012) Depression in medically ill patients. Psychiatr Clin N Am 35:231–247. https://doi.org/10.1016/j.psc.2011.11.001

Russo A, Capogrosso P, Ventimiglia E, La Croce G, Boeri L, Montorsi F et al (2016) Efficacy and safety of dapoxetine in treatment of premature ejaculation: an evidence-based review. Int J Clin Pract 70:723–733. https://doi.org/10.1111/ijcp.12843

Severus E, Taylor MJ, Sauer C, Pfennig A, Ritter P, Bauer M et al (2014) Lithium for prevention of mood episodes in bipolar disorders: systematic review and meta-analysis. Int J Bipolar Disord 2:15. https://doi.org/10.1186/s40345-014-0015-8

Singh AN, Matson JL, Cooper CL, Dixon D, Sturmey P (2005) The use of risperidone among individuals with mental retardation: clinically supported or not? Res Dev Disabil 26:203–218. https://doi.org/10.1016/j.ridd.2004.07.001

Teslyar P, Stock VM, Wilk CM, Camsari U, Ehrenreich MJ, Himelhoch S (2013) Prophylaxis with antipsychotic medication reduces the risk of post-operative delirium in elderly patients: a meta-analysis. Psychosomatics 54:124–131. https://doi.org/10.1016/j.psym.2012.12.004

Tsoi DT, Porwal M, Webster AC (2010) Efficacy and safety of bupropion for smoking cessation and reduction in schizophrenia: systematic review and meta-analysis. Br J Psychiatry 196:346–353. https://doi.org/10.1192/bjp.bp.109.066019

Wong CW (2016) Pharmacotherapy for dementia: a practical approach to the use of cholinesterase inhibitors and memantine. Drugs Aging 33:451–460. https://doi.org/10.1007/s40266-016-0372-3

Yeung WF, Chung KF, Yung KP, Ng TH (2015) Doxepin for insomnia: a systematic review of randomized placebo-controlled trials. Sleep Med Rev 19:75–83. https://doi.org/10.1016/j.smrv.2014.06.001

Zhou X, Ravindran AV, Qin B, Del Giovane C, Li Q, Bauer M et al (2015) Comparative efficacy, acceptability, and tolerability of augmentation agents in treatment-resistant depression: systematic review and network meta-analysis. J Clin Psychiatry 76:e487–e498. https://doi.org/10.4088/JCP.14r09204

Rehabilitation for Persons with Severe Mental Illness in Lower- and Middle-Income Countries

21

9

Jagadisha Thirthalli, Thanapal Sivakumar, and
Chethan Basavarajappa

Contents

Abstract

Psychiatric conditions are a leading cause of disability across the globe. Majority of persons experiencing disability due to mental illness live in rural areas of low- and middle-income countries. The fact that a plethora of efficacious rehabilitation interventions are available to help these persons makes little difference in their lives, given the fact that the delivery of these interventions is nearly non-existent in most countries. There have been major changes in the conceptualization of disability and rehabilitation in the recent decades. Disability is viewed as a product of interaction of a person's impairment with his/her environment. Hence, rehabilitation, as a concept, has moved away from a purely medical model. Person-centered,

J. Thirthalli (✉) · T. Sivakumar · C. Basavarajappa
Psychiatric Rehabilitation Services, Department of Psychiatry, National Institute of Mental Health and Neurosciences, Bangalore, Karnataka, India
e-mail: jagatth@yahoo.com; drt.sivakumar@yahoo.co.in; drt.sivakumar@outlook.com; drchethanraj@gmail.com

© Springer Nature Singapore Pte Ltd. 2020
S. Chaturvedi (ed.), *Mental Health and Illness in the Rural World*, Mental Health and Illness Worldwide, https://doi.org/10.1007/978-981-10-2345-3_19

rights-based comprehensive rehabilitation, addressing health, education, livelihood, social, and empowerment needs of the patients, has become the standard of care. There is serious dearth of literature demonstrating successful and sustainable models of rehabilitation meeting these standards. World Health Organization's Community-Based Rehabilitation (WHO-CBR) provides a roadmap to develop locally applicable, comprehensive rehabilitation for persons with disability (PwD) due to any health condition. It consists of a bottom-up approach, in which all stakeholders in the community take part in managing all aspects of rehabilitation of PwD. While a few case studies show promise, there is need for systematic studies of feasibility and effectiveness of WHO-CBR.

Keywords
Rural · Rehabilitation · Lower- and middle-income countries · Severe mental illness

Introduction

In this chapter, we discuss the challenges and opportunities for rehabilitation of persons experiencing severe mental illnesses, with special focus on rural areas of lower- and middle-income (LAMI) countries. We start our chapter with an overview of the concept of rehabilitation with reference to persons with mental illness; this would be followed by a brief discussion on evidence-based rehabilitation interventions for persons with disability due to mental illness (PwDMI); we then discuss how rehabilitation of PwDMI in LAMI countries differs from that in high-income countries; and we follow this up with unique features of rural communities insofar as rehabilitation of PwDMI is considered. We end the chapter with a detailed discussion on World Health Organization's Community-Based Rehabilitation (WHO-CBR) and conclusive remarks with suggestions for services and research in the future.

Evolution of the Concept of Rehabilitation

The term rehabilitation is used in many contexts. Here, we confine ourselves to rehabilitation in the context of persons with disabilities (PwD). Definitions of rehabilitation abound. The one that the World Health Organization (WHO) uses is arguably the most comprehensive. The WHO defines rehabilitation of people with disabilities as a process aimed at enabling them to reach and maintain their optimal physical, sensory, intellectual, psychological, and social functional levels. Rehabilitation helps PwD attain independence and self-determination (World Health Organization 2017). Clearly, rehabilitation is person-centered and individualized; "one-size-fits-all" approaches are not suitable. This becomes more evident when one closely follows the evolution of the term disability.

Historically, disability was considered to be an individual's deficit. In that case, society's responses were restricted to one of two paths: individuals can be "fixed" through medicine or surgery by professionals (*medical model*) or they can be cared for, through charity or welfare programs (*charity model*). According to these models, the lives of PwD are handed over to others who make decisions for them. Over the past few decades, however, there has been a shift in focus from the individual to the society. According to the *social model*, disability is recognized as the consequence of the interaction of the individual with an environment that does not accommodate that individual's differences and limits or impedes the individual's participation in society (Bhargava et al. 2012). This is reflected in the definition of persons with disability given by the United Nations Convention on Rights of Persons with Disability (UNCRPD), "persons with disabilities include those who have long-term physical, mental, intellectual or sensory impairments which, in interaction with various barriers, may hinder their full and effective participation in society on an equal basis with others" (United Nations General Assembly 2007). It follows that rehabilitation of PwD is as much to do with contextual factors, particularly, societal factors, as it is with the impairment per se. To understand this better, one should pay attention to different terms used by the WHO International Classification of Functioning and Health (ICF) (see Box 1, see Fig. 1) (World Health Organization 2001). In summary, persons with similar impairments may have vastly different levels of disability depending on a number of contextual factors. These contextual factors are more influential in determining disability in the context of PwDMI.

Box 1 Conceptualization in ICF
Terms used in WHO ICF:

Body functions *are the physiological functions of body systems (including psychological functions).*

Body structures *are anatomical parts of the body such as organs, limbs, and their components.*

Impairments *are problems in body function or structure such as a significant deviation or loss.*

Activity *is the execution of a task or action by an individual.*

Participation *is involvement in a life situation.*

Activity limitations *are difficulties an individual may have in executing activities.*

Participation restrictions *are problems an individual may experience in involvement in life situations.*

Environmental factors *make up the physical, social, and attitudinal environment in which people live and conduct their lives.*

Consequent to the change in the conceptualization of disability, the concept of rehabilitation has moved a long way from being a purely a medical method to a set

Health condition
(disorder / disease)

Body functions & structures ⟷ Activity ⟷ Participation

Environmental factors   Personal factors

Contextual factors

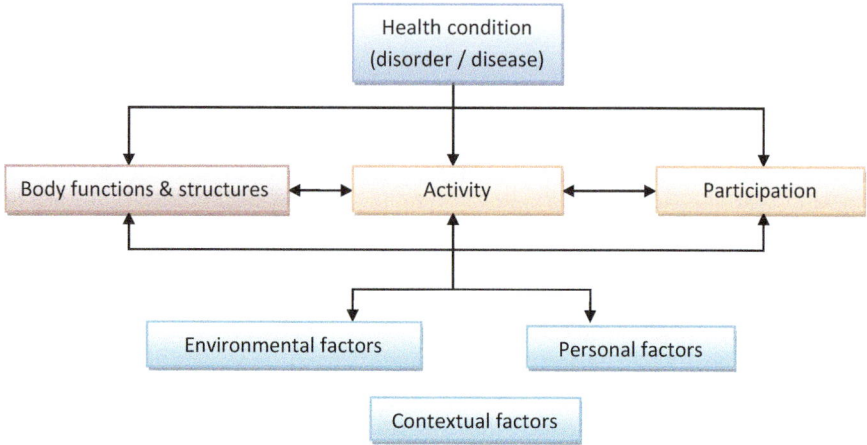

Fig. 1 Conceptualization of Disability in ICF (World Health Organization 2001)

of comprehensive and highly individualized approaches. Moreover, rehabilitation has ceased to be imposed upon a PwD; for all the right reasons, now it is the PwD, who takes a decisive role in defining and his/her own rehabilitation. The slogan "nothing about us without us" aptly summarizes this spirit: programs or plans for rehabilitation of PwD would have to necessarily include their views.

It is customary to differentiate treatment and rehabilitation. According to William Anthony, *treatment* is actions designed to cure a disease or reduce its symptoms and *Rehabilitation* is actions intended to reduce negative effects of the disease on the person's life (Anthony 2013). It is apparent that there is considerable degree of overlap between these – the goals of treatment (symptom reduction) and rehabilitation (improve role functioning) are complementary and they go hand in hand.

Unique Features in Rehabilitation of Persons with Mental Illness

The current concepts of disability and rehabilitation transcend the type of impairment – disability is measured in terms of the disadvantages the PwD experiences, irrespective of which bodily structure and/or function is affected. However, the unique features of PwDMI cannot be ignored. PwDMI differ from those with disability due to physical impairments in several ways (Thirthalli and Basavarajappa 2014): First, unlike in most other disabilities, majority of PwDMI needs continued medical care – treatment and rehabilitation happen in parallel, and not in sequential order. Second, unlike in the case of disabilities due to most physical causes, where rehabilitation seldom affects the impairment (e.g., braille books would not improve visual ability), rehabilitation of PwDMI may substantially reduce the impairment itself (e.g., vocational rehabilitation improves depressive symptoms in a person with schizophrenia), and thus rehabilitation becomes essential part of treatment of the impairment. Third, the level of impairment due to mental illness may fluctuate widely – this is seldom the case in most physical disabilities. Finally, sizable proportion of PwDMI suffers from lack of insight into their disability, unlike those with other disabilities. This may prevent them from seeking or accepting rehabilitation services that may substantially improve the quality of their life. This is in contrast to those with other disabilities, where the PwDs are proactive in realizing their rightful rehabilitation needs. Programs and plans for rehabilitation of PwDMI need to be cognizant of these realities.

Course and Outcome of Severe Mental Illnesses in LAMI Countries

A surprising but consistently replicated finding of several comparative follow-up studies is that the proportion of schizophrenia patients with favorable outcome is significantly higher in many LAMI countries like India (Jablensky and Sartorius 2008). At this stage, the reasons for this finding are largely conjectural. Social and cultural factors including family support (Lauriello et al. 2005), lower expressed emotions (Cohen et al. 2008; Wig et al. 1987), external causal models of illness (Desjarlais et al. 1995), availability of informal work (Desjarlais et al. 1995), absence of formal social security benefits and consequent pressure to work for survival (Thara 2012), and psychiatric pluralism (Halliburton 2004) are hypothesized to contribute to a less severe course of illness in developing countries (Desjarlais et al. 1995). These very factors, often referred to as the "sociocultural black-box," form important components of the contextual factors that are mentioned in the ICF (see above) (Jablensky and Sartorius 2008).

Evidence-Based Rehabilitation Interventions for Severe Mental Illnesses

PwDMI suffer from a number of challenges and require rehabilitative inputs to overcome these. Important among them are persistent negative symptoms; cognitive deficits; social skill deficits; unemployment; residual positive, affective, and anxiety symptoms; poor treatment adherence; different levels of stigma, prejudice, and discrimination in the society etc. Several rehabilitative interventions are available for overcoming these challenges. Evidence-based practice (EBP) is the contemporary gold standard for clinical practice, and many of these interventions have substantial evidence base for their efficacy. These include supported education/employment/housing, family interventions, cognitive behavioral therapy (CBT), social skills training, etc. (Bond and Campbell 2008; Dixon et al. 2010; National Institute for Health and Care Excellence 2014).

While these are exciting developments in the field of psychiatric rehabilitation, challenges galore while implementing them at the ground level. Availability of optimally trained therapists, clinical reality of each individual experiencing multiple challenges at the same time, shifting goalposts of rehabilitation needs in the same individuals at different stages of their lives, absence of evidence for durability of effects of interventions, inter-therapist variability, cost of treatment, etc. are challenges even in high-income countries (Thirthalli 2016). In the case of LAMI countries, these challenges are only compounded. Added to these are genuine questions regarding the relevance and acceptability of most of these interventions, which are primarily developed in western sociocultural contexts. Finally, given that families are involved in the care of PwDMI, in most LAMI countries, interventions, which do not consider their role in them, would have questionable value in day-to-day practice.

Complexities Within LAMI Countries and the Need to Find Local Solutions

When compared to high-income countries, LAMI countries have modest living conditions and low standard of living, generate more revenue from service sector than industries, and have lower per capita income (Surbhi 2015) and gross domestic product (GDP) (27 trillion vs. 48.4 trillion) (The World Bank 2017). The difference is evident even in health sector where there is low life expectancy (70 vs. 81 years), high infant mortality rate, birth rate, and death rate (Surbhi 2015). Ostensibly, such differences exist even within the LAMI countries.

Within each LAMI country, the contextual factors differ between rural and urban areas. Generally, life is relatively simple in rural areas; there is little need for having to constantly adapt to fast-changing life. There is less scope for occupational mobility and there are limited numbers of employment options. Family plays considerably greater role in the day-to-day functioning of an individual and social support tends to be good (Chand 2016). Access to healthcare delivery systems, which are crucial cogs in the wheel of rehabilitation of PwDMI, is starkly different between urban and rural areas. Rehabilitation interventions for PwDMI would thus differ a lot between rural and urban areas within each LAMI countries.

Finally, rural areas can seldom be stereotyped. Vast differences exist across rural areas even within small regions of LAMI countries. Most rural areas depend on agrarian economy. A rural area with fertile soil and ample water supply is likely to have better finances and consequently higher literacy, better living amenities and concentration of resources including health infrastructure and health professionals. People are also more likely to be empowered and more likely to claim their rights. In contrast, if the area is dry and drought-prone, there is higher likelihood of poor literacy rates and poor living conditions. During droughts, its inhabitants face issues related to poverty, change of livelihood, and migration for livelihood. People in such places are likely to be less empowered. In such impoverished conditions, the likelihood of well-qualified people to stay and serve the area decreases. This affects the quality of health services available which may further enhance rates of disability setting up a vicious circle. Often, they are neglected, discriminated against and excluded from mainstream development initiatives, and find it difficult to access health, education, housing, and livelihood opportunities. This results in greater poverty or chronic poverty, isolation, and even premature death. The costs of medical treatment, physical rehabilitation, and assistive devices also contribute to the poverty cycle of many people with disabilities. It is evident that rehabilitation initiatives for PWD in general and PwDMI in particular should conform to the local realities.

Literature on Rehabilitation of PwDMI in Rural Areas of LAMI Countries

Literature on rehabilitation of PwDMI in rural areas of LAMI countries is sparse, but is encouraging. A series of papers from a rural cohort of persons with schizophrenia in South India suggests that continued treatment with low-dose, low-cost antipsychotic medications along with low-intensity psychosocial interventions delivered by relatively less-qualified staff makes a meaningful difference in the disability suffered by these individuals (Thirthalli et al. 2009, 2010; Suresh et al. 2012). PwDMI seem to benefit from support from families, NGOs, government welfare programs meant for the general public, etc., to obtain income-generation opportunities (Kumar et al. 2016). Programs which are more structured, yet delivered by relatively less-qualified community-level health workers, have also shown good promise in reducing disability associated with schizophrenia in rural areas of LAMI countries (Chatterjee et al. 2009, 2014). Group interpersonal therapy for treatment of depression in Uganda (Bolton et al. 2007), CBT for postnatal depression by Lady Health Workers in Pakistan (Rahman et al. 2008), culturally adapted carer-supervised CBT for depression in Pakistan (Naeem et al. 2014), task shifting in treatment of depression in Chile (Araya et al. 2003), and other services and challenges in implementation and opportunities have also been documented (Rathod et al. 2017). Community-based rehabilitation for schizophrenia is also being conducted in Ethiopia (Asher et al. 2015, 2016) and India (ongoing project by the authors).

World Health Organization's Community-Based Rehabilitation (WHO-CBR): Philosophy and Scope

Initiation of treatment of the mental illness forms an important first step in the rehabilitation of PwDMI. Universal access to mental healthcare thus becomes the cornerstone of rehabilitation. Integration of mental healthcare with primary healthcare is a natural choice of providing such access. Efforts at providing such integrated healthcare have met with limited success in some of the LAMI countries. A number of reasons have been proposed for this (Jacob 2001). Many of these reasons stem from the fact that these programs did not involve different levels of stakeholders, most importantly, the PwDMI themselves and their families. A bottom-up approach is fundamental to the success of such programs. This is one of the most important components of the WHO-CBR.

The WHO-CBR is a multi-sectoral bottom-up strategy within general community development for the rehabilitation, poverty reduction, equalization of opportunities, and social inclusion of all people with disabilities (World Health Organization 2010a). CBR makes optimum use of primary healthcare and community resources, to bring primary healthcare and rehabilitation services closer to people with disabilities, especially in low-income countries. Poverty is both a cause and consequence of disability. Addressing disability is a concrete step to reducing the risk of poverty in any country. At the same time, addressing poverty also reduces disability. Poverty must be eliminated to achieve a better quality of life for people with disabilities; hence, one of the main objectives of any CBR program is to reduce poverty by ensuring that health, education, and livelihood opportunities are accessible to people with disabilities. Since most of the PwDMI have financial difficulties, making consultation and medications available free of cost/as a part of national programs would improve adherence and prevent disability to a considerable extent.

Among various disabilities, disability due to mental illnesses has been low on priority on the development agenda and for society in general. There is lack of knowledge about mental illness, with widespread stigma, prejudice, and discrimination. In every community, there are people living with mental illness who are likely to be isolated, stigmatized, and deprived of their fundamental human rights. People with mental illness have extremely limited access to support and health services particularly in LAMI countries and have historically been excluded from CBR programs (World Health Organization 2010b). A community-based rehabilitation approach facilitates community reintegration of the concerned individual. When community sees a person with mental illness regain valued social roles and contribute to society, it shatters many myths and misconceptions about mental illness and helps in holistic care of persons with psychiatric disorders.

CBR typically uses the management cycle (see Fig. 2) (Khasnabis and Motsch 2010). Importantly, all stakeholders should be part of the management cycle. PwDMI and their family form the core of stakeholders; their community, community leaders, teachers, health and community workers, etc., form the next level of care; local governments, NGOs, and disability groups form the next level; and federal government, political leaders, policymakers, media, etc. become the final layer of

Fig. 2 CBR management cycle

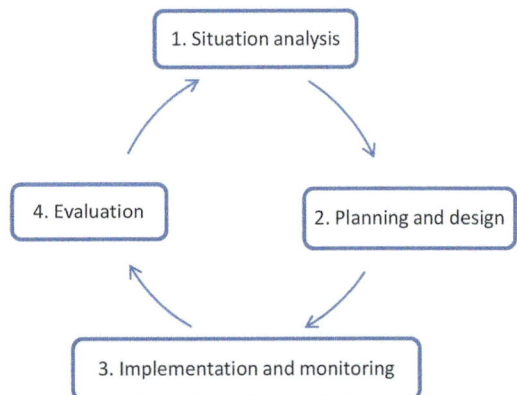

stakeholders. The involvement of stakeholders at all levels would ensure that the program responds to the needs of the community and that the community helps to sustain the program in the long run. Each stakeholder comes with his/her own unique resources and limitations. Government-led programs or government-supported programs provide more resources and have a larger reach and better sustainability. Involvement of the community tends to make CBR more appropriate, make it work in difficult situations, and ensure better community participation, as it instils a sense of ownership. CBR has been most successful where there is government support and where it is sensitive to local factors, such as culture, finances, human resources, and support from stakeholders, including local authorities and disabled people's organizations (Khasnabis and Motsch 2010). Where possible, funds from community should be preferred, as it helps in long-term sustainability of programs. Staff for CBR program should desirably be selected from the local community as they would have knowledge of the local culture and language and better access to community members. CBR programs should also be strongly committed to recruiting people with disabilities or family members of disabled people, as this contributes to their empowerment and send positive signals about their ability to community.

The matrix (see Fig. 3) consists of five key components reflecting multi-sectoral focus of CBR (consistent with the contextual factors of disability mentioned above): health, education, livelihood, social, and empowerment. The matrix has been designed to allow programs to select options which best meet their local needs, priorities, and resources. The programs are not expected to implement every component and element of the matrix. Due to this, each CBR program will be different from each other as it is influenced by a wide range of factors, viz., physical, socioeconomic, cultural, and political factors.

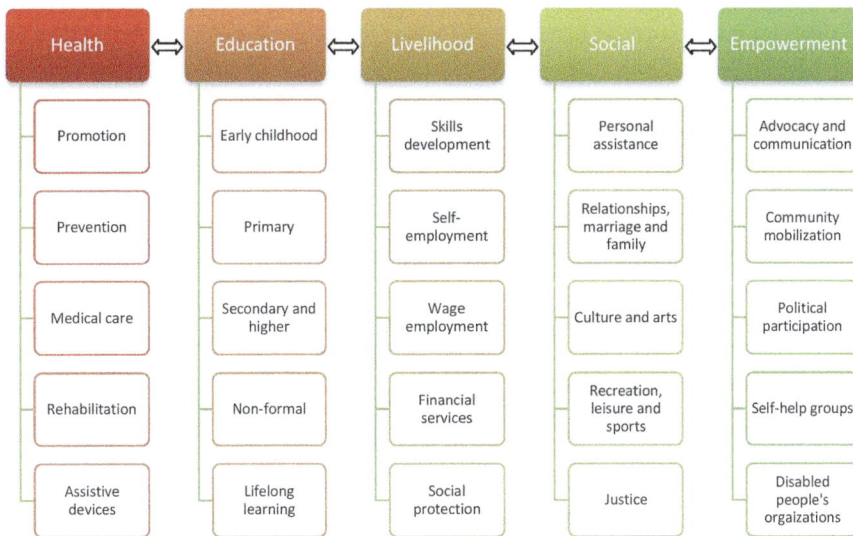

Health	Education	Livelihood	Social	Empowerment
Promotion	Early childhood	Skills development	Personal assistance	Advocacy and communication
Prevention	Primary	Self-employment	Relationships, marriage and family	Community mobilization
Medical care	Secondary and higher	Wage employment	Culture and arts	Political participation
Rehabilitation	Non-formal	Financial services	Recreation, leisure and sports	Self-help groups
Assistive devices	Lifelong learning	Social protection	Justice	Disabled people's orgaizations

Fig. 3 WHO CBR matrix

The Role of Healthcare and Rehabilitation Staff in CBR

After the Alma-Ata declaration in 1978 (World Health Organization 1978), the LAMI countries have introduced primary healthcare (PHC) system in the public healthcare system. Under this system, there exist a number of cadre to deliver healthcare in the rural communities. Typically, the existing staff at the rural/community level deliver mental healthcare and rehabilitation services as well in some LAMI countries; the extent of their involvement varies widely. The Accredited Social Health Activist (ASHA) worker and Village Rehabilitation Workers (VRW) in India, Lady Health Workers in Pakistan (Rahman et al. 2008), and volunteers are the community-level health workers in the LAMI countries. As they hail from local community, they are aware about the local realities and can act as a link between PwDMI and the government health system. Unfortunately, there are multiple challenges in the form of lack of awareness about mental illness, competing priorities, poor remuneration, and absence of emphasis on mental illness in the PHC system. The staff is generally overburdened with other national-level health programs. However, these programs offer excellent opportunities to integrate mental healthcare with general healthcare. Such integrated programs would not cause as much additional burden on the cadre as stand-alone mental health programs would.

Mental health professionals can gain trust of local community and grassroot workers by interacting with them at the grassroot level, sensitizing them about mental illness, understanding the local realities, being approachable, offering quality services with accessible and affordable medications, and showcasing successful rehabilitation of PwDMI. It is important to "walk the talk" by offering services rather than confine themselves to sensitization programs. When communities see PwDMI regain valued social roles and contribute to society, it shatters many myths and misconceptions about mental illness through principles of "social contact." Once the community is convinced about the effect of medications on a PwDMI, they extend support in terms of spreading the word in locality and relatives encouraging them to come for treatment (snowballing), exerting pressure on some families who are reluctant to take treatment due to various factors, and accepting the PwDMI in the community and offering livelihood options when they are recovered.

Case Vignettes of Patients Benefited by CBR

1. Ms. P, a 45-year-old mother of two daughters, has been psychotic and dysfunc-tional for last 10 years. Her husband had to take care of household responsibilities and earning a livelihood. The family found it difficult to take her to nearest district headquarters which offers treatment for mental illness. Even if they started early in the first bus, there was no guarantee that they would reach the Outpatient department (OPD) on time before noon to get a consultation and free medications. They dropped out after two visits as they observed little difference in her condition with treatment and also because, besides incurring travel expenses,

they would be deprived of a day's earning, which her husband could use to feed the family. The eldest daughter managed to pass her 12th standard in a nearby government school despite troubled home atmosphere. The family could not send her for higher education due to difficult family situation. She chose to stay at home, taking care of her sick mother and household responsibilities. The family learned about the CBR program for persons with mental illness in their locality and brought Ms. P for treatment. Antipsychotic medications were initiated and the psychotic symptoms remitted within 3 months. The family environment improved and Ms. P started to take care of the family. Through the CBR program, she obtained a disability certificate, disability pension, and health insurance scheme for PwDMI. The family brings her for treatment to the nearest PHC where mental health camps are conducted (which is much nearer than the district hospital) and free medications are provided. The peaceful domestic circumstances had a positive impact on the family. Her husband could focus on work and earn better; there was added financial support from the disability pension that Ms. P was provided with. Their youngest daughter could clear her 12th standard with good scores and could pursue her higher education. Father was in a position to pay for her education. She will be the first graduate of the family. The eldest daughter, in the meanwhile, is learning tailoring and family is searching for a prospective bridegroom for her.

This demonstrates that CBR has ripple effect well beyond the welfare of the PwDMI – the position of the entire family had a dramatic positive shift with simple measures of providing comprehensive medical and social benefits at an accessible place.

2. Mr. S, a 50-year-old gentleman, known case of schizophrenia with epilepsy. His wife works as agriculture laborer, and his son has shifted to a nearby city and works in a hotel. Patient remains at home due to psychotic symptoms and fear of fits. Medications were started and symptoms are under control. He attended government vocational training certificate course on sheep rearing for unemployed people. NGO helped him get disability certificate and disability pension, open a bank account, and get a loan. Recently, he got Rs. 35,000 loan from a local bank and has bought 11 sheep. They plan to rear it and sell it in another 6 months. The family will make a profit of Rs. 5000 per sheep. They can repay the principal amount to the bank within 1 year. On prompt repayment, they also become eligible for higher loans in the future.

For additional case vignettes, one can refer to the supplementary booklet of WHO-CBR guidelines (Khasnabis and Motsch 2010).

Conclusive Remarks

The concept of disability and rehabilitation has seen tectonic shift from medical, charity-based approaches to person-centered, rights-based ones. This distinctly positive development faces several challenges in the context of disability due to mental illnesses. Past few decades have seen exciting developments in the field of rehabilitation of persons with disability due to mental illnesses. However, as there are several challenges in making these available to millions who need them, there is urgent need to develop and test models of rehabilitation, which are accessible and acceptable across different settings and cultures. The WHO-CBR model holds promise in this context, as, at least in theory, it has several features which make it practical and applicable across diverse settings and cultures. Several case studies kindle hopes that programs based on this model can make substantial change in the lives of persons with disabilities. However, there is need for systematic studies with focus on implementation, sustenance, and comprehensive impact of such programs. Governments and organizations across the globe should focus on funding implementation research in this seriously neglected field.

References

Anthony W (2013) The experience of mental illness: an introduction to psychiatric rehabilitation. In: Pratt CW, Gill KJ, Barrett NM, Roberts MM (eds) Psychiatric rehabilitation. Academic, Waltham, p 9

Araya R, Rojas G, Fritsch R, Gaete J, Rojas M, Simon G et al (2003) Treating depression in primary care in low-income women in Santiago, Chile: a randomised controlled trial. Lancet 361(9362):995–1000. https://doi.org/10.1016/S0140-6736(03)12825-5

Asher L, Fekadu A, Hanlon C, Mideksa G, Eaton J, Patel V et al (2015) Development of a community-based Rehabilitation Intervention for People with Schizophrenia in Ethiopia. PLoS One 10(11):e0143572. https://doi.org/10.1371/journal.pone.0143572

Asher L, De Silva M, Hanlon C, Weiss HA, Birhane R, Ejigu DA et al (2016) Community-based Rehabilitation Intervention for people with Schizophrenia in Ethiopia (RISE): study protocol for a cluster randomised controlled trial. Trials 17(1):299. https://doi.org/10.1186/s13063-016-1427-9

Bhargava R, Thanapal S, Rozatkar A (2012) Persons with disability act. In: Chavan B, Nitin G, Priti A, Ajeet S, Sushrut J (eds) Community mental health in India, 1st edn. Jaypee Brothers Medical Publishers (P), New Delhi

Bolton P, Bass J, Betancourt T, Speelman L, Onyango G, Clougherty KF et al (2007) Interventions for depression symptoms among adolescent survivors of war and displacement in Northern Uganda: a randomized controlled trial. JAMA 298(5):519–527. https://doi.org/10.1001/jama.298.5.519

Bond GR, Campbell K (2008) Evidence-based practices for individuals with severe mental illness. J Rehabil 74(2):33–44

Chand S (2016) 10 Major differences between rural and urban societies. Available via Your Article Library. http://www.yourarticlelibrary.com/society/10-major-differences-between-rural-and-urban-societies/23390/. Accessed 02 July 2017

Chatterjee S, Pillai A, Jain S, Cohen A, Patel V (2009) Outcomes of people with psychotic disorders in a community-based rehabilitation programme in rural India. Br J Psychiatry 195(5):433–439. https://doi.org/10.1192/bjp.bp.108.057596

Chatterjee S, Naik S, John S, Dabholkar H, Balaji M, Koschorke M et al (2014) Effectiveness of a community-based intervention for people with schizophrenia and their caregivers in India (COPSI): a randomised controlled trial. Lancet 383(9926):1385–1394. https://doi.org/10.1016/S0140-6736(13)62629-X

Cohen A, Patel V, Thara R, Gureje O (2008) Questioning an axiom: better prognosis for schizophrenia in the developing world? Schizophr Bull 34(2):229–244

Desjarlais R, Eisenberg L, Good B, Kleinman A (1995) World mental health: problems and priorities in low-income countries. Oxford University Press, New York

Dixon LB, Dickerson F, Bellack AS, Bennett M, Dickinson D, Goldberg RW et al (2010) The 2009 schizophrenia PORT psychosocial treatment recommendations and summary statements. Schizophr Bull 36(1):48–70. https://doi.org/10.1093/schbul/sbp115

Halliburton M (2004) Finding a fit: psychiatric pluralism in south India and its implications for WHO studies of mental disorder. Transcult Psychiatry 41(1):80–98. https://doi.org/10.1177/1363461504041355

Jablensky A, Sartorius N (2008) What did the WHO studies really find? Schizophr Bull 34(2):253–255. https://doi.org/10.1093/schbul/sbm151

Jacob KS (2001) Community care for people with mental disorders in developing countries. Br J Psychiatry 178(4):296–298. https://doi.org/10.1192/bjp.178.4.296

Khasnabis C, Motsch KH (eds) (2010) Community-based rehabilitation: CBR guidelines. World Health Organization, Geneva

Kumar CN, Suresha KK, Arunachala U, Thirthalli J (2016) Role of self-help groups and income generation programs towards real world functioning of persons with schizophrenia: experience from the Thirthahalli cohort. Workability Asia, Bangalore

Lauriello J, Bustillo JR, Keith SJ (2005) Schizophrenia: scope of the problem. In: Sadock BJ, Sadock VA (eds) Kaplan & Sadock's comprehensive textbook of psychiatry, 8th edn. Lippincott Williams & Wilkins, Baltimore, pp 1346–1354

Naeem F, Sarhandi I, Gul M, Khalid M, Aslam M, Anbrin A et al (2014) A multicentre randomised controlled trial of a carer supervised Culturally adapted CBT (CaCBT) based self-help for depression in Pakistan. J Affect Disord 156:224–227. https://doi.org/10.1016/j.jad.2013.10.051

National Institute for Health and Care Excellence (2014) Psychosis and schizophrenia in adults: prevention and management. In: Psychosis and schizophrenia. NICE guidance. Available via NICE. https://www.nice.org.uk/guidance/cg178. Accessed 20 Apr 2016

Rahman A, Malik A, Sikander S, Roberts C, Creed F (2008) Cognitive behaviour therapy-based intervention by community health workers for mothers with depression and their infants in rural Pakistan: a cluster-randomised controlled trial. Lancet 372(9642):902–909. https://doi.org/10.1016/S0140-6736(08)61400-2

Rathod S, Pinninti N, Irfan M, Gorczynski P, Rathod P, Gega L et al (2017) Mental health service provision in low- and middle-income countries. Health Serv Insights 10:1–7. https://doi.org/10.1177/1178632917694350

Surbhi S (2015) Difference between developed countries and developing countries. Available via Key Differences. http://keydifferences.com/difference-between-developed-countries-and-developing-countries.html. Accessed 23 June 2017

Suresh KK, Kumar CN, Thirthalli J, Bijjal S, Venkatesh BK, Arunachala U et al (2012) Work functioning of schizophrenia patients in a rural south Indian community: status at 4-year follow-up. Soc Psychiatry Psychiatr Epidemiol 47(11):1865–1871. https://doi.org/10.1007/s00127-012-0495-8

Thara R (2012) Twenty-five years of schizophrenia: the Madras longitudinal study. Indian J Psychiatry 54(2):134–137. https://doi.org/10.4103/0019-5545.99531

The World Bank (2017) Countries and economies. Available via The World Bank Group. http://data.worldbank.org/country. Accessed 23 June 2017

Thirthalli J (2016) Evidence-based Psychosocial Intervention for Schizophrenia: many a barrier between the bench and the bedside. J Psychosoc Rehabil Ment Health 3(1):27–30. https://doi.org/10.1007/s40737-016-0052-y

Thirthalli J, Basavarajappa C (2014) Distinctiveness of disability due to psychiatric conditions. J Psychosoc Rehabil Ment Health 1(2):85–86. https://doi.org/10.1007/s40737-014-0015-0

Thirthalli J, Venkatesh BK, Kishorekumar KV, Arunachala U, Venkatasubramanian G, Subbakrishna DK et al (2009) Prospective comparison of course of disability in antipsychotic-treated and untreated schizophrenia patients. Acta Psychiatr Scand 119(3):209–217. https://doi.org/10.1111/j.1600-0447.2008.01299.x

Thirthalli J, Venkatesh BK, Naveen MN, Venkatasubramanian G, Arunachala U, Kishorekumar KV et al (2010) Do antipsychotics limit disability in schizophrenia? A naturalistic comparative study in the community. Indian J Psychiatry 52(1):37–41. https://doi.org/10.4103/0019-5545.58893

United Nations General Assembly (2007) Convention on the rights of persons with disabilities. https://www.un.org/development/desa/disabilities/convention-on-the-rights-of-persons-with-disabilities.html. Accessed 02 July 2017

Wig NN, Menon DK, Bedi H, Leff J, Kuipers L, Ghosh A et al (1987) Expressed emotion and schizophrenia in north India. II. Distribution of expressed emotion components among relatives of schizophrenic patients in Aarhus and Chandigarh. Br J Psychiatry 151:160–165

World Health Organization (1978) Declaration of Alma-Ata. World Health Organization, Alma-Ata. Retrieved from http://www.who.int/publications/almaata_declaration_en.pdf. Accessed 10 Apr 2017

World Health Organization (2001) International classification of functioning, disability and health. World Health Organization, Geneva

World Health Organization (2010a) Best practices: mental health service development. In: Mental health policy, planning & service development. http://www.who.int/mental_health/policy/services/mh_bestpractices_servdevlpt_2010_en.pdf?ua=1. Accessed 03 July 2017

World Health Organization (2010b) mhGAP intervention guide for mental, neurological and substance use disorders in non-specialized health settings: Mental Health Gap Action Programme (mhGAP). World Health Organization, Geneva

World Health Organization (2017) Health topics rehabilitation. Available via World Health Organization. http://www.who.int/topics/rehabilitation/en/. Accessed 02 July 2017

Future of Rural Living and Impact on Mental Health

22

Dinesh Bhugra and Antonio Ventriglio

Contents

Abstract

As is evident from this volume, practice of psychiatry in rural areas carries with it certain imperatives. Firstly, there is clear variation in incidence and prevalence of various psychiatric disorders in comparison with urban data. Secondly, the availability of human resources and services in rural areas is often poor with difficulties to access services. Thus, the duration of untreated symptoms is likely to be high leading to poor outcomes. Another complicating factor is massive rise in urban sprawls with people moving from rural areas to gain employment, thereby often leaving older individuals behind and isolated. Future aspects of improving practice of rural psychiatry must take into account

D. Bhugra (✉)
Mental Health and Cultural Psychiatry, Institute of Psychiatry (KCL), London, UK

Department of Psychosis Studies, King's College London, London, UK
e-mail: dinesh.bhugra@kcl.ac.uk

A. Ventriglio
University of Foggia, Foggia, Italy

© Springer Nature Singapore Pte Ltd. 2020
S. Chaturvedi (ed.), *Mental Health and Illness in the Rural World*, Mental Health and Illness Worldwide, https://doi.org/10.1007/978-981-10-2345-3_25

digital psychiatry, integrated psychiatric care between primary and secondary care. These changes are also important factors in developing training modules and packages for the next generation of psychiatrists.

Keywords
Rural psychiatry · Integration · Primary care · Secondary care · Training

Introduction

There is no doubt that epidemiological data over the past few decades have illustrated that rates and patterns of psychiatric disorders vary between urban and rural areas. This volume is a tribute to pulling these arguments together. There is clear evidence that urban sprawls are appearing rapidly and increasingly across the globe and within the next couple of decades for the first time there are likely to be more people living in urban areas than in rural areas. There is thus every likelihood that as social changes occur, people's perceptions of mental health and mental disorders will change too. There are several issues that need to be taken on board by the policymakers.

With increased globalization and rapid industrialization in many parts of the world, there is massive internal migration to urban areas which brings with it specific issues related to mental ill-health and changing needs which require changes in service development and delivery. Rural mental health remains of paramount importance due to specific issues and needs related to demographic changes as people move away and those who are left behind may find it difficult to access services or get social support as family members may well have moved to urban areas. The variations in rates of various psychiatric disorders have been described in this volume as well as potential etiological factors. For example, rural communities who across the world face higher rates of suicide than the general population require more targeted interventions need to be specific and well designed in order to reduce risk including social protective factors, healthcare, and policy changes. Therefore, any changes that need to be made to health provisions in rural settings to adequately provide mental health care for this at-risk population have to be incorporated into any policy developments.

There is increasing and compelling mountain of evidence that suggests that social determinants of mental illness such as poverty, overcrowding, and unemployment cause and contribute to psychiatric disorders. Furthermore, accessibility and appropriate services often are available only in urban areas as most of the training centers are based in urban areas and most clinicians prefer to work and live in urban areas. Consequently, not surprisingly rural psychiatry is ignored in training as well as often in practice not deliberately perhaps but accidentally. Another possible factor is the type of training provided which focuses on urban environment. Trainees of younger age groups learn and function in very different ways and that gives a major opportunity for trainers and teachers to use tele-mental health and e-mental health for training, supervision, as well as for providing services. These problems are obvious in high-income countries, but only, of late, low- and middle-income countries are trying to come to terms with these challenges. Reasons why practice of psychiatry in rural areas is challenging is not only because of poor resources, professional isolation, lack of multidisciplinary professionals, but also due to per-ceived or real fewer opportunities for professional development. Another key factor which is linked with lower levels of education which may very much focus on explanatory models of illness which rely on traditional and folk causation thereby people may seek non-medical or folk interventions which in turn may be looked down upon by medical professionals. Of course, many countries around the world are thinking in innovative and creative manner for recruitment of mental health professionals but also working with faith and folk healers to provide services which are accessible and acceptable to patients and their carers and families. However, simply placing trainees and training centers in rural areas is not likely to improve recruitment and retention unless other measures are put in place. There are often many erroneous assumptions about the needs of rural populations and their needs.

It has been argued in this volume that stress and rural are difficult words to define conceptually. Where does urban areas end, and rural areas begin? What are levels of stress that are more dangerous than others? Rural stress is conceptualized and seen as type of stress which may be considered unique to rural settings or are more pertinent to the rural areas. Of course, the very same factors, which define rural areas, may sometimes work as perpetuating factors for stress in rural population. It is important to recognize that seasonal variations in farming output are likely to influence prevalence and expression of stress through physical and psychological symptoms and will be mediated by individual's resilience and support systems. It is important to carry out studies in rural areas to understand the burden of stress and its role in the precipitation of psychiatric disorders so that appropriate interven-tions can be set in place.

Current Challenges

Often epidemiological and population studies tend to ignore rural areas as often the population is spread over a large geographical area making the task of data collection more onerous. Thus, the epidemiological data looking at incidence and prevalence of various psychiatric disorders are often scarce making it critical that research is carried out with appropriate and culturally sensitive assessment tools. A further problem is with the likely interpretation of the data were these available. Furthermore, as mentioned earlier, the access to clinical services is problematic. Not only primary care services are often poor, but access to social care is neglected often leaving the families to struggle by themselves without adequate support. In rural settings, the infra-structure may be very basic which can put people off from seeking help. Types of psychiatric disorders seen in rural areas are not dissimilar to those seen in urban areas although presentations may vary. For some groups such as the elderly, LGBT individuals, or those with intellectual disabilities, there are additional problems in seeking and receiving appropriate help.

The mental health treatment gap is high across the globe, but the variation depends upon countries, resources, and type of psychiatric disorders. It is inevitable that the gap will be worse for rural areas and population. The reasons for this treatment gap are many and often complex. Stigma, low investment in mental health, low mental health resources, poor literacy levels, and nonmedical explanatory models will influence treatment seeking and will vary from the actual need.

A further complicating factor is prevalence of stigmatizing attitudes and behaviors against those with psychiatric disorders. In rural settings often, communities are small and familiar with other members. There is no doubt that this stigma against people with mental illness can affect help-seeking and therapeutic alliance. Stigma is defined as the assignation to a person of a label that marks them out, tarnishes them, and leads to their exclusion by society through negative attitudes and negative behaviors. Broadly this may influence funding for research and services which in turn will increase stigma further as poor services and clinical settings play on to the image that those with psychiatric disorders are not worthy of care. Observations from many parts of the world indicate that most of the community psychiatric services such as community mental health centers and teams are placed in urban areas. Traditionally, psychiatric asylums and institutions have been on the peripheries of urban areas often in villages where these may be the sole source of employment leading to further isolation and a sense of alienation but also as a source of fear contributing to further isolation and stigma. Despite multiple anti-stigma campaigns, in many parts of the world stigma has not gone down and this may be because often stigma has been shown to play a role in creation of "the Other" validating one's own identity, as well as a lack of adequate knowledge leading to negative attitudes and behaviors. Surprisingly, there appear to be some differences in attitudes towards people with mental illness between rural and urban populations. These need further exploration and explanation and planning to decrease and eliminate stigma.

It has been argued in this volume that somatoform as well as dissociative disorders are common presentations in rural health settings – whether this is universal, needs further exploration. These disorders have been described conceptually as a method of expression of emotional distress or problems according to cultural parameters and values. In low- and middle-income rural settings, presentations of somatic complaints to the health professionals are equally common across rural and urban areas although data from high-income countries are not very clear. These symptoms might be a result of stress leading also to underlying depression or anxiety. In many cultures, these may be seen as a more acceptable form of seeking and accepting help. Integrated care between primary care physicians and secondary care specialists may be an optimal form of management where both physicians can work jointly for better outcomes.

Although called common mental disorders anxiety, depressions, phobias, and other conditions under this rubric can be crippling and associated with significant disability and might not be recognized in the primary health settings which is another reason for integrating services. Integration of mental health at primary care level is the only way forward to address this unmet need. Primary care health workers may well lack appropriate knowledge and skills to detect and manage mental disorders. There have been many initiatives affecting engagement of primary care workers. These have included training volunteers, religious leaders and healers, school teachers, as well as children who can all help identify needs and then refer them to appropriate services. Some recent initiatives have shown considerable promise in serving rural communities.

Common mental disorders are influenced by and associated with multiple social and cultural factors which influence the prevalence, presentation, and help-seeking. These clinical presentations are strongly affected by cultural explanatory models and clinicians need to be sensitive to the type and variety of presentation and help-seeking. Any treatment of these disorders also needs to be culturally sensitive and appropriate in that the patient's explanatory models are incorporated in any therapeutic interventions in managing these common conditions.

There is significant research evidence that women in general show much higher rates of various psychiatric disorders and in addition they are also likely to be carers for those who are ill. In rural settings, it is also likely that they will be facing further disadvantages related to gender, poverty, poor physical health, roles of care-giving, and being women farmers although these may vary across different countries. Besides gender disadvantage, exposure to intimate partner violence (IPV) as well as alcohol abuse by men folk may contribute to many psychiatric disorders including common mental disorders. Seeking help may well be more problematic for women in rural settings because of factors such as an absence of women-friendly mental health care services, lack of trained mental health professionals in rural areas or those who understand the unique needs of rural culture, stigma associated with help seeking, poor knowledge about treatment facilities, and lack of accessibility to transport to urban facilities. Women may be more vulnerable to psychiatric disorders due to a number of reasons including older age, widowhood, poverty, and living in areas of armed conflict. Comorbid infections and physical illnesses can play a major role in adding to health burden. In common with urban women, rural women's well-being is strongly associated with their physical health, inner well-being, economic security, rural identity, household and family well-being, and community relations. It has been shown that rural women are more accepting of help from their own community level workers and peer volunteers, which means that any mental health services must respond to cultural values and tailored to the realities of rural women.

As mentioned earlier, corresponding to the population, the level of health care workers in rural areas is abysmally poor. This gets further complicated when there are shortages of health care workers. Substance abuse rates are higher in many rural settings due to a number of social factors, and the treatment gap is considerably high. It is therefore inevitable that varying from population dynamics both in terms of aging and movement to implementation of interventions, differences between rural and urban populations remain across a range of psychiatric disorders.

It is important to bear in mind that whenever services are planned, organized, and delivered, stakeholders including local population and patient groups with their carers and families are involved from the very first step. This will enable the planners to understand and measure the needs and models of explanation, their views, and perception of mental disorders. It is also critical that any public mental health campaigning is culturally sensitive and culturally appropriate through which issues such as stigma and discrimination, early interventions, and public education can be tackled. For all of these to happen, adequate funding and investment in human resources is of paramount importance. Giving up Cartesian dichotomy is difficult but integrating mental healthcare with physical healthcare is crucial. This can be achieved by training primary care frontline health workers such as health visitors, community midwives, and others in managing mental disorders, supported by specialists through upscaling services according to need. E-mental health and tele-mental health can play a significant role in reaching out to individuals in their homes through Skype, mobile devices, and the web especially if the healthcare professionals are well supported by mental health specialists but issues related to ethics and confidentiality must be addressed.

Inhabitants of rural areas live under specific physical conditions which make them possibly more vulnerable to adverse economic, social and natural situations and the lack of services may well add to the delay in treatment and creating longer duration of untreated periods making outcomes potentially worse. Some of the advantages in rural settings in many low and middle-income countries is perhaps better income with less competitive environment who may also have better connections and links with other groups such as community leaders, religious leaders and healers and other resources. As mentioned above, among many disadvantages of managing mental problems in rural areas include stigma, isolation, limited resources for interventions, and limited opportunities for training and supervision and support from other colleagues. Workforce planning needs to ensure that human and financial resources meet the local needs, but training also needs to take into account personal and professional development.

On the one hand, often rural living is described as idyllic and peaceful, but reality may vary; especially in many countries, rural communities have high rates of gun ownership with possibility of increased rates of violence and violent crimes. Higher rates of alcohol abuse may play a role in influencing high rates of violence and deaths.

Medications remain as a core intervention for many psychiatric conditions in most centers around the world, with or without psychological interventions. Drug treatments have indeed helped many patients with psychiatric disorders. However, there remain major challenges in accepting medication as many cultures tend to prefer injections or capsules rather than tablets. Furthermore, some cultures would prefer big tablets as they are seen as more potent, whereas others will prefer small tablets as they are seen as having more concentrated of medication and therefore more potent. Some cultures prefer red tablets, whereas others prefer blue or green tablets depending upon whether these are seen as stimulating or calming. It is theoretically possible that such types of cultural differences may be seen across rural/urban divide. Thus, there are significant limitations to current medications and psychological approaches that make persons with psychiatric disorders seek nonpharmacological treatments as add-on or independent therapies, also known as traditional methods of healing or by the term "complementary and alternative medicine" (CAM). This is even more pronounced in the developing countries, especially rural populations, as often herbs and alternative medicines are seen as natural products and therefore less toxic, whereas reality may be quite different. The wide use of CAM approaches by patients has led clinicians and researchers to also explore the efficacy and evidence base for these interventions. The CAM approaches include whole alternative medical systems, mind-body therapies, biologically based therapies, energy-based therapies, and body-based manipulative therapies.

Delivery of treatments such as medication as well as psychotherapies in rural areas may well be affected by a number of factors, as has been highlighted in this volume; the practice of psychopharmacology in rural settings may well be constrained by a lack of availability, accessibility, and affordability of the medications. Hence, in some cases, rather than first-line medications, second-line medications may be used compromising the outcome. It is, therefore, important to recognize that evidence-based medicine may need to be tailored to suit to these ground realities. Rehabilitation in rural settings brings with it additional problems. Majority of persons experiencing disability due to mental illness live in rural areas especially in low- and middle-income countries with limited access to rehabilitation facilities. The concept of disability is viewed as a product of interaction of a person's impairment with his/her environment. Hence, it makes sense that the concepts of rehabilitation have moved away from a purely medical model. Concepts of rehabilitation must include a person-focused approach in culturally appropriate and sensitive manner with clear focus on human rights of the patient in mind. Any rehabilitation must, therefore, be fully comprehensive addressing health, education, livelihood, social, and empowerment needs of the patients. These standards of care must be aspired to as well as delivered. Although World Health Organization's Community-Based Rehabilitation (WHO-CBR) provides a roadmap to develop locally applicable, comprehensive rehabilitation for persons with disability (PwD) due to any health condition variation in resources can play havoc with actual delivery of rehabilitation services. In theory, the approach is a bottom-up one, where all stakeholders in the community take part in managing all aspects of rehabilitation of PwD; reality on the ground can be very different. However, such an approach needs a careful evaluation in varying conditions and countries.

In managing psychiatric disorders no matter what setting is, mind-body therapies have come into their own. With increasing evidence base suggesting that these interventions are efficacious in several mental disorders, these therapies have become accepted and indeed popular among not only patients but also their carers and healthcare professionals. In many parts of the world, these therapies have become much more popular. For example, in India, meditation and yoga-based therapies have emerged as important adjunct therapeutic options. Clinical trials with yoga have been conducted in many psychiatric conditions such as depression, anxiety, and schizophrenia, with promising results. While the evidence is preliminary, the available data suggest that yoga has potential distinct benefits in patients with these disorders. Furthermore, as described in this volume, there is research evidence that these effects are mediated by certain neurobiological processes that may have been deranged in the disease, and yoga may correct such an imbalance and is therefore of potential therapeutic importance. Future work must explore the impact of these methods in different settings to accurately measure the usefulness of such interventions.

As experience from Bangladesh suggest, integration of mental health service into primary care with community-based approaches and effective referral systems and scaling up of services can be used.

Future

Following steps may enable to deliver better mental healthcare in rural settings bearing in mind that not all rural areas will have the same needs or resources available.

Integration: Whether in rural areas or in urban areas, psychiatric services are often fragmented not only between physical healthcare and psychiatric healthcare but also between primary and secondary care, sometimes with great barriers between the two. Furthermore, super-specialization in psychiatry makes the risks of fragmented services even greater. There is no doubt that patients require holistic care for mental illness including physical care, but they also require social care, employment decent housing, employment, and educational opportunities, so there must be an integrated approach.

In secondary care, different services for different needs must be developed. For example, triage for 48/72 h and then appropriate placements, crisis centers, day hospitals, etc. should be available and accessible. Primary care deals with a majority of psychiatric disorders and therefore should be well supported. Within primary care, there needs to be a clear shift towards managing long-term conditions. By placing psychiatric services in primary care, patients can have better access to counseling, to community psychiatric nurses, and specialist skills with a strong possibility of reducing stigma. Circles of support in their communities through third sector and/or community well-being hubs or centers: day care, respite care, carer support, family support, resource centers, pharmacists, physiotherapists, counselors can all provide extra services to meet the needs of psychiatric patients. In addition, integration can provide trust, better communication, and patient-centeredness – all important aspects of producing better care. Integration will enable to reduce stigma and will also increase likelihood of working with the media to improve education.

Prevention and Health Promotion

But more emphasis needs to be placed on preventive and promotive service on mental health. Besides, for expansion of service, new allocation of resources and capacity development are essential. Awareness and anti-stigma campaign is important to successful service development. Development of organized and effective rural mental health service should be prioritized for better future of the nation. Mental health needs to be treated as a *public health issue*, so that it will be normal to look after our mental health starting early. Better recognition and understanding of prodromal symptoms will help support people to take an active role in their own care and self-management. Families and other sources such as teachers can be involved. Community engagement through education and helping build resilience and well-being should be a major priority. Working across disciplines and medical specialties such as medicine for managing alcohol use and abuse is a must.

Life Span Issues

Services should look at broader age span system-wide change rather than simply creating artificial age-related boundaries. This is perhaps more relevant in rural areas where the movement of people may be slower. Training in developmental aspects should be lifelong.

Workforce Development

Services need to manage the workforce issues with clear skill-mix. Skills that are needed include: excellent communication; flexible 24-hour approach; more working across professional boundaries and hospital/community boundaries; certain competencies and skills; closer liaison with the third sector (nongovernmental and voluntary organizations); ability to scale up; and task shifting – providing training and support and care.

Research

It is important that service providers also boost research in social aspects and especially translational research as well as research into the usefulness and impact of alternative and complementary treatments; the role of spiritual healers/elders/priests; the use of social media among others.

Digital Psychiatry

Digital psychiatry offers psychiatry the potential for radical change in terms of service delivery particularly in rural areas and the development as well as use of new treatments. Novel research methods, transparency standards, clinical evidence, and care delivery models must be created in collaboration with a wide range of stakeholders. However, ethical issues must be clearly identified. Using web-based assessment or treatment methods carry with them challenges related to accepting treatments.

Training the Psychiatrist of the Future

Psychiatrists of the future need to deliver services in a different way. They also have different attitudes and ways of learning, communicating, and functioning. Rapid scientific advance and evolving models of healthcare delivery need to be incorporated into training, but trainees and patients must also adapt to a changing landscape.

Psychiatry in general, but rural psychiatry in particular, is at a very significant point in the development. From psychopharmacogenomics to tele-mental health, these are indeed exciting times and we as clinicians must adapt and grab the opportunities.

Index

© Springer Nature Singapore Pte Ltd. 2020
S. Chaturvedi (ed.), *Mental Health and Illness in the Rural World*, Mental Health and
Illness Worldwide, https://doi.org/10.1007/978-981-10-2345-3